EVOLUTION OF HYDROTHERMAL ECOSYSTEMS ON EARTH (AND MARS?)

The Ciba Foundation is an international scientific and educational charity (Registered Charity No. 313574). It was established in 1947 by the Swiss chemical and pharmaceutical company of CIBA Limited — now Ciba-Geigy Limited. The Foundation operates independently in London under English trust law.

The Ciba Foundation exists to promote international cooperation in biological, medical and chemical research. It organizes about eight international multidisciplinary symposia each year on topics that seem ready for discussion by a small group of research workers. The papers and discussions are published in the Ciba Foundation symposium series. The Foundation also holds many shorter meetings (not published), organized by the Foundation itself or by outside scientific organizations. The staff always welcome suggestions for future meetings.

The Foundation's house at 41 Portland Place, London W1N 4BN, provides facilities for meetings of all kinds. Its Media Resource Service supplies information to journalists on all scientific and technological topics. The library, open five days a week to any graduate in science or medicine, also provides information on scientific meetings throughout the world and answers general enquiries on biomedical and chemical subjects. Scientists from any part of the world may stay in the house during working visits to London.

Ciba Foundation Symposium 202

EVOLUTION OF HYDROTHERMAL ECOSYSTEMS ON EARTH (AND MARS?)

1996

JOHN WILEY & SONS

Chichester · New York · Brisbane · Toronto · Singapore

Baffins Lane, Chichester,
West Sussex PO19 1UD, England

National 01243 779777
International (+44) 1243 779777
e-mail (for orders and customer service enquiries): cs-books@wiley.co.uk
Visit our Home Page on http://www.wiley.co.uk
or http://www.wiley.com

Other Wiley Editorial Offices

John Wiley & Sons, Inc., 605 Third Avenue,
New York, NY 10158-0012, USA

Jacaranda Wiley Ltd, 33 Park Road, Milton,
Queensland 4064, Australia

John Wiley & Sons (Canada) Ltd, 22 Worcester Road,
Rexdale, Ontario M9W 1L1, Canada

John Wiley & Sons (Asia) Pte Ltd, 2 Clementi Loop #02-01,
Jin Xing Distripark, Singapore 0512

Ciba Foundation Symposium 202
xii+335 pages, 77 figures, 19 tables

Library of Congress Cataloging-in-Publication Data

Evolution of hydrothermal ecosystems on Earth (and Mars?)/[edited by] Gregory R. Bock, Jamie A. Goode.
p. cm.—(Ciba Foundation Symposium; 202)
"Symposium on Evolution of Hydrothermal Ecosystems on Earth (and Mars?), held at the Ciba Foundation, London, January 30–February 1, 1996"—P.
Includes bibliographical references and indexes.
ISBN 0 471 96509 X (alk. paper)
1. Life—Origin—Congresses. 2. Hot spring ecology—Congresses. 3. Exobiology—Congresses. I. Bock, Gregory. II. Goode, Jamie. III. Symposium on Evolution of Hydrothermal Ecosystems on Earth (and Mars?) (1996: Ciba Foundation) IV. Series.
QH325.F963 1996 96-31351
574.5′263—dc20 CIP

British Library Cataloguing in Publication Data

A catalogue record for this book is available from the British Library

ISBN 0 471 96509 X

Typeset in 10/12pt Garamond by Dobbie Typesetting Limited, Tavistock, Devon.
Printed and bound in Great Britain by Biddles Ltd, Guildford.
This book is printed on acid-free paper responsibly manufactured from sustainable forestation, for which at least two trees are planted for each one used for paper production.

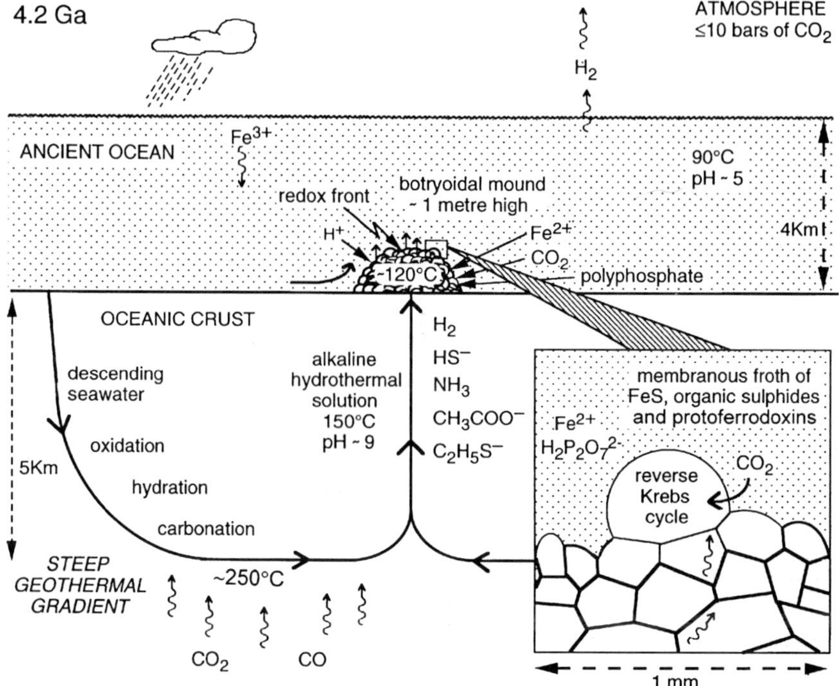

Molecular phylogenetic studies show that the deepest branches of the 'universal tree of life' are occupied by hyperthermophiles. Either life arose at high temperatures (one possible model for this is shown above), or only the hyperthermophiles survived the 'late heavy bombardment' during the final stages of accretion of the Earth. Thermal springs were abundant on the early Earth and their deposits are well known to mineral explorers because of the contained gold, copper and other valuable metals; they are a source of palaeobiological information that has yet to be tapped. All the indications are that such springs also would have been abundant on Mars early in its history; their deposits are a prime target for the exploration for former life on that planet.

The figure is modified from M. J. Russell and A. J. Hall (1996) The emergence of life from iron monosulphide bubbles at a submarine hydrothermal redox and pH front. Journal of the Geological Society of London (in press).

Contents

Symposium on Evolution of hydrothermal ecosystems on Earth (and Mars?), held at the Ciba Foundation, London, January 30–February 1 1996

Editors: Gregory R. Bock (Organizer) and Jamie A. Goode

This symposium was based on a proposal made by Malcolm Walter, David Des Marais, Jack Farmer and Nancy Hinman

Participants

S. M. Barns Environmental Molecular Biology, M888, Life Sciences Division, Los Alamos National Laboratory, Los Alamos, NM 87545, USA

S. L. Cady NASA Ames Research Center, MS 239-4, Moffett Field, CA 94035-1000, USA

M. H. Carr US Geological Survey, MS-975, 345 Middlefield Road, Menlo Park, CA 94025, USA

D. Cowan Department of Biochemistry and Molecular Biology, University College London, Gower Street, London WC1E 6BT, UK

P. C. W. Davies Department of Physics and Mathematical Physics, The University of Adelaide, Adelaide, SA 5005, Australia

D. J. Des Marais Space Science Division, NASA Ames Research Center, MS 239-4, Moffett Field, CA 94035-1000, USA

J. D. Farmer NASA Ames Research Center, MS 239-4, Moffett Field, CA 94035-1000, USA

W. F. Giggenbach Institute of Geological and Nuclear Sciences, PO Box 31312, Lower Hutt, New Zealand

R. W. Henley Etheridge Henley Williams, PO Box 250, Deakin West, ACT 2600, Australia

R. A. Horn INCO Limited, 2060 Flavelle Boulevard, Missisauga, Ontario, Canada L5K 1Z9

J. F. Huntington CSIRO Exploration and Mining, Mineral Mapping Technologies Group, PO Box 136, North Ryde, NSW 2113, Australia

B. M. Jakosky Laboratory for Atmospheric and Space Physics and Department of Geological Science, University of Colorado, Boulder, CO 80309-0392, USA

A. H. Knoll Botanical Museum, Harvard University, 26 Oxford Street, Cambridge, MA 02138, USA

R. Kuzmin Vernadsky Institute of Geochemistry and Analytical Chemistry, Russian Academy of Sciences, Kosygin Street 19, Moscow 177975, Russia

E. G. Nisbet Department of Geology, Royal Holloway, University of London, Egham, Surrey, TW20 0EX, UK

R. J. Parkes Department of Geology, University of Bristol, Bristol, BS8 1RJ, UK

A. Pentecost Division of Life Sciences, King's College London, Campden Hill Road, London W8 7AH, UK

C. L. Powell Department of Geology and Petroleum Geology, Meston Building, Kings College, Aberdeen, AB9 2UE, UK

M. J. Russell Department of Geology and Applied Geology, University of Glasgow, Glasgow G12 8QQ, UK

H. Sakai 1-4-7-1508 Seishim-cho, Edogawa-ku, Tokyo, Japan

E. L. Shock Department of Earth and Planetary Sciences and McDonnell Center for the Space Sciences, Washington University, St Louis, MO 63130, USA

B. R. T. Simoneit College of Oceanic and Atmospheric Sciences, Oregon State University, Corvallis, OR 97331-5503, USA

K. O. Stetter Lehrstuhl für Mikrobiologie, Universität Regensburg, Universitätstrasse 31, D-93053 Regensburg, Germany

R. E. Summons Australian Geological Survey Organisation, GPO Box 378, Canberra, ACT 2601, Australia

N. H. Trewin Department of Geology and Petroleum Geology, Meston Building, Kings College, University of Aberdeen, Aberdeen AB24 3UE, UK

M. R. Walter School of Earth Sciences, Macquarie University, North Ryde, NSW 2109 and Rix & Walter Pty Ltd, 265 Murramarang Road, Bawley Point, NSW 2539, Australia

Preface

The purpose of the meeting which lead to the production of this book was to bring together a diverse group of specialists who have complementary skills in the search for the earliest life on Earth, for its equivalents on Mars, and ultimately for the origin of life.

Biologists write about 'the universal tree of life', and believe that they can recognise the deepest roots of that tree, and that these are thermophilic microorganisms. Many geochemists think that this makes sense in terms of what geologists know about the abundance of hydrothermal systems on the early Earth and in terms of the potent chemical reactions possible in such systems. Geologists have the opportunity and the techniques to test these hypotheses. Despite what is sometimes thought by biologists, there are ways for geologists to find a fossil record of the different domains or super-kingdoms of life thought to lie near the base of the universal tree, and thus both to test the phylogenetic hypotheses and to calibrate them against time. Well preserved rock successions 3.5 Ga old have yielded the oldest evidence of life on Earth, and even older successions are now known. The search for evidence of life in these successions has barely begun, and essentially no attention has yet been paid to the hydrothermal systems that were the focus of the meeting.

Going beyond the Earth, one of the greatest exploration efforts of modern times is the search for life elsewhere in the Universe. The SETI program, of searching for radio signals from other civilizations, has to contend with enormous odds, given the vastness of the Universe and the uncertainties of recognizing biogenic signals. The odds may well be more in favour of discovering whether there once was, and maybe still is, life on Mars.

The discovery of evidence of former life on Mars could be interpreted in one of two ways: the first is that life arose on Mars independently of that on Earth. That to me seems every bit as significant as the discovery of life in the more distant realms of the Universe. The second possibility, in some ways even more startling, would be that early life on Mars could be found to be indistinguishable from that on Earth. That would re-ignite interest in the Panspermia hypothesis of Hoyle and Wikramisinge, or give a great boost to research on self-organizing systems (which might drive evolution in particular directions).

In a petition published in *Science* in 1982, Carl Sagan wrote, *inter alia*, 'We are unanimous in our conviction that the only significant test of the existence of extraterrestrial intelligence is an experimental one. No a priori arguments on this subject can be compelling or should be used as a substitute for an observational program'. That was a petition in support of the SETI program, but it has equal force in the search for former life on Mars, and in the quest for the origin of life on Earth. Exploring the Solar System is a grand venture, comparable to the great European voyages of discovery of earlier centuries. Unlike those voyages, it is now a cooperative international program. The prospects are immediate: there is a defined series of missions to Mars beginning with three launches this year (1996), two American and one Russian. Late in 1995 the Administrator of NASA announced that a sample return mission to Mars will be launched in 2005; sample return missions almost certainly will be required ultimately to determine whether there once was life on Mars. But we shall have to wait to see if the US Congress shares NASA's vision.

Exploration costs a lot of money, and we are never going to be able to collect all the information we would hope for. So if we are to have any hope of answering the question as to whether there was once life on Mars we will have to finesse our way through a forest of uncertainties and inadequate information. Exploration geologists make a profession of doing just that, in their search for mineral deposits and petroleum accumulations. Using their experience and techniques, we can help design precise experiments and define precise targets on the surface of Mars (and indeed on Earth, in our search for earliest life here). This symposium attempted to harness that expertise.

The quests for earliest life on Earth and Mars are great intellectual ventures to which the meeting and this book will, I hope, make a contribution.

Malcolm Walter
Chairman of the symposium

Hyperthermophiles in the history of life

Karl O. Stetter

Lehrstuhl für Mikrobiologie, Universität Regensburg, Universitätsstrasse 31, D-93053 Regensburg, Germany

Abstract. Prokaryotes requiring extremely high growth temperatures (optimum 80–110 °C) have recently been isolated from water-containing terrestrial, subterranean and submarine high temperature environments. These hyperthermophiles consist of primary producers and consumers of organic matter, forming unique high temperature ecosystems. Surprisingly, within the 16S rRNA-based phylogenetic tree, hyperthermophiles occupy all the shortest and deepest branches closest to the root. Therefore, they appear to be the most primitive extant organisms. Most of them (the primary producers) are able to grow chemolithoautotrophically, using CO_2 as sole carbon source and inorganic energy sources, suggesting a hyperthermophilic autotrophic common ancestor. They gain energy from various kinds of respiration. Molecular hydrogen and reduced sulfur compounds serve as electron donors while CO_2, oxidized sulfur compounds, NO_3^- and O_2 (only rarely) serve as electron acceptors. Growth demands of hyperthermophiles fit the scenario of a hot volcanism-dominated primitive Earth. Similar anaerobic chemolithoautotrophic hyperthermophiles, completely independent of a sun, could even exist on other planets provided that active volcanism and liquid water were present.

1996 Evolution of hydrothermal ecosystems on Earth (and Mars?). Wiley, Chichester (Ciba Foundation Symposium 202) p 1–18

On Earth, most known life forms are adapted to ambient temperatures within the mesophilic temperature range (15–45 °C). Among microorganisms, thermophiles growing optimally (fastest) at temperatures between 45 and 70 °C have been known for some time. As a rule, these regular thermophiles (sometimes rather anthropocentrically described as 'extreme') have close relatives growing within the mesophilic temperature range. In 1972, from a terrestrial hot spring in Yellowstone National Park, T. D. Brock isolated *Sulfolobus acidocaldarius*, an aerobic acidophilic prokaryote growing optimally at pH 3 and 75 °C, and exhibiting a maximum growth temperature of 85 °C (Brock et al 1972). *S. acidocaldarius* was found to be a facultative chemolithoautotroph, gaining energy by oxidation of organic matter or elemental sulfur. 16S rRNA studies showed that *Sulfolobus* represents a major phylogenetic branch within the archaea. Archaea, originally named 'archaebacteria', together with bacteria and eucarya, form the three domains of all life (Woese & Fox 1977), and are described later. Due to the low solubility of oxygen at high temperatures and the presence of reducing gases, boiling hydrothermal systems are mainly anaerobic. In our attempts to take samples from these ecosystems we took special care to avoid contamination with oxygen, which might be toxic to the organisms growing there (Stetter 1982, Balch et al 1979). By this strategy, we isolated novel anaerobic bacteria and archaea exhibiting unprecedented optimal growth temperatures in excess of 80 °C (Blöchl et al 1995). In contrast to the more usual (moderate) thermophiles, we designated them as hyperthermophiles (Stetter et al 1990). As a rule, they grow fastest between 80 and 105 °C (upper temperature border of *Pyrolobus fumarius*: 113 °C; E. Blöchl, H. W. Jannasch & K. O. Stetter, unpublished results) and are unable to propagate at 50 °C or below. For *Pyrodictium*, *Pyrolobus* and *Methanopyrus*, 80 °C is still too low to support growth. In this paper, I give an insight into the modes of life, biotopes and phylogeny of hyperthermophiles, and show evidence for their primitiveness and their probable existence since the dawn of life in the early Archaean age.

Biotopes of hyperthermophiles

Hyperthermophiles are found in natural and artificial environments. On land, volcanic emissions from deep magma chambers heat up soils and surface waters, forming sulfur-containing solfataric fields and neutral to slightly alkaline hot springs. The salinity of such terrestrial hydrothermal systems (e.g. the Krafla solfataras in Iceland, the solfatara crater in Pozzuoli, Italy, and the Yellowstone National Park, USA) is usually low. The surface of solfataric soils is rich in sulfate and exhibits an acidic pH (pH 0.5–6). It is oxidized by diffusing oxygen and has an ochre colour due to the presence of ferric iron. Deeper down, solfataric fields are less acidic (pH 5–7) and are anaerobic, exhibiting a blackish-blue colour due to the presence of ferrous sulfide. As a rule, solfataric fields contain large amounts of elemental sulfur which may be formed by oxidation from hydrogen sulfide at the surface. In addition, sulfate is present at the same concentration in solfataric fields as it is in sea water (about 30 mM/l).

Artificial biotopes include smouldering coal refuse piles (e.g. Ronneburg, Thuringen, Germany; Fuchs et al 1996) and hot outflows from geothermal and atomic power plants. Deep subterranean non-volcanic geothermally heated biotopes were discovered recently about 3500 m below the bottom of the North Sea and below the Alaskan North Slope permafrost soil (Stetter et al 1993), where *in situ* temperatures are approximately 100 °C. These Jurassic oil-bearing sandstone and limestone formations harbour hyperthermophilic communities as indicated by hydrogen sulfide formation ('reservoir souring') and a mixture of about 10^6 viable cells of various species of hyperthermophiles per litre of extracted fluids (Stetter et al 1993). Similar organisms exist in a reservoir in France (L'Haridon et al 1995). Marine biotopes may be shallow (e.g. at the beach of Vulcano Island, Italy) or deep hot sediments and hydrothermal systems. Most impressive are the deep sea 'smoker' vents, where mineral-laden hydrothermal fluids with temperatures of up to 400 °C build huge rock chimneys. Although these hot fluids are sterile, the surrounding rock material with a much lower temperature is teeming with hyperthermophiles (e.g. 10^8 cells of *Methanopyrus* spp. per g of rock at the Mid-Atlantic Snake Pit vent). A further type of submarine high temperature environment is provided by the active seamounts (e.g. Teahicya and Macdonald seamounts, close to Tahiti). When the Macdonald seamount erupted, about 10^4 viable cells of hyperthermophiles per litre were detected within its submarine plume (Huber et al 1990). Marine hydrothermal systems usually contain the high salt concentration of sea water (about 3%) and pHs ranging from slightly acidic to alkaline (pH 5–8.5). Volcanic emissions usually contain large amounts of steam, carbon dioxide, hydrogen sulfide, variable quantities of carbon monoxide, hydrogen, methane and nitrogen, and traces of ammonia. In addition, hydrogen may be created from ferrous iron and hydrogen sulfide by pyrite formation (Drobner et al 1990).

Although unable to grow at the low ambient temperatures, hyperthermophiles are able to survive under these conditions for several years. This is true for strict anaerobes even in the presence of oxygen, when they are kept cold in the laboratory. Not surprisingly, they can even be isolated from cold sea water after enrichment by ultrafiltration (Stetter et al 1993).

Physiological properties and energy-yielding reactions

So far, about 50 species of hyperthermophilic bacteria and archaea have been described: these are grouped into 23 genera in 10 taxonomic orders (Blöchl et al 1995). Hyperthermophiles are well adapted to growing in extremes of temperature, pH, redox potential and salinity (Table 1). Within their habitats, they form complex ecosystems consisting of both primary producers and consumers of organic matter (Table 2). Members of the extremely acidophilic genera *Stygiolobus*, *Acidianus*, *Metallosphaera* and *Sulfolobus* (Table 1) are coccoid-shaped and are found in acidic hot solfataric fields and coal refuse piles. They are aerobic, facultative aerobic and anaerobic chemolithoautotrophs gaining energy through the use of H_2, S^{2-} and S^0 as electron donors and S^0 and O_2 as electron acceptors (Table 2). Several isolates are facultative or obligate heterotrophs, growing on organic material such as cell extracts, sugars and amino acids. Slightly acidic to alkaline terrestrial hydrothermal systems contain members of the rod-shaped archaeal genera *Thermoproteus* (Fig. 1A), *Thermofilum*, *Pyrobaculum* and the coccoid-shaped *Desulfurococcus*, which are chemolithoautotrophs, and facultative and obligate heterotrophs depending on the species (Tables 1 and 2). Most of them are anaerobes, usually growing by sulfur respiration. As an exception, *Pyrobaculum aerophilum* is able to grow by respiration of nitrate or (low concentrations of) oxygen. Terrestrial hot springs in the Kerlingarfjoll mountains in the southwest of Iceland harbour *Methanothermus,* an ultra-aerophobic rod-shaped methanogen which most likely represents a genus endemic to this area (Lauerer et al 1986). Neutrophilic terrestrial communities may also contain *Thermotoga,* a carbohydrate-fermenting rod-shaped bacterium exhibiting a characteristic sheath-like structure ('toga') overballooning the ends. The highest growth temperatures observed occur among communities of marine hyperthermophiles which consist of members of the bacterial genera *Aquifex* and *Thermotoga* and the archaeal genera *Pyrobaculum*, *Staphylothermus*, *Pyrodictium*,

TABLE 1 Some basic features of hyperthermophiles

	Growth conditions							
	Temp (°C)							
Species	*Minimum*	*Optimum*	*Maximum*	*pH*	*Aerobic (ae) or anaerobic (an)*	*Biotope (marine [m] or terrestrial [t])*	*DNA G+C (mol%)*	*Morphology*
Thermotoga maritima	55	80	90	5–9	an	m	46	Rods with a 'toga'
Aquifex pyrophilus	67	85	95	5–7	ae	m	40	Rods
Sulfolobus acidocaldarius	60	75	85	1–5	ae	t	37	Lobed cocci
Metallosphaera sedula	50	75	80	1–4	ae	t	45	Cocci
Acidianus infernus	60	88	95	2–5	ae/an	t	31	Lobed cocci
Stygiolobus azoricus	57	80	89	1–5	an	t	38	Lobed cocci
Thermoproteus tenax	70	88	97	3–6	an	t	56	Regular rods
Pyrobaculum islandicum	74	100	103	5–7	an	t	46	Regular rods
Pyrobaculum aerophilum	75	100	104	6–9	ae/an	m	52	Regular rods
Thermofilum pendens	70	88	95	4–6	an	t	57	Slender regular rods
Desulfurococcus mobilis	70	85	95	5–7	an	t	51	Cocci
Staphylothermus marinus	65	92	98	5–8	an	m	35	Cocci in aggregates
Pyrodictium occultum	82	105	110	5–7	an	m	62	Disks with tubules
Thermodiscus maritimus	75	88	98	5–7	an	m	49	Disks
Pyrococcus furiosus	70	100	105	5–9	an	m	38	Cocci
Archaeoglobus fulgidus	60	83	95	5–7	an	m	46	Irregular cocci
Methanothermus sociabilis	65	88	97	5–7	an	t	33	Rods in clusters
Methanopyrus kandleri	84	98	110	5–7	an	m	60	Rods in chains
Methanococcus igneus	45	88	91	5–7	an	m	31	Irregular cocci

TABLE 2 Energy-yielding reactions in chemolithoautotrophic hyperthermophiles

Energy-yielding reaction	*Genera (examples)*
$2H_2+O_2 \rightarrow 2H_2O$	*Pyrobaculum,*[a] *Sulfolobus,*[a] *Acidianus,*[a] *Metallosphaera,*[a] *Aquifex*
$4H_2S+2O_2 \rightarrow 4H_2SO_4$	*Sulfolobus,*[a] *Acidianus,*[a] *Metallosphaera,*[a] *Aquifex*
$2S^0+3O_2+2H_2O \rightarrow 2H_2SO_4$	*Sulfolobus,*[a] *Acidianus,*[a] *Metallosphaera,*[a] *Aquifex*
$H_2+NO_3^- \rightarrow NO_2^-+H_2O$	*Pyrobaculum,*[a] *Aquifex*
$H_2+S^0 \rightarrow H_2S$	*Acidianus,*[a] *Stygiolobus, Pyrodictium,*[a] *Thermoproteus,*[a] *Pyrobaculum*[a]
$4H_2+H_2SO_4 \rightarrow H_2S+4H_2O$	*Archaeoglobus*
$4H_2+CO_2 \rightarrow CH_4+2H_2O$	*Methanopyrus, Methanothermus, Methanococcus*

[a]Facultative heterotrophs.

Thermodiscus, *Thermococcus*, *Pyrococcus*, *Archaeoglobus*, *Methanopyrus* and *Methanococcus* (Table 1). They are found within shallow or deep submarine hydrothermal systems. *Aquifex* spp. are rod-shaped obligate chemolithoautotrophs, achieving the highest growth temperatures within the bacteria (Tables 1 and 2). They gain energy by oxidizing H_2 or S^0 under microaerobic conditions. In the absence of O_2, *Aquifex* spp. can grow by nitrate reduction (Table 2). The organisms with the highest growth temperatures, 110 °C, are of the genera *Pyrodictium* and *Methanopyrus*. *Pyrodictium* is an anaerobic facultative chemolithoautotrophic sulfur respirer. It forms disk-shaped cells which connect to one another through unique networks of hollow tubules about 30 nm in diameter (Fig. 1B). Although resting forms, such as spores, have never been observed, cultures of *Pyrodictium occultum* grown at 110 °C exhibit an extraordinary heat resistance and even survive autoclaving for 1 h at 121 °C (Table 3). Under these conditions, about 70% of the soluble protein consists of the thermosome (Phipps et al 1993). *Methanopyrus* is a rod-shaped methanogen that grows optimally at 100 °C with a population doubling time of 50 min. In contrast to all other archaea, it contains 2,3-di-*O*-geranylgeranyl-*sn*-glycerol (Fig. 2) as the dominating membrane lipid (Hafenbradl et al 1993). This ether lipid represents the unsaturated precursor molecule of the 2,3-di-*O*-phytanyl-*sn*-glycerol lipid that is common among the other members of archaea. Since it is generally assumed that biochemical pathways reflect their own evolution, the use of a precursor molecule by an organism can be seen as a primitive feature. This is in line with the phylogenetic position of *Methanopyrus* (see below). *Archaeoglobus* spp. are coccoid-shaped facultatively lithoautotrophic archaeal sulfate reducers with a broad spectrum of substrates. They share the coenzymes F_{420}, methanopterin, tetrahydromethanopterin and methanofuran with methanogens. Consequently, under the UV microscope at 420 nm excitation they exhibit the blue–green fluorescence otherwise characteristic of methanogens (Stetter et al 1987). In addition to the chemolithoautotrophic producers of organic material, submarine hydrothermal systems harbour many strictly heterotrophic consumers, which gain energy either by sulfur respiration or by fermentation. They are coccoid-shaped and belong to the genera *Staphylothermus*,

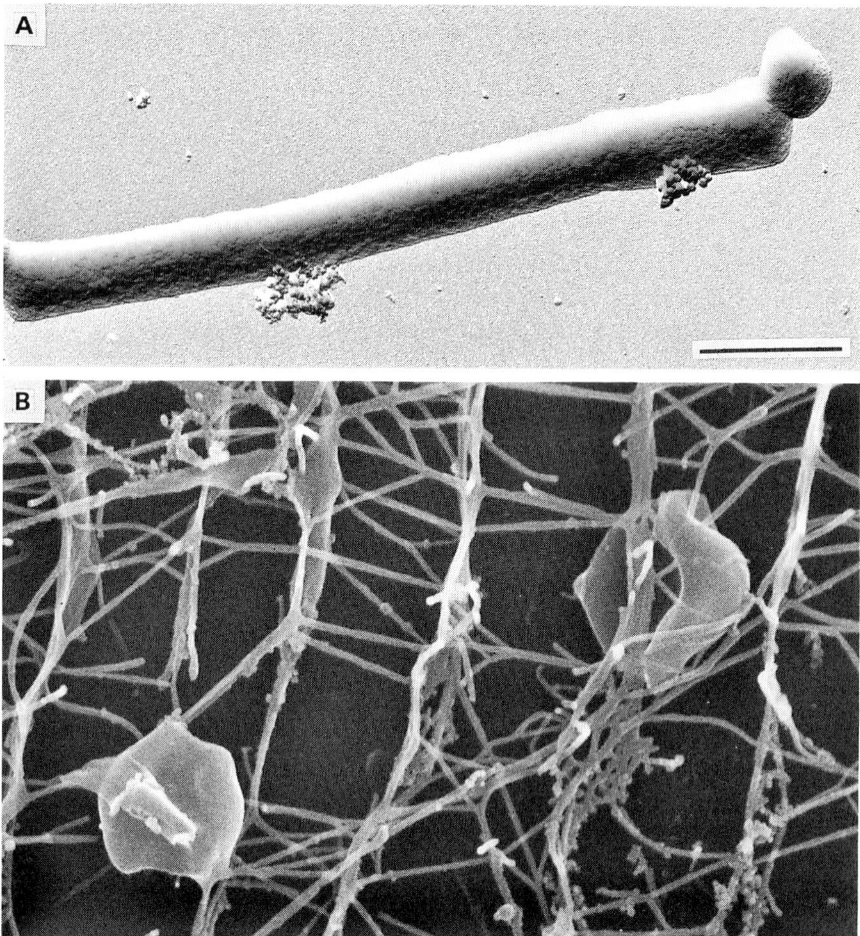

FIG. 1. Electron micrographs of hyperthermophilic archaea. (A) Golf club-shaped cell of *Thermoproteus tenax*, platinum shadowed (scale bar = 1 μm). (B) Network of hollow tubules connecting disk-shaped cells of *Pyrodictium abyssi*. Scanning electron micrograph (Rieger et al 1995).

Thermodiscus, *Hyperthermus*, *Thermococcus* and *Pyrococcus* (Tables 1 and 2; Blöchl et al 1995). Deep subterranean oil reservoirs contain species of hyperthermophiles common to submarine hydrothermal systems. So far, *Archaeoglobus fulgidus*, *Archaeoglobus lithotrophicus* and *Archaeoglobus profundus*, *Thermococcus celer*, *Thermococcus litoralis* and new species of *Pyrococcus* and *Thermotoga* have been isolated from them (Stetter et al 1993).

TABLE 3 ***Pyrodictium occultum*: autoclave survival after heat pretreatment**

	Heat treatment		
Experiment no.	*3 h 110 °C*	*1 h 121 °C*	*Survivors (viable cells/ml)*
1	–	+	0
2	+	+	10^3
3	+	–	10^7

Cultures pre-grown overnight at 102 °C.

Phylogenetic tree of hyperthermophiles

The 16S rRNA-based universal phylogenetic tree (Fig. 3; Woese & Fox 1977, Woese et al 1990) is rooted by reference to phylogenetic trees of duplicated genes of ATPase subunits and elongation factors Tu and G (Iwabe et al 1989). Deep branches in the tree are evidence for early separation. For example, the separation of the bacteria from the stem common to archaea and eucarya represents the deepest and earliest branching point within the phylogenetic tree (Fig. 3). Short phylogenetic branches indicate a rather slow rate of evolution. Surprisingly, hyperthermophiles are represented among all the deepest and shortest prokaryotic lineages (Fig. 3, bold lines). The deepest and shortest phylogenetic branches are represented by *Aquifex* and *Thermotoga* within the bacteria and *Methanopyrus* and *Pyrodictium* within the archaea. On the other hand, mesophilic and moderately thermophilic members of bacteria and archaea represent long lineages within the phylogenetic tree and, therefore had a fast rate of evolution (e.g. proteobacteria, Gram-positives, *Halobacterium*, *Methanosarcina*; Fig. 3).

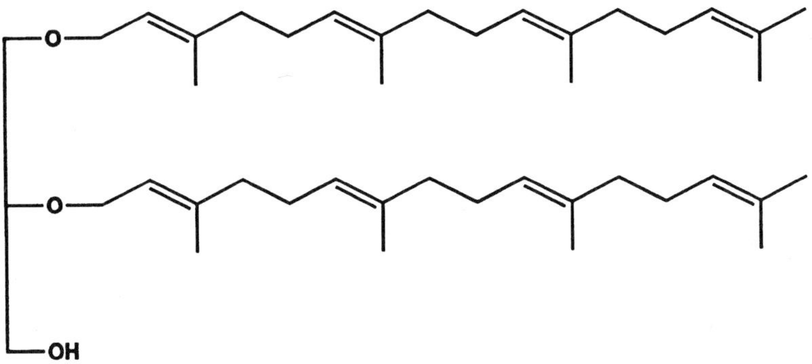

FIG. 2. Structure of the *Methanopyrus kandleri* lipid 2,3-di-*O*-geranylgeranyl-*sn*-glycerol.

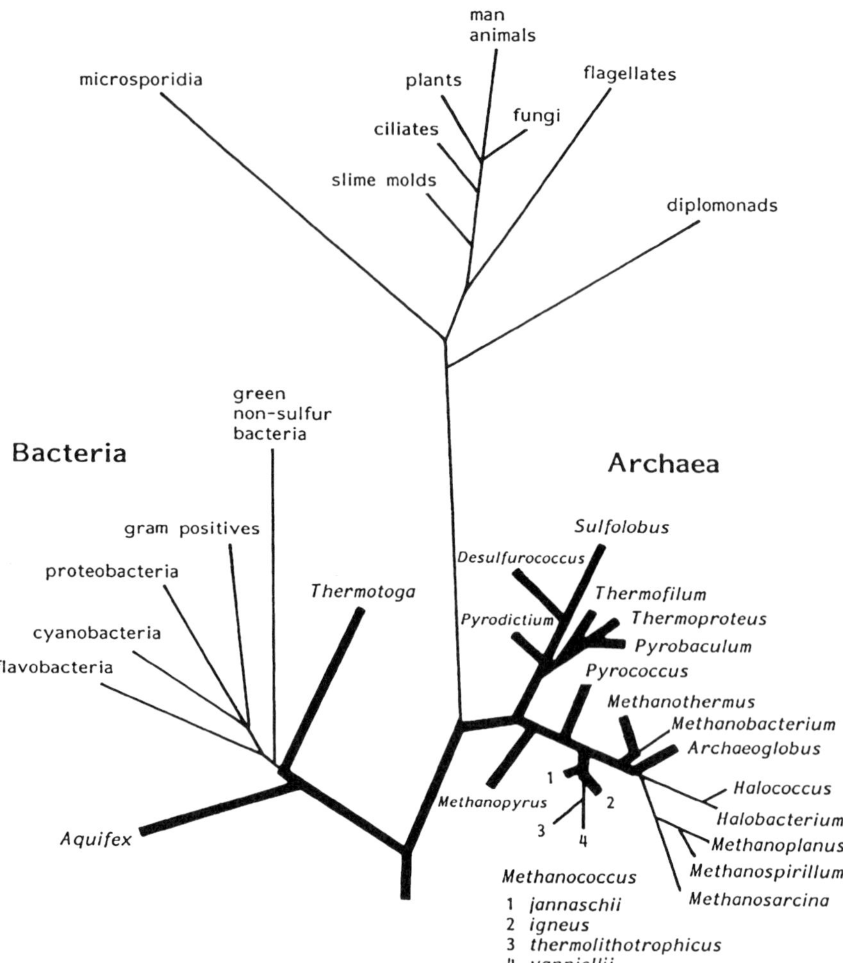

FIG. 3. Hyperthermophiles within the 16S rRNA-based phylogenetic tree. Redrawn and modified (based on C. R. Woese's model; Woese et al 1990, Blöchl et al 1995).

Conclusions: testimony of hyperthermophiles

On Earth, complex ecosystems exist at high temperatures of about 100 °C within volcanically, geochemically or artificially heated rocks, sediments and waters. They comprise communities of hyperthermophilic archaea and bacteria which are primary producers and consumers of organic matter. Hyperthermophilic primary producers are chemolithoautotrophs exhibiting very simple growth requirements: they are able to use various mixtures of oxidized and reduced minerals and gases as energy sources and to assimilate carbon from CO_2 — in addition, they just need liquid water and heat for growth. Hydrogen and sulfurous compounds are widely used substrates which mainly occur within volcanic environments. Most hyperthermophiles do not require or even tolerate oxygen for growth and are, therefore, completely independent of photosynthesis, and hence sunlight. In addition, the deep subterranean hyperthermophilic communities known so far may depend mainly on organic crude oil compounds as an energy source. They are able to grow at high pressure several kilometres below the soil surface. Most organisms isolated from this source are already known from submarine and terrestrial hydrothermal systems. Therefore, they may have entered the reservoirs comparatively recently through natural faults, oil seeps and during drilling and flushing while recovering oil (Stetter et al 1993). Moreover, there is a good chance that still-unknown hot microcosms exist deep in the Earth's crust which could gain energy using inorganic redox reactions (e.g. pyrite formation; Drobner et al 1990).

Within the 16S rRNA-based universal phylogenetic tree of life, hyperthermophiles form a cluster around the root, occupying all the shortest and deepest phylogenetic branches. This is true for both the archaeal and the bacterial domains (Fig. 3). As a rule, members of the deepest and shortest lineages are chemolithoautotrophs exhibiting the highest growth temperatures ('the deeper, the hotter'; e.g. *Pyrodictium*, *Methanopyrus*, *Aquifex*). From their 16S rRNA, these slowly evolving organisms appear to be the most primitive ones still existing. The presence of a primitive precursor lipid in *Methanopyrus* supports this conclusion. In view of a possible close similarity of recent hyperthermophiles to ancestral life, a chemolithoautotrophic hyperthermophilic common ancestor appears most probable. In agreement, growth conditions of recent hyperthermophiles fit well to our view of the primitive Earth about 3.9 Ga ago, when life could have originated: the atmosphere was overall reducing and on the basis of the evidence of a thin brittle crust, with much stronger volcanism and radioactive decay, Earth was much hotter (Ernst 1983). Therefore, ancestral hyperthermophiles should even have dominated in the Early Archaean age. Because of their ability to survive for long times in the cold, recent hyperthermophiles can even spread in cold conditions in a kind of dormant state. Therefore, hyperthermophiles could have successfully inoculated other planets, such as Mars, whenever growth conditions had been favourable there. To accomplish this, cells present in aerosols and volcanic dust could had been picked up by meteorites which passed Earth's atmosphere and finally carried them to other planets which may have had hydrothermal systems. Currently, Mars' surface is too cold and neither liquid water nor active volcanism has been be detected. Therefore, hyperthermophiles could not grow there at present. However, deep below the surface, assuming that water and heat are present, there may exist active microcosms of hyperthermophiles which could have survived since the Archaean and which await discovery.

Acknowledgements

I thank John Parkes for critically reading the manuscript and Reinhard Rachel for electron microscopy. The work presented from my laboratory was supported by grants of the DFG (Leibniz Award), the BMFT, the EEC and the Fonds der Chemischen Industrie.

References

Balch WE, Fox GE, Magrum LJ, Woese CR, Wolfe RS 1979 Methanogens: re-evaluation of a unique biological group. Microbiol Rev 43:260–296

Blöchl E, Burggraf S, Fiala G et al 1995 Isolation, taxonomy and phylogeny of hyperthermophilic microorganisms. World J Microbiol Biotechnol 11:9–16

Brock TD, Brock KM, Belly RT, Weiss RL 1972 *Sulfolobus:* a new genus of sulfur-oxidising bacteria living at low pH and high temperature. Arch Microbiol 84:54–68

Drobner E, Huber H, Wächtershäuser G, Rose D, Stetter KO 1990 Pyrite formation linked with hydrogen evolution under anaerobic conditions. Nature 346:742–744

Ernst WG 1983 The early Earth and the Archaean age rock record. In: Schopf JW (ed) Earth's earliest biosphere, its origin and evolution. Princeton University Press, Princeton, NJ, p 41–52

Fuchs T, Huber H, Teiner K, Burggraf S, Stetter KO 1996 *Metallosphaera prunae*, sp nov, a novel metal-mobilizing thermoacidophilic archaeum, isolated from a uranium mine in Germany. Syst Appl Microbiol 18:560–566

Hafenbradl D, Keller M Thiericke R, Stetter KO 1993 A novel unsaturated archaeal ether core lipid from the hyperthermophile *Methanopyrus kandleri.* Syst Appl Microbiol 16:165–169

Huber R, Stoffers P, Cheminee JL, Richnow HH, Stetter KO 1990 Hyperthermophilic archaebacteria within the crater and open-sea plume of erupting Macdonald Seamount. Nature 345:179–182

Iwabe N, Kuma K, Hasegawa M, Osawa S, Miyata T 1989 Evolutionary relationship of archaebacteria, eubacteria, and eucaryotes inferred from phylogenetic trees of duplicated genes. Proc Natl Acad Sci USA 86:9355–9359

L'Haridon S, Reysenbach A-L, Glénat P, Prieur D, Jeanthon C 1995 Hot subterranean biosphere in a continental oil reservoir. Nature 377:223–224

Lauerer G, Kristjansson JK, Langworthy TA, Konig H, Stetter KO 1986 *Methanothermus sociabillis* sp nov, a second species within the *Methanothermaceae* growing at 97 °C. Syst Appl Microbiol 8:100–105

Phipps BM, Typke D, Hegerl R et al 1993 Structure of a molecular chaperone from a thermophilic archaebacterium. Nature 361:475–477

Rieger G, Rachel R, Hermann R, Stetter KO 1995 Ultrastructure of the hyperthermophilic archaeon *Pyrodictium abyssi.* J Struct Biol 115:78–87

Stetter KO 1982 Ultrathin mycelia-forming organisms from submarine volcanic areas having an optimum growth temperature of 105 °C. Nature 300:258–260

Stetter KO, Lauerer G, Thomm M, Neuner A 1987 Isolation of extremely thermophilic sulfate reducers: evidence for a novel branch of archaebacteria. Science 236:822–824

Stetter KO, Fiala G, Huber G, Huber R, Segerer A 1990 Hyperthermophilic microorganisms. FEMS Microbiol Rev 75:117–124

Stetter KO, Huber R, Blöchl E et al 1993 Hyperthermophilic archaea are thriving in deep North Sea and Alaskan oil reservoirs. Nature 365:743–745

Woese CR, Fox GE 1977 Phylogenetic structure of the prokaryotic domain: the primary kingdoms. Proc Natl Acad Sci USA 74:5088–5090

Woese CR, Kandler O, Wheelis ML 1990 Towards a natural system of organisms: proposal for the domains Archaea, Bacteria, and Eucarya. Proc Natl Acad Sci USA 87:4576–4579

DISCUSSION

Carr: How is it possible to work out the root of the 'universal tree of life'?

Barns: We root the tree by looking at a gene that was duplicated and became two functionally separate genes within or before the last common ancestor of all life, i.e. a gene that underwent gene duplication before the three major lineages radiated from each other (Doolittle & Brown 1994). In this way, the sequence of one gene can be used as an outgroup for the others.

Carr: This issue is pretty fundamental to what we're talking about at this meeting. Is there a consensus that the root actually is where it is indicated in Karl Stetter's diagram?

Barns: All the data that we have so far from duplicated genes — there are three that have enough data to be believable — give the same result (Iwabe et al 1989, Gogarten et al 1989, Brown & Doolittle 1995, Baldauf et al 1996).

Stetter: Even if the root were to be proven wrong, which I don't think is likely, it wouldn't change the different domains. The hyperthermophiles would still occupy the deepest branches. The only question concerns whether it was the archaea or the bacteria that first branched off on the way to the eukarya. Further confirmation of the tree as I depicted it is provided by the complete sequences of the genomes of some of the archaeal hyperthermophiles that have only recently been obtained. I have access to some of these data, which show clearly that many of the proteins show a close relationship to their equivalents in higher eukaryotes.

Davies: Are you saying that we are more closely related to these archaea than to bacteria?

Stetter: Yes.

Davies: When you look for these hyperthermophiles in different hotspots, do you find the same species in widely separated regions?

Stetter: We have had access only to a few of these high temperature environments, so it is difficult to generalize, but for example we find the same species of *Methanopyrus kandleri* in the Pacific and the mid-Atlantic ridge, and we find the same species of *Pyrodictium* in the south sea on the Macdonald Seamount as we do at the beach of Vulcano, Italy.

Davies: It seems to me that there are then two possibilities — one is that these are extremely primitive organisms that have not evolved over billions of years; the other is that there is some hidden continuity there.

Stetter: A kind of evolution is indicated by the existence of closely related species at the same and at distant locations.

Davies: But if they have remained isolated for billions of years, it is astonishing that they have not evolved into separate species. Could they be connected through the rocks?

Stetter: I have a much simpler explanation: they have travelled by air.

Davies: Wouldn't it be too cold for them? Surely they would die?

Stetter: This is a very important point: at low temperatures hyperthermophiles are unable to propagate, but they can survive for many years. Our oldest samples in the laboratory still contain viable hyperthermophiles after 15 years at lower temperatures. In this respect they are very different from other bacteria.

Walter: The possibility that these organisms may travel through rocks is potentially very interesting. You made the point that there are abundant bacteria within the thermal plumes that come out of black smokers. The inference is that these bacteria are coming from the subterranean hydrothermal environment. Are you or others inferring that there is in fact a subterranean microbiota associated with the black smokers? If so, perhaps there is a plumbing system that connects them all.

Stetter: The problem with that idea is that the plume temperatures of black smokers are in the range of 400 °C — any organisms present would die within milliseconds. Holger Jannasch from Woods Hole has a nice 'smoker-poker' facility with which he can collect samples of plume fluids; he has shown that these are sterile.

Parkes: My group works on the Ocean Drilling Program, looking at the bacterial populations and processes in deep marine sediments. About two years ago we had the opportunity to do some deep drilling off Juan de Fuca Ridge hydrothermal system (Cragg & Parkes 1994). We did a transect from a control site into the high temperature region and found that the bacterial populations decreased more rapidly with sediment depth as the temperature started to rise. Amazingly, when we were a few metres away from an active chimney, where there is horizontal hydrothermal fluid flow at depth, bacterial populations actually increased (Fig. 1 [*Parkes*]). These results were from direct microscopy alone, so they need to be confirmed. But there is an indication that at about 169 °C we had an elevation in bacterial populations which was associated with hydrothermal fluid flow. I think there is a distinct possibility of the existence of an even higher-temperature and more extensive microbial system in the deep lateral fluid flow systems within these hydrothermal vent systems than that found previously in hydrothermal plumes and chimney walls.

Stetter: Life at 160 °C could be possible. However, this is only on the basis of microscopical evidence. Several years ago, on the basis of similar evidence, John Baross claimed that there is life at 250 °C (Baross et al 1983) — data which were never reproduced and which even he doesn't believe anymore. But it's an exciting idea.

Jakosky: How will you confirm these results?

Parkes: We were really looking for bacterial populations actually being decreased by high temperatures, so we only took preserved samples. We intend to go back and use a variety of techniques to demonstrate those populations.

Farmer: An interesting and fundamental problem about this tree is interpreting what it means with regard to the origin of life. The tendency is to assume that life originated at high temperatures. However, an alternative scenario concerns the role of large impact events, as Sleep et al (1989) have discussed. This throws a different light on the problem. What we might be seeing here is a relic of these early events where the biosphere was 'bottlenecked' through high temperature conditions at different points. I would be interested in hearing what others think about this.

Shock: The idea you are referring to is that just because the phylogenetic tree is rooted in hyperthermophilic organisms, it doesn't mean that there's necessarily a

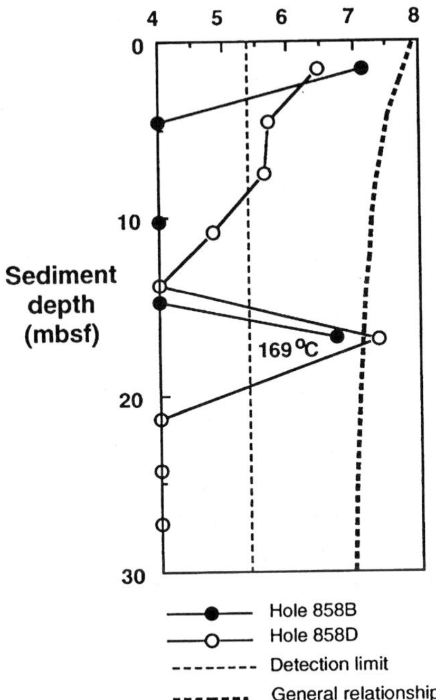

FIG. 1. *(Parkes)* Bacterial populations in sediments close to active hydrothermal vents at Juan de Fuca Ridge, N. E. Pacific. mbsf, metres below sea floor. (From Cragg & Parkes 1994.)

connection to the origin of life—these could simply be the organisms that survived some large impact event that would heat up the surface of the planet and more-or-less obliterate life in all other environments. Hydrothermal systems are among the places where life would be most likely to survive. It's certainly an interesting idea. There is ample evidence from studies of the moon and other planets of what the impact record was, but as I'll try to describe in my paper (Shock 1996, this volume), there are good reasons to believe that simple organic synthesis could get started in hydrothermal systems. To postulate that life originated elsewhere and then radiated and invaded hydrothermal systems is an added and perhaps unnecessary complexity.

Nisbet: One of the strong arguments in support of the assertion that hyperthermophiles are very primitive is the importance of the heat-shock proteins. In modern living organisms, they are vital in an incredible variety of processes, such as photosynthesis. They help fold other proteins, and seem to be of the greatest antiquity, dating back perhaps to the last common ancestor. Their original role may have been to repair damage. Why do all the most deeply-rooted organisms need heat-shock proteins? To a geologist heat shock proteins look like a zeolite mineral — a mineral with a cavity in it. They are almost organic equivalents of zeolites, which various people have suggested may have been hosts of pre-biotic chemicals in settings where life could have started. Also, some of the metal proteins incorporate very hydrothermal-looking metals: Cn, Zn, Mo. This is surely circumstantial evidence that the hyperthermophiles did start in hydrothermal systems. Perhaps the hydrothermal setting was more than a bottleneck our ancestors passed through: perhaps it was our birthplace.

Jakosky: Jack Farmer and Everett Shock have both referred to the possible role of impacts in the early earth history as providing a filter through which species will have to go through, effectively selecting for the hyperthermophiles. That is not clearly required. The uncertainties in the impact rate are such that the impacts can die off early enough that you can still form life in whatever type of environment you want, without forcing it through the hyperthermophiles.

Pentecost: You mentioned that the upper temperature for the hyperthermophiles so far discovered is 113 °C. Do we have any theoretical basis for an absolute upper temperature for life?

Stetter: We and others have found that macromolecules are no longer stable at temperatures of 200–250 °C (Trent et al 1984, Bernhardt et al 1984). As a guess, I would say that 150–160 °C might be possible. However, organisms living at these temperatures wouldn't be able to use small molecules such as ATP, which have easily hydrolysed phosphates, and would have to find alternatives. In principle, macromolecules are stable to 180–200 °C. There really is no way an organism could exist above that temperature range.

Cowan: Just to build on that, the stability of enzymes and proteins, although critical, is by no means the only important effect. One can take already stable hyperthermophilic proteins and enzymes and enhance stability by site-directed mutagenesis or immobilization. In this way, they can be stabilized to 130–140 °C and probably rather higher. However, this is conformational stabilization, and one inevitably strikes the problem of chemical stability. This problem occurs at very much lower temperatures when you start dealing with small molecules, intermediates and co-factors. Anyone who has ever carried out an NADPH-dependent assay at 85 °C in a cuvette knows very well that you're pushing the upper limit of chemical stability.

Barns: That is true, but although we tend to think of the insides of cells as being like cuvettes, they're not; they are very highly concentrated with respect to solutes, macromolecules and so forth. Consequently, probably none of these molecules exists free in solution, but rather they are all in complexes with proteins and nucleic acids. It has been proposed that small molecules such as NADH and ATP may be stabilized by being maintained in such complexes.

Shock: I would like to emphasize that nature and the laboratory can be very different. An experiment you can conduct in the lab might not accurately represent the conditions of a hydrothermal system. If you manage to cause macromolecules or other molecules to decompose at an elevated temperature, you must be absolutely convinced that you have controlled everything in your experiment to represent what can go on in a hydrothermal system. This has rarely if ever been done in any experimental work. Consequently, it is rather early to be claiming or advocating any particular upper temperature for these sorts of reactions. I would also make the same claim about calculations: you have to be just as careful looking at somebody's calculation of the temperature at which something is or isn't stable, and ask yourself if the calculation includes all the aspects of the real natural system that are quantifiable.

Henley: Some 10 years ago, Roger Summons and I opened some fluid inclusions that are formed naturally in veins at temperatures from 400–550 °C. Astonishingly, we found high molecular weight steroids and hopanes. So we have the evidence that these things can be preserved. The veins themselves were formed 1.8 Ga ago in the outer margins of granites. We felt that what was happening there was that, because the residence times of the hydrocarbons in the vicinity of the granite were short, the hydrocarbons were able to remain intact.

A frequent problem with calculations is that they're made on the assumption that the chemical system rushes to equilibrium. That is not the case, otherwise we wouldn't be here today.

Davies: How relevant is pressure as a parameter here? In a very high pressure system, can you inhibit the molecular dissociation pattern? That is, can we imagine that some of these organisms might exist at 200 °C by living sufficiently deep, so that the pressure stabilizes the dissociation?

Stetter: This was one of the first things we studied with our cultured hyperthermophiles. Pressure seems to make no difference to the upper temperature boundary. If anything, the upper temperature limit decreases a little at higher pressures.

Cowan: Extrapolating that to studies done *in vitro*, and taking into account all the caveats concerning extrapolating *in vitro* experiments to *in vivo* situations, quite a lot of work has been done on pressure effects on macromolecules. At intermediate pressures you can get significant stabilization. The same destabilization effect that Professor Stetter just referred to has recently been demonstrated by Cavaille & Combes (1995), so I don't think that pressure is necessarily a feasible mechanism for stabilizing macromolecules to very high temperatures.

Russell: Are iron–sulfur proteins always present in the membranes of hyperthermophiles?

Stetter: I'm not an expert on iron–sulfur proteins, but I know from other investigators that they are present in many of these organisms.

Russell: Do they use Fe(III) complexes as electron acceptors, rather than nitrate and sulfate?

Stetter: We have tried to find evidence for this, so far without success. Now we have a new hyperthermophile that is an anaerobic iron oxidizer, but this requires nitrate as an acceptor and not ferric iron.

Summons: I'm interested in molecular fossils. You showed the distinctive 'golf club' morphology of *Thermoproteus tenax* (Fig. 1A, p 6), which seems to be one of the few structures you could recognize in a fossil. What about fossil lipids? Is there anything distinctive about these organisms that would leave a long-lasting chemical record?

Stetter: They have phytanyl ether lipids. A special feature of *Methanopyrus* occurring in the deepest phylogenetic branch of archaea is that they have 2,3-di-*O*-geranylgeranyl-sn-glycerol in the membrane.

Summons: Are these lipids also common to methanogens that live at normal temperatures?

Stetter: No. They have phytanyl ethers, but never the non-reduced form with double bonds intact. This unsaturated form is unique to the methanogens that grow in black smokers.

Knoll: My question isn't really about the origin of life, but something that might be just as important, which is the long-term maintenance of life. Ecosystems on Earth today are characterized by complementary metabolisms that end up cycling important elements. Have you identified complementary physiologies among hyperthermophiles that would allow you to construct complete biogeochemical cycles for carbon, nitrogen and sulfur?

Stetter: There are complete carbon and sulfur cycles. However, for nitrogen we have nitrate reducers all the way through to ammonia, but nitrification is missing. This seems to have been invented much later.

Walter: Karl Stetter, looking at your tree of life, where are the hyperthermophilic stem eukaryotes?

Stetter: That is a question I'm concerned about, of course. It is a very long lineage to the eukaryotes, and no deep branches are so far known. I'm also very interested in finding the upper temperature border for eukaryotes; I'm currently hunting for them in high temperature environments. Our results are not very spectacular so far, but we have an amoeba from Yellowstone park which grows at about 60 °C which we can cultivate in the laboratory. I have also brought up some rocks from black smokers which were very hot on the surface, roughly between 90–100 °C. When I examined these under the microscope they contained amoeba-like organisms and possibly flagellates, but we couldn't culture them.

Barns: It is important to point out that all of the deepest branches in the eukaryotic tree that we know of are occupied by pathogens such as *Giardia* and *Trichomonas*. We only know of these organisms because they make us sick, so we probably haven't sampled widely enough to know if deeply branching, hyperthermophilic eukaryotes exist. Microbiologists have been looking for eukaryotes in high temperature environments mostly using cultivation methods, and so far the upper temperature limit is 63 °C for a fungus that Mike Tansey isolated from Yellowstone 20 years ago. This organism is neither a true hyperthermophile nor a deeply branching lineage. My thesis project was basically to use rRNA-based molecular techniques to look for hyperthermophilic eukaryotes, but I didn't find any in my study. This may be due to sampling the wrong environments, using the wrong molecular techniques, or just not looking hard enough. Or perhaps ancient hyperthermophilic eukaryotes, if they ever existed, may have been outcompeted by prokaryotes in high temperature environments and become extinct.

Alternatively—and this is almost pure speculation on my part—it may be that the modern archaea *are* the thermophilic eukaryotes. One of the striking things about the 16S rRNA tree is the difference in the lengths of the lineages that go to the different domains: the line that leads to the eukaryotes is very long, while those among the thermophilic archaea are quite short. The lineages of extant mesophilic (low temperature) organisms that are derived from hyperthermophilic lineages in the archaea also tend to be quite long, suggesting that life at lower temperatures promotes (or allows) rapid sequence evolution. Perhaps this is what has caused the long line segment leading to the eukaryotes: that is, the seminal event in shooting off the eukaryotes from the ancestral line leading from the bacteria to archaea was adaptation to lower temperature environments. Thus, if this speculation is true, we won't find any hyperthermophilic lineages along the eukaryotic line of descent.

Nisbet: Supplementing that, the heat shock proteins in the eukaryotic nucleus have affinities to heat shock proteins in the archaea, whereas the heat shock proteins in mitochondria and chloroplasts are very directly related to those of bacteria.

Shock: How many strictly anaerobic eukarya are there?

Barns: Only a few are known. Interestingly, these occupy the deepest branches in the tree.

Shock: In the absence of an oxygenated atmosphere, all conditions above about 60 °C are going to be highly anaerobic in any submarine hydrothermal systems because of the mixing of the fluids. *Sulfolobus* and similar organisms are now found in continental settings because the highly oxygenated ground waters bring oxygen in and they can actually do oxidation. But in a hydrothermal system in the absence of an oxygenated atmosphere, or even in the modern submarine system separated from the atmosphere, there simply cannot be hyperthermophilic oxidation using O_2—there's not enough to make it an energy efficient system. In order to find the most primitive eukarya you would have to follow up strictly anaerobic organisms; things which even the slightest whiff of oxygen will send way below detection limits.

Stetter: Among the eukaryotes there are strict anaerobes without mitochondria which are named *archezoa* and which are known from the intestines of higher organisms (Cavalier-Smith 1993). They have solved the problem of getting rid of hydrogen formed during metabolism by employing endosymbionts. These may be methanogens located either in the cytoplasm or around or even inside the nucleus and which form methane from H_2 and CO_2.

Shock: Do all these strictly anaerobic eukaryotes rely on methanogens or other symbiotic organisms?

Stetter: This has been poorly studied.

References

Baldauf SL, Doolittle WF, Palmer JD 1996 The root of the universal tree and the origin of eukaryotes based on elongation factor phylogeny. Proc Natl Acad Sci USA, in press

Baross JA, Deming JW 1983 Growth of 'black smoker' bacteria at temperatures of at least 250 °C. Nature 303:423–426

Bernhardt G, Luedemann H-D, Jaenicke R, Koenig H, Stetter KO 1984 Biomolecules are unstable under black smoker conditions. Naturwissenschafeten 71:583–586

Brown JR, Doolittle WF 1995 Root of the universal tree of life based on ancient aminoacyl-transfer RNA synthetase gene duplication. Proc Natl Acad Sci USA 92:2441–2445

Cavaille D, Combes D 1995 Effect of temperature and pressure on yeast invertase stability: a kinetic and conformational study. J Biotech 43:221–228

Cavalier-Smith T 1993 Kingdom Protozoa and its 18 phyla. Microbiol Rev 57:953–994

Cragg BA, Parkes RJ 1994 Bacterial profiles in hydrothermally active deep sediment layers from Middle Valley (N. E. Pacific) Sites 857 and 858. Proc Ocean Drilling Prog, Sci Res 139:509–516

Doolittle WF, Brown JR 1994 Tempo, mode, the progenote and the universal root. Proc Natl Acad Sci USA 91:6721–6728

Gogarten JP, Kibak H, Dittrich P et al 1989 Evolution of the vacuolar H^+-ATPase: implications for the origin of eukaryotes. Proc Natl Acad Sci USA 86:6661–6665

Iwabe N, Kuma K, Hasegawa M, Osawa S, Miyata T 1989 Evolutionary relationship of archaebacteria, eubacteria, and eukaryotes inferred from phylogenetic trees of duplicated genes. Proc Natl Acad Sci USA 86:9355–9359

Shock EL 1996 Hydrothermal systems as environments for the emergence of life. In: Evolution of hydrothermal ecosystems on Earth (and Mars?). Wiley, Chichester (Ciba Found Symp 202) p 40–60

Sleep NH, Zahnle K, Kasting JF, Morowitz HJ 1989 Annihilation of ecosystems by large asteroid impacts on the early Earth. Nature 342:139–142

Trent JD, Chastian RA, Yayanos AA 1984 Possible artefactual basis for apparent bacterial growth at 250 °C. Nature 307:737–740

General discussion I

'Beehive' diffusers

Russell: I would like to recall Prigogine & Stengers' (1984) famous statement about life being the supreme expression of self-organizing processes. Life is a dissipative structure generated in far-from-equilibrium conditions. Black smokers are also far from equilibrium. Therefore we might expect to find other types of dissipative structures associated with them as well. After all, dissipative structures are normally coupled to each other. I would like to describe one example, the 'beehive' diffuser, that may be tangentially significant.

Beehive structures occur on the Mid-Atlantic Ridge, at 23 °N (Fouquet et al 1993). I would like to show Rickard et al's (1994) model of how they work. They are self-organizing structures that may be significant in the origin of life. I don't think they are, personally, but I think we should remind ourselves of them. Black smokers attain temperatures of 350 °C or more and their chimney formation depends on the inverse solubility of calcium sulfate in sea water. They are a kind of chemical and thermal exchanger — Rickard et al (1994) call them diffusers — and as the sea water is entrained into the black smoker via the *venturi* effect, calcium sulfate precipitates, and the escaping black smoker water precipitates out sulfides and oxides. It's just conceivable (although the oceans were probably too hot and too reducing, and therefore had too little sulfate to allow beehive growth in the Hadean) that life could have emerged at the interface between relatively oxidized and highly reduced minerals and, as Everett Shock is going to show in his paper, mixing of hydrothermal solution with sea water affords a large amount of energy. Be that as it may, these could be places to look for bacteria.

Cowan: In my laboratory we have started looking in deep sea hydrothermal vent systems for one of the less exciting thermal niches — the aerobic thermophilic niche. One of our target sites is the diffuser structures, because the entrainment of highly oxygenated water and mixing with the anaerobic heated water yields temperature ranges which are appropriate for mid-range thermophiles, about 50–80 °C. We have also found a biotope where similar entrainment of cold oxygenated water occurs in the diffuse flow areas around the periphery of black smokers. These regions support substantial populations of aerobic thermophiles.

Stetter: Three years ago I had the chance to sample these 'beehives' at both 'TAG' and 'snakepit' sites at the mid-Atlantic Ridge. They are full of hyperthermophiles of the type we find in the walls of the black smokers. The temperatures are as high as 100 °C in several areas. The beehives harbour very complex ecosystems of aerobic and anaerobic hyperthermophiles.

Cady: Assuming that you have continuous outward growth via mineralization of the beehive structures, have you observed any preserved microbiota that may have existed at the mineral–water interface?

Russell: No, beehive diffusers are not found preserved in ore bodies because, as we might expect for dissipative structures, they are ephemeral. As soon as they cool down the calcium sulfate dissolves and they fall apart.

How much oxygen was there in the Archaean atmosphere?

Nisbet: I'd like to go back to the oxygen problem. In the Mid Archaean there is moderate consensus that oxygen levels in the atmosphere were very low. But in the Hadean to the earliest Archaean, there's a substantial chance there was massive loss of hydrogen from the top of the early atmosphere. If you look at the work of Yung et al (1989) on D/H ratios, possibly a kilometre or so of the ocean might have been lost. That leaves you with a kilometre of oxygen — 100 bars — to get rid of. Yet the record of life seems to be telling us that oxygen levels were pretty low by the time life started. How much oxygen did the most deeply rooted organisms need?

Stetter: Kenneth Towe (1994) thinks that there could have been oxygen locally, at least in the range of 100 ppm. There's no doubt that the atmosphere was generally reducing, but locally there could have been limited amounts of free oxygen available. Oxygen could have been formed permanently by the splitting of water molecules into H_2 and O_2 by high energy radiation. Therefore, oxygen-using microorganisms could have evolved on Earth already rather early. Then, much later (about 2 Ga ago) when the big 'oxygen explosion' occurred, most of the oxygen users known today evolved.

Nisbet: But several processes are occurring here. After photosynthesis arrives (presumably moderately late on) there is production of oxygen, and you can get local concentrations of oxygen. But the atmospheric lifetime of oxygen today is 10–20 Ma: if we shut-off all photosynthesis it will probably be gone in about 20 Ma. In the Hadean and very earliest Archaean, say about 4.3–4.0 Ga ago, there was the possibility of making oxygen by losing hydrogen into space, but this presumably declined as the planet cooled. Somewhere around 4 Ga ago we've got the problem of no obvious source of oxygen, and yet some of these biological processes do seem to require a little bit of oxygen.

Giggenbach: Why does H_2 loss produce elemental oxygen?

Nisbet: Well, probably OH: I didn't specify. If the atmosphere is warm enough to put water vapour up into the high atmosphere, photodissociation occurs, hydrogen is lost to space and you are left with a surplus of oxygen. As I said, the D/H ratio evidence suggests that up to a kilometre of ocean might have been lost. It would have produced a supply of oxygen to the air, which would then eventually have passed, via the oceans, to the subduction zones. That would also tally with the oxidation state of the mantle, which is perhaps a little bit higher than it might have been.

Sakai: The D/H ratio of the present ocean is 0‰, which represents fairly well that of the whole crustal water, whereas the hydrogen from the upper mantle as measured in mid-ocean ridge basalt, for instance, shows much lighter D/H ratios of −60‰ or so. This discrepancy has been interpreted as due to the preferential loss of light hydrogen into space, as mentioned by Euan Nisbet. However, some workers are now claiming that the basaltic value does not represent the original value but is instead the result of preferential degassing of heavy water from ascending magma. If they are correct, the D/H ratios of the mantle hydrogen could be much closer to zero, and the amount of water lost into space could have been smaller than previously thought.

Des Marais: The issue of hydrogen lost to space is very important. My view is that the origin of life created a tremendous sink for hydrogen. As soon as the biosphere was present, I think that the hydrogen partial pressure was very low. Possibly some of this hydrogen might have been converted to methane, which would then have lost its hydrogen to space. It would be important to have an estimate from the microbiological research community as to the percentage of hydrogen going into these hyperthermophilic ecosystems that enters the atmosphere as methane versus hydrogen sequestered in the crust in organic material and other non-volatile products of biosynthesis. This would give the atmospheric modellers a constraint about this hydrogen loss to space, which is very important for interpreting the history of the redox balance in the crust and mantle.

Jakosky: I thought that even before the biosphere arose, photodissociation of CO_2 would have produced small amounts of oxygen.

Henley: One of the problems that I face whenever there's a discussion about levels of oxygen is that I'm not quite sure what levels of oxygen people actually want to believe. Moreover, if there is a need to express the reducing nature of an environment, it is more useful to indicate this as a concentration of hydrogen.

Nisbet: The old geology text books all say that there was a reducing atmosphere in the early Archaean. But the support for this idea is actually pretty thin. The idea is biologically driven, by the notion that we needed these conditions as a precursor for life, as in the Miller experiment. We haven't really got any decent geological handle on the amount of oxygen that was around. The only potential new lead on the problem might be to look at what the most deeply rooted bacteria seem to be able to tolerate.

Cowan: The most deeply rooted examples are all obligate anaerobes, to many of which oxygen is extremely toxic. From the biological point of view, for a simple biochemical system a compound as toxic as oxygen poses problems. It becomes necessary to evolve complex mechanisms to try and protect sensitive biochemistry from that toxicity.

Nisbet: But then you run up against the loss of hydrogen to space: maybe this toxic stuff was indeed around?

Stetter: Not all of the deepest branches of hyperthermophiles are obligate anaerobes. We have recently discovered *Pyrobaculum aerophilum*, an aerobic hydrogen oxidizer closely related to other species of *Pyrobaculum* which are strictly anaerobic sulfur reducers (Völkl et al 1993). Our latest isolate is *Pyrolobus*, a relative of the strict anaerobe *Pyrodictium*, that is able to use oxygen in the ppm range.

Parkes: You've got the same problem, in the sense that the anaerobes may have been the ancient ancestors, and a few strains have been exposed to oxygen, tolerate it, and probably get some small energetic gain from using it.

Stetter: They use it, but at incredibly low concentrations. With routine procedures you would not detect this and you would classify the organism as a strict anaerobe.

Des Marais: All organisms today are highly evolved. Even in these lineages of archaea you might expect that some organisms had evolved the ability to use oxygen sometime later in history.

Shock: It is important to remember that in a hydrothermal system, the hydrogen content of the fluid is controlled by reactions between water and minerals. It doesn't necessarily have any connection to the atmospheric composition. The important reaction is the disproportionation of water to oxygen and hydrogen. The hydrogen stays in the fluid, and the oxygen gets buried into iron oxide and silicate minerals.

Knoll: Given the nature of the Archaean geological record and the sensitivities to oxygen of some of the minerals that have been used to try to gauge early oxygen levels, it may be that our best sense of what oxygen was like at the time of early evolution will come from biology. What we think about the nature of early life based on these phylogenetic trees is highly dependent on how well we think we've sampled the tree as it really exists, rather than the tree that we draw. Are we confident that we have a good enough sampling of archaea and bacteria, so that we really know where organisms like *Aquifex* sit in the phylogeny of their domains?

Stetter: That is a good question. The 16S rRNA sequences used to plot these phylogenic relationships represent only about one-thousandth of the whole genome of these organisms. The assumption is made that the 16S rRNA genes, which have 1500 base pairs, represent the evolution of the organisms, which is not self-evident and therefore other control experiments are important. Total genome sequencing for one of these organisms takes only three to four months, and this should give a final answer to the question of phylogeny. Despite this, I'm rather confident that the phylogeny based on 16S rRNA is fairly accurate at least in most cases. For example, *Archaeoglobus* is an archaeal sulfate reducer. It was surprising that Woese et al (1991) found that its 16S rRNA sequence placed it in the midst of the methanogens. However, new data have shown that it is 'almost' a methanogen — it contains most of the unique co-enzyme equipment typical of methanogens, although it has a different end product, H_2S (Möller-Zinkhan et al 1989). Consequently, I think that using 16S rRNA is a good approach, but there could be exceptions where it does not accurately reflect real relationships; we'll have to wait and see.

Barns: Even though 16S rRNA is a single molecule and may not reflect the evolution of the entire organism, the fact that it was present in the last common ancestor of life and has evolved in the separate lineages of life gives us an estimate of life's diversification. It is a molecular signature for extant organisms. Comparison of the rRNA sequences that we have obtained from organisms that are available in culture with the diversity of sequences that we can pull from organisms in the natural environment strongly indicates that we so far have only a very small sampling of the diversity of microorganisms. As we get more of these sequences, we will fill out the tree better. When we get these organisms into culture we will have a better idea of their common properties and that will give us a better handle on the nature of the last common ancestor.

References

Fouquet Y, Wafik A, Cambon P, Mevel C, Meyer G, Gente P 1993 Tectonic setting and mineralogical and geochemical zonation in the Snake Pit sulfide deposit (MidAtlantic Ridge at 23° N). Econ Geol 88:2018–2036

Möller-Zinkhan D, Börner G, Thauer RK 1989 Function of methanofuran, tetrahydromethanopterin, and coenzyme F_{420} in *Archaeoglobus fulgidus*. Arch Microbiol 152:362–368

Prigogine I, Stengers I 1984 Order out of chaos. London, Heinemann.

Rickard D, Knott R, Duckworth R, Murton B 1994 Organ pipes, beehive diffusers and chimneys at the Broken Spur hydrothermal sulphide deposits, 29° N MAR. Mineralog Mag 58:774–775

Towe KM 1994 Earth's early atmosphere: constraints and opportunities for early evolution. In: Bengston S (ed) Early life on Earth. Columbia University Press, New York (Nobel Symp 84) p 36–47

Völkl P, Huber R, Drobner E et al 1993 *Pyrobaculum aerophilum* sp. nov., a novel nitrate-reducing hyperthermophilic archaeum. Appl Environ Microbiol 59:2918–2926

Woese CR, Achenbach L, Rouviere P, Mandelco L 1991 Archaeal phylogeny: reexamination of the phylogenetic position of *Archaeoglobus fulgidus* in light of certain composition-induced artifacts. System Appl Microbiol 14:364–371

Yung YL, Wen J-S, Moses JI, Landry BM, Allen M 1989 Hydrogen and deuterium loss from the terrestrial atmosphere: a quantitative assessment of non-thermal escape fluxes. J Geophysical Res 94:14971–14989

Phylogenetic perspective on microbial life in hydrothermal ecosystems, past and present

Susan M. Barns*, Charles F. Delwiche, Jeffrey D. Palmer, Scott C. Dawson, Karen L. Hershberger and Norman R. Pace

Department of Biology and Institute for Molecular and Cellular Biology, Indiana University, Bloomington, IN 47405, USA

Abstract. Understanding hydrothermal ecosystems, both past and present, requires basic information on the types of organisms present. Traditional methods, which require cultivation of microorganisms, fail to detect many taxa. We have used phylogenetic analyses of small subunit rRNA sequences obtained from microorganisms of a hot spring in Yellowstone National Park to explore the archaeal (archaebacterial) diversity present. Analysis of these sequences reveals several novel groups of archaea, greatly expanding our conception of the diversity of high temperature microorganisms, and demonstrating that hydrothermal systems harbour a rich variety of life. Many of these groups diverged from the archaeal line of descent early during evolution, and an understanding of their common properties may assist in inference of the nature of the last common ancestor of all life. The data also show a specific relationship between low-temperature marine archaea and some hot spring archaea, consistent with a thermophilic origin of life. Future use of rRNA-sequence-based techniques in exploration of hydrothermal systems should greatly facilitate study of modern thermophiles and give us insight into the activities of extinct communities as well.

1996 Evolution of hydrothermal ecosystems on Earth (and Mars?). Wiley, Chichester (Ciba Foundation Symposium 202) p 24–39

Understanding the evolution of hydrothermal ecosystems requires information on the diversity and activities of extant thermophilic microorganisms. Although a limited fossil record does exist for early microbes, it is of little help in identifying organisms in terms of their relatedness to present-day species. This is because, unlike macroscopic organisms, morphological features are generally uninformative regarding genetic relationships between microbes. Since information on the basic identities of ancient organisms is lacking, it is problematic to reconstruct their activities in ancient communities. Therefore, it is useful to turn our attention to the study of modern thermophilic

*Present address: Environmental Molecular Biology, M888, Life Sciences Division, Los Alamos National Laboratory, Los Alamos, NM 87545, USA.

microorganisms in an attempt to understand better the ecology and diversity of microbial life in hydrothermal systems and the evolutionary processes that have led to it.

Unfortunately, our current understanding of microbial life at high temperatures—or indeed in any type of environment—is quite limited. This is largely due to the traditional requirement for laboratory cultivation for the discovery or characterization of microorganisms. As was recognized even by the earliest microbiologists (summarized in Ward et al 1992), only a small fraction of species present in any given environment are recovered in routine laboratory cultures, presumably due to the difficulty of reproducing the complex physical and nutritional requirements of the organisms in the laboratory. Those species whose growth needs are not met are not obtained for study and thus remain unknown to us. In order to overcome this impediment to discovery and investigation, our laboratory and others have worked at developing methods for identifying and characterizing microorganisms in the environment without the need for laboratory cultivation. These techniques revolve about the use of ribosomal RNA (rRNA) sequences obtained directly from microbes in environmental samples without first cultivating the organisms. The sequences can be used in (at least) two ways to gain information regarding the organisms from which they are obtained: for phylogenetic inferences, and as identifiers for the organisms.

Ribosomal RNAs are present in all known organisms, and hence were present in the last common ancestor of all life. Thus, the history of the molecule (as recorded in the changes in its sequence) can be used to trace the history of the organism. Comparison of the rRNA sequences derived from different organisms gives a measure of the evolutionary relatedness of those organisms: closely related organisms have more similar sequences than do distantly related species. Such sequence comparisons can be used to construct relatedness (phylogenetic) trees which are maps of evolutionary relationships between organisms. When an rRNA sequence obtained from an otherwise unknown organism in an environmental sample is found to be highly similar to that of a previously studied species, it is likely that the two organisms will share many other properties as well. Conversely, if an environmental rRNA sequence has no known close relative, it derived from a novel organism with possibly unique biochemistry. In addition, rRNA sequences retrieved from uncultivated organisms, when analysed with other available sequences, help to complete the picture of the phylogenetic diversity of all life, giving us a more comprehensive understanding of the course of evolution.

In a second application, rRNA sequences can serve as 'molecular signatures' for organisms. Through the use of hybridization probes specific for particular rRNA sequences, organisms can be identified in environmental samples and laboratory cultures. Such probes can be used to study the activities and interactions of microbes in their natural settings, allowing *in situ* study of their morphology, interactions, movement in response to environmental variables, mineral deposition, etc. (Amman et al 1990, Poulsen et al 1993, Reysenbach et al 1994, Stahl et al 1988). Probes have also proven useful in the laboratory cultivation of microorganisms (Huber et al 1995, Kane et al 1993), facilitating the isolation of organisms to permit in-depth study of their physiology.

We have analysed small subunit rRNA sequences obtained from the archaeal (archaebacterial) microbial community of Obsidian Pool hot spring (75–93 °C) in Yellowstone National Park. Phylogenetic trees constructed with these sequences indicate the presence of a surprising array of previously unknown organisms, indicating that life in hydrothermal systems may be quite evolutionarily and physiologically diverse. Comparison of these sequences with those obtained from archaea of lower temperature environments supports the theory of a thermophilic origin of prokaryotes. In addition, many of the Obsidian Pool sequences branch more deeply in the archaeal phylogenetic tree than do those of most previously known organisms. Now that the existence of these organisms has been revealed, further study of their activities and common properties should shed light on the nature of Earth's earliest thermophilic organisms and their communities.

Methods

Methods for obtaining and analysing small subunit rRNA sequences from Obsidian Pool have been described previously (Barns et al 1994). Briefly, cells were collected in sediment samples and treated to liberate their DNA. This mixed-community DNA was then purified from contaminating sediment compounds in order to allow subsequent enzymic steps to be performed. Polymerase chain reaction (PCR) amplification, employing one primer specific for the sequences of archaeal (and eukaryotic) small subunit rRNA genes (rDNAs) and one universal primer capable of amplifying rRNA genes from all known organisms, was then used to selectively copy rRNA genes from the DNA mixture. This step allowed exponential amplification of the target genes from small ($< 10^{-6}$ g) amounts of mixed starting DNA, to produce usable amounts of DNA highly enriched for archaeal rRNA genes of the original microbial community. Individual rRNA gene copies were then separated from the mixture by cloning into plasmid vectors and transformation into bacterial host strains, to produce cell lines each of which contained an rRNA gene copy from just one organism in the starting population. This process creates a clone 'library' of the rDNAs of the organisms in the Obsidian Pool community. These rDNAs were then sequenced, and the sequences used in phylogenetic comparative analyses to produce relatedness trees between the unknown organisms and previously characterized ones. Analysis of Obsidian Pool sequences and those of some cultivated archaeal species was used to infer the tree of Fig. 1. Because many of the recovered sequences did not appear to be closely related to any previously known species, a more extensive phylogenetic analysis was performed on eight representative Obsidian Pool sequences together with those from 56 additional archaeal, bacterial and eukaryotic taxa to produce a universal tree of life (Fig. 2).

Results and discussion

Analysis of Obsidian Pool rRNA sequences reveals a wide diversity of archaea

Screening of 239 clones by partial sequencing produced a collection of 32 distinct archaeal rDNA sequence types ('phylotypes'). Phylogenetic analysis of these sequences

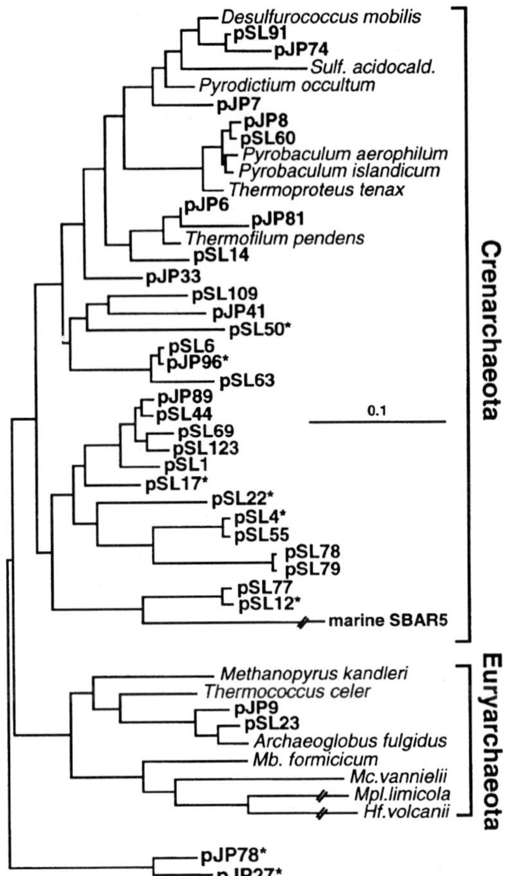

FIG. 1. Phylogenetic tree of rDNA clones from Obsidian Pool, illustrating the wide diversity of archaea present in the community. Sequences recovered from sediment organisms (designated 'pJP' and 'pSL') were analysed with sequences from cultivated archaeal species (in italics) obtained from the Ribosomal Database Project database (Maidak et al 1994). Both full (approx. 1300 nucleotide) and partial (approx. 500 nucleotide) sequences were used in the analysis. 'Marine SBAR5' sequence is that of clone SBAR 5 recovered from Pacific marine bacterioplankton by DeLong (1992). Tree was inferred by maximum likelihood analysis and is arbitrarily rooted in the pJP27/pJP78 lineage. The subdivision of the archaeal domain into the kingdoms Crenarchaeota and Euryarchaeota is indicated. Scale bar represents 10 mutations per 100 nucleotide sequence positions. Sequences of clones designated with asterisks were used in the three domain analysis of Fig. 2. Trees presented in Figs 1 and 2 are based on analyses by Barns et al (1996).

together with those from previously described species revealed a surprising diversity of novel phylotypes (designated by 'pJP' and 'pSL' in Fig. 1). Remarkably, approximately 41% of the 32 phylotypes recovered were present in only one clone each out of the 239 clones screened. This indicates that considerably greater diversity of phylotypes is present in this hot spring community than was detected by this analysis. More

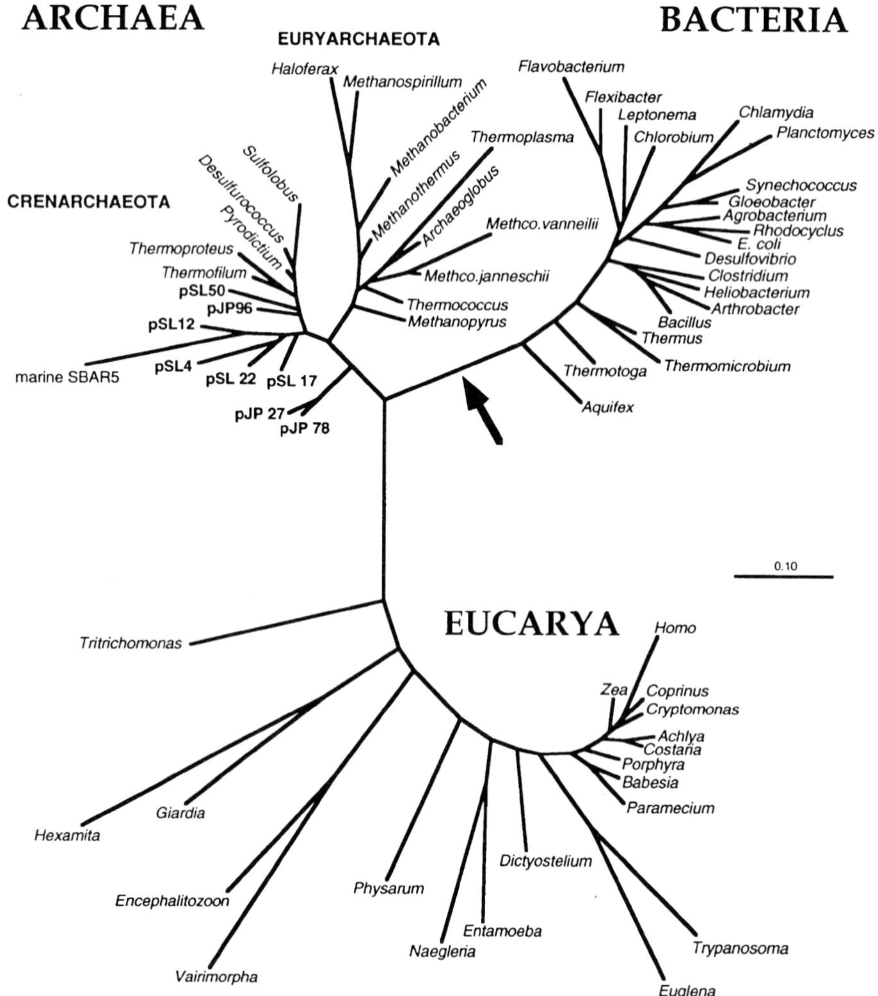

FIG. 2. 'Universal' unrooted phylogenetic tree showing positions of major groups of sequences recovered from the Obsidian Pool community. Tree was inferred by maximum likelihood analysis of 922 nucleotides of rDNA sequence from each organism or clone. 64 taxa from the domains archaea, bacteria and eucarya were analysed. The root of the tree, as inferred by analysis of duplicated gene sequences (Iwabe et al 1989, Gogarten et al 1989), is indicated with an arrow. Scale bar represents 10 mutations per 100 nucleotide sequence positions.

extensive screening of clones probably would reveal many additional sequences of previously unknown organisms.

Previous analyses of rRNA sequences of cultivated species indicated that the archaeal domain could be phylogenetically subdivided into two groups, designated the kingdoms Crenarchaeota and Euryarchaeota (Woese et al 1990). As can be seen in

Fig. 1, most of the sequences obtained from Obsidian Pool affiliate with the crenarchaeal kingdom. Representatives of all previously known orders of Crenarchaeota are present in this hot spring. Sequences affiliated with the *Desulfurococcus*/*Pyrodictium* clade (pJP7, pJP74, pSL91), *Pyrobaculum* sp. (pJP8, pSL60) and *Thermofilum pendens* (pJP6, pJP81) were recovered, but none was identical to any available sequence from a cultivated organism. In addition, two sequences of euryarchaeal affiliation (pJP9 and pSL23) were obtained, both highly similar to that of *Archaeoglobus fulgidus*, a genus previously known only from isolates of marine origin (Burggraf et al 1990).

Most of the sequences obtained from Obsidian Pool, however, did not affiliate closely with any previously studied organism (sequences pSL63 to pSL12, Fig. 1). Addition of these sequences to the archaeal phylogenetic tree changes our conception of the phylogeny and diversity of that domain. Based on previous sequence analysis of cultivated species, the crenarchaeal kingdom was thought to comprise a distinct, closely-related group of a few genera, separated by a large phylogenetic distance from their closest relatives in the Euryarchaeota (Woese et al 1990). The addition of the many deeply-branching lineages recovered from this hot spring substantially fills in that distance, blurring the distinction between the kingdoms (Figs 1 and 2). These sequences, presumably derived from thermophilic microbes, also greatly expand the known evolutionary diversity of high temperature organisms, demonstrating that hydrothermal environments may be as rich in diversity as lower temperature ones.

Analysis of low temperature Crenarchaeota supports a thermophilic origin of life

This study also identified close, probably thermophilic relatives (pSL12 and pSL77) of recently discovered, abundant Crenarchaeota from diverse coastal and open ocean sites in the North Pacific and Antarctic (this group is represented by sequence 'marine SBAR5' in Figs 1 and 2; DeLong 1992, DeLong et al 1994, Fuhrman et al 1992). The cold-water source of these marine sequences, together with features of their rDNAs, indicates that the organisms contributing these sequences, in remarkable contrast to all other known Crenarchaeota, are probably not thermophilic. Previous analyses were unable to resolve the phylogenetic position of these marine rRNA sequences within the Crenarchaeota, but indicated that they constitute a deeply branching group. The present analyses show that they group specifically with Obsidian Pool sequences pSL12 and pSL77. The affiliation of the marine SBAR5 sequence with the hot spring sequences, together with their nested position within other thermophilic lineages, imply that the low temperature marine archaea are descendants of ancestral thermophiles. Preliminary analyses of additional crenarchaeal rDNA sequences obtained from low temperature freshwater lake sediments place these sequences also as peripheral relatives of thermophilic organisms (analyses not shown). The uniform occurrence of thermophilic lineages near the base of the archaeal tree clearly implies that the common ancestor of the archaea was thermophilic, and that lower-temperature species derive from it. Placement of hyperthermophilic bacteria (*Aquifex*, *Thermotoga*, *Thermus*, *Thermomicrobium*) as the deepest branches of that domain in rRNA analyses

(Fig. 2) also is consistent with the theory of a high-temperature origin of life (Woese 1987).

Deep branches in the tree of life may help us understand primordial thermophiles

In an effort to better resolve the placement of some of the deeply-branching Obsidian Pool sequences, we included eight of them in a phylogenetic analysis of sequences from all three domains of life, archaea, bacteria and eucarya (Fig. 2). As was seen in Fig. 1, all of these sequences branch more deeply in the archaeal tree than do those of cultivated Crenarchaeota. Two of the sequences, pJP27 and pJP78, are seen to branch before the bifurcation of the Crenarchaeota and Euryarchaeota kingdoms under some conditions of analysis (Fig. 2), and may therefore constitute yet a third, previously unknown kingdom of archaea (Barns et al 1996). Phylogenetic analyses of duplicated genes indicate that the root of the universal tree of life, the position of the last common ancestor of all life, lies on the bacterial lineage (indicated by arrow in Fig. 2; Iwabe et al 1989, Gogarten et al 1989, Brown & Doolittle 1995, Baldauf et al 1996). If this is the case, these Obsidian Pool sequences constitute some of the earliest-branching lineages yet discovered in the tree of life. Preliminary phylogenetic analyses of sequences obtained from the bacterial community of Obsidian Pool indicate that deeply branching thermophilic bacteria also are present in this hot spring (P. Hugenholtz, unpublished work 1995). Efforts are currently underway to cultivate these novel Obsidian Pool organisms, using rRNA-sequence based probes to assist isolation (Huber et al 1995). Once cultivated, further study of the common physiological properties of these anciently-diverged lineages of bacteria and archaea should help us to infer characteristics of Earth's earliest thermophilic organisms.

Microbial life in modern hydrothermal systems is clearly phylogenetically diverse, and this flowering of diversity probably began quite early during evolution. Although currently we are only at the stage of discovering and exploring this diversity, molecular sequence-based tools now available will greatly enhance our ability to detect and study the tremendous variety of organisms present in high temperature environments. Phylogenetic analysis of sequences recovered from such organisms will tell us about the evolutionary relationships between organisms present and can be used to refine our understanding of the history of all life. And sequence-based probes can be used to study organisms in laboratory cultures and the environment, permitting in-depth research on their physiology and community interactions. As we gain insight regarding the nature of life at high temperatures today, we come closer to understanding such life in the past.

Acknowledgements

We gratefully acknowledge Ed DeLong, Mitchell Sogin and Robin Gutell for providing aligned sequence data; Carl Woese, Gary Olsen, Joe Felsenstein, Pete Lockhart, Sandie Baldauf, Ford Doolittle, Michael Gray, David Spencer, Mitchell Sogin, Charles Marshall, Séan Turner, Walter Fitch and David Hillis for extensive helpful comments on phylogenetic analyses; and the staff of

Yellowstone National Park, especially Bob Lindstrom, for their enthusiastic cooperation. This material was based on work supported by a National Science Foundation Graduate Fellowship to S. M. B., a Department of Energy grant (DE-FG02–93ER-20088) and National Institutes of Health grant (NIH GM34527) to N. R. P., and a National Institutes of Health grant (GM-35087) to J. D. P.

References

Amann RI, Krumholz L, Stahl DA 1990 Fluorescent-oligonucleotide probing of whole cells for determinative, phylogenetic, and environmental studies in microbiology. J Bacteriol 172: 762–770

Baldauf SL, Doolittle WF, Palmer JD 1996 The root of the universal tree and the origin of eukaryotes based on elongation factor phylogeny. Proc Natl Acad Sci USA, in press

Barns SM, Fundyga RE, Jeffries MW, Pace NR 1994 Remarkable archaeal diversity detected in a Yellowstone National Park hot spring environment. Proc Natl Acad Sci USA 91:1609–1613

Barns SM, Delwiche CF, Palmer JD, Pace NR 1996 A third kingdom of Archaea? Perspectives on archaeal diversity, thermophily and monophyly from environmental rRNA sequences. Proc Natl Acad Sci USA 93:9188–9193

Brown JR, Doolittle WF 1995 Root of the universal tree of life based on ancient aminoacyl-transfer RNA synthetase gene duplications. Proc Natl Acad Sci USA 92:2441–2445

Burggraf S, Jannasch HW, Nicolaus B, Stetter KO 1990 *Archaeoglobus profundus* sp-nov represents a new species within the sulfate-reducing Archaeobacteria. System Appl Microbiol 13:24–28

DeLong EF 1992 Archaea in coastal marine environments. Proc Natl Acad Sci USA 89:5685–5689

DeLong EF, Wu KY, Prezelin BB, Jovine RVM 1994 High abundance of Archaea in Antarctic marine Picoplankton. Nature 371:695–697

Fuhrman JA, McCallum K, Davis AA 1992 Novel marine archaebacterial group from marine plankton. Nature 356:148–149

Gogarten JP, Kibak H, Dittrich P et al 1989 Evolution of the vacuolar H^+-ATPase: implications for the origin of eukaryotes. Proc Natl Acad Sci USA 86:6661–6665

Huber R, Burggraf S, Mayer T, Barns SM, Rossnagel P, Stetter KO 1995 Isolation of a hyperthermophilic Archaeum predicted by *in situ* RNA analysis. Nature 376:57–58

Iwabe N, Kuma K, Hasegawa M, Osawa S, Miyata T 1989 Evolutionary relationship of archaebacteria, eubacteria, and eukaryotes inferred from phylogenetic trees of duplicated genes. Proc Natl Acad Sci USA 86:9355–9359

Kane MD, Poulsen LK, Stahl DA 1993 Monitoring the enrichment and isolation of sulfate-reducing bacteria by using oligonucleotide hybridization probes designed from environmentally derived 16S ribosomal RNA sequences. Appl Environ Microbiol 59:682–686

Maidak BL, Larsen N, McCaughey MJ et al 1994 The Ribosomal Database Project. Nucleic Acids Res 22:3485–3487

Poulsen LK, Ballard G, Stahl DA 1993 Use of ribosomal RNA fluorescence *in situ* hybridization for measuring the activity of single cells in young and established biofilms. Appl Environ Microbiol 59:1354–1360

Reysenbach A-L, Wickham GS, Pace NR 1994 Phylogenetic analysis of the hyperthermophilic pink filament community in Octopus Spring, Yellowstone National Park. Appl Environ Microbiol 60:2113–2119

Stahl DA, Flesher B, Mansfield HR, Montgomery L 1988 Use of phylogenetically based hybridization probes for studies of ruminal microbial ecology. Appl Environ Microbiol 54:1079–1084

Ward DM, Bateson MM, Weller R, Ruff-Roberts AL 1992 Ribosomal RNA analysis of microorganisms as they occur in nature. Adv Microb Ecol 12:219–286

Woese CR 1987 Bacterial evolution. Microbiol Rev 51:221–271

Woese CR, Kandler O, Wheelis ML 1990 Towards a natural system of organisms: proposal for the domains Archaea, Bacteria, and Eucarya. Proc Natl Acad Sci USA 87:4576–4579

DISCUSSION

Stetter: In your wonderful experiments you have shown this incredible diversity exists. However, one point I am unsure of is that you do not find the same sequences again or at another site. We have found the opposite: we have looked at many populations with the technique pioneered by Norman Pace and yourself, often finding 100% sequence identity between organisms from distant places, for example from Italy, Iceland and the Yellowstone National Park.

Barns: This probably has something to do with the different sampling methods we use. Because you are looking for a specific type of organism, you have already narrowed down the diversity of sequences that you obtain by isolating those particular organisms in pure culture. Whereas if we analyse an environment that contains a great variety of different types of organisms, such as Obsidian Pool, the chance of us picking up that particular type is lower.

Stetter: There may also be a site-dependent *in situ* enrichment. The chemistry of the springs may differ greatly from site to site. This is a kind of selection of the population of organisms you will find there. Some types may dominate in one site, and in another one they're far below the detection level.

Davies: In the construction of these phylogenetic trees, by making certain assumptions about the rates of mutations of certain genes, is it possible for you to date the divergence of each branch?

Barns: It is problematic to do so for rRNA-derived trees, because different lineages have different rates of evolution — the clock is not constant across the universal tree. The rate probably is also not constant *within* lineages.

Davies: People have done it for hominids.

Barns: I believe that those studies were done with genes other than those for rRNA; I'm not familiar enough with phylogenetic analysis of these genes to comment on the validity of this approach. Any assumptions one could make about the rate of evolution of rRNA genes probably wouldn't be valid across all lineages.

Davies: Given that these hot springs are so isolated, one would expect to see very different organisms in each of them.

Shock: Why?

Davies: Because of the difficulty of getting from one to another.

Shock: The surface of the planet is moving constantly and there is water everywhere.

Davies: Are you suggesting that there is a continuity between them?

Shock: Hydrothermal systems have been around for at least 4 Ga.

Davies: This gives you some angle on how far back you go, and that relates to the first part of my question, which concerns the degree of diversity between the different sites, as opposed to within a given site. I am having a lot of difficulty understanding how these things work.

Barns: Because rRNA sequences do not change in a clock-like way, we can't infer how long ago these types were distributed or over what period they have evolved.

Davies: Perhaps you can get your clock from what we know about the formation of these hydrothermal sites.

Des Marais: You can put a bucket of sterilized seawater in the middle of the North American continent, and, within a few weeks or months, you get marine bacteria in the bucket.

Davies: I'm prepared to accept that as an explanation. It seems to me that this is an absolutely crucial issue.

Nisbet: Marine aerosols get everywhere, so there is no reason they shouldn't take these archaea around the world.

Can I ask a dumb geologist's question? Isn't it highly significant that you have just discovered eight or nine new phyla and a new kingdom?

Barns: What is significant is our current ignorance of the diversity of the microbial world. With the further use of molecular sequence-based techniques in microbiology, new phyla are going to fall like apples off a tree in the next few years.

Henley: What's the relative distribution of these things in the total biomass of the Earth?

Barns: Although I can't give you a figure, I believe that the microbes outweigh the macrobes, simply because they probably exist deep into the crust. The thin layer of macrobial life on the top probably doesn't compare.

Davies: Tommy Gold has estimated that the majority of the biomass is under our feet rather than on the surface.

Cady: Why is it that 16S rRNA sequencing techniques can be used to identify organisms but not their properties?

Barns: rRNA is a molecule that is essential to all organisms and evolves very slowly, which is why we can use it to look at all of life. But most of the environmentally important properties — the things that the cells eat and excrete — are more evolutionarily volatile. The genes for these properties change rapidly and, as far as we know, independently of those for rRNA. So we generally can't use one to predict the other.

Cady: How long does it take you to do an analysis?

Barns: The Obsidian Pool analysis shown in Fig. 1 took about two years, but these days it could probably be done in 2–4 weeks. This reflects the current availability of automated sequencers and different screening techniques. It is getting faster all the time.

Farmer: Once you've identified that these species exist in the communities you are dealing with, are there any techniques for actually isolating specific organisms? And once you have isolated them, can you then work out the metabolism?

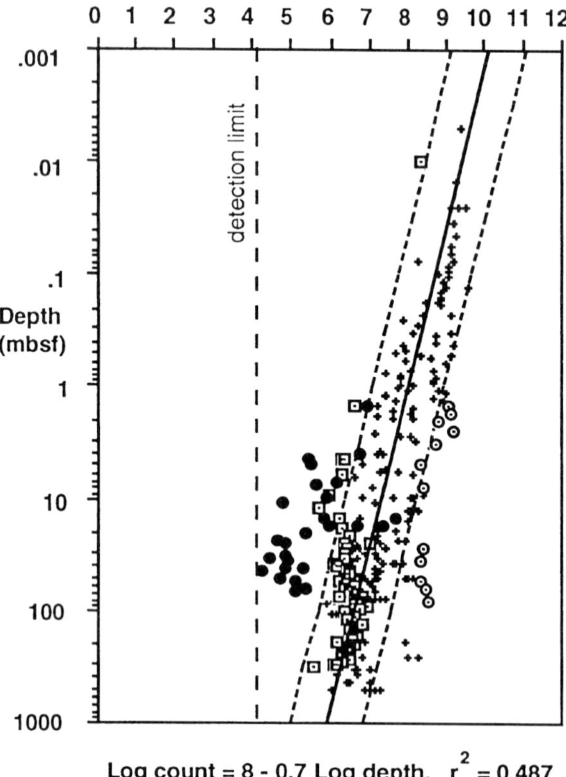

FIG. 1. *(Parkes)* Distribution of deep sediment bacteria in five sites in the Pacific Ocean. Three sites are highlighted: shallow, high productivity, Peru Margin (■); low productivity Eastern Equatorial Pacific (○); and Juan de Fuca Ridge hydrothermal system (●). Other sites are combined (+): Japan Sea, Lua Basin, Juan de Fuca Ridge non-hydrothermal sites and deep Peru Margin sites. (From Parkes et al 1994.)

Barns: There are some elegant techniques that Karl Stetter's group has devised, involving fluorescently labelling the cells to identify the morphology, and then isolating them in pure culture (Huber et al 1995).

Stetter: Due to Susan Barns' work it is now simple to isolate specific archaea. If the 16S rRNA sequence is known, we can design a specific fluorescent oligonucleotide probe against it. Then we can identify these cells in enrichment cultures which we have already in the laboratory. This step identifies the shape of the organism. We then go back to the original culture and use a step which was developed in my laboratory, involving optical tweezers. Here we use an infrared laser under the microscope, with which we catch a single cell and transfer it through a capillary about 10 cm long, within

6 min. This leaves us with a single cell, which we put into a multisubstrate medium. If we are lucky the cell multiplies, and we then characterize the resulting population's 16S rRNA to check it matches the original sequence we used.

Nisbet: In all of the various depictions of the tree of life, we see the last common ancestor—a sort of 'Queen Victoria' of all life. She passed her genes into the population of monarchs; this ancestor represents the last common ancestral population. Could you describe this beast? Obviously it has ribosomes and rRNA, it's a DNA-based organism, it has heat shock protection, but what other features does it have?

Barns: There is so much in common between all known forms of life that it is clear that much of what makes up modern organisms was present in that ancestor. It already had sophisticated machinery for synthesis of DNA, RNA and proteins.

Stetter: It looks like it could have been a chemolithoautotroph, since all the deepest and shortest branches within the phylogenetic tree are obligate and facultative chemolithoautotrophs. In addition, the surprising thing is that all of life is extremely similar on a molecular level. All the basic mechanisms are the same and the proteins are so closely related.

Pentecost: Viruses haven't been mentioned at all. Have they an evolution? Do they evolve separately?

Barns: They certainly have an evolution. They probably have been present since nucleic acids arose. Many viruses have obtained portions of their genome from their host organisms, so they are chimaeras of genetic information. Unfortunately, those we have now are evolving so rapidly that it's very difficult to draw any kind of deep genealogy for them.

Stetter: Viruses occur in all three domains of life. The archaea have very unusual viruses (Zillig et al 1986).

Parkes: If there is still life on Mars, the prospect is that this will be deep below the surface. But is a deep biosphere a realistic prospect, and could this be independent of surface biospheres fuelled by sunlight? Surely, an important consideration has to be whether there is a deep bacterial biosphere on Earth and, if so, what is it like and can this be used as a model for exploring Mars? We haven't really discussed this, and therefore I would like to present some of the results of our studies as part of the Ocean Drilling Program (ODP) with regard to bacterial populations in deep marine sediments.

We have studied sediments from 10 sites in the Atlantic and Pacific Oceans and the Mediterranean Sea, and have made approximately 700 direct microscopic determinations of bacterial populations within the sediment. Bacterial populations were present in all samples—our deepest samples are currently from 748 m (unpublished results) and our oldest are from > 10 Ma. Bacterial distributions are remarkably similar at all sites (e.g. Fig. 1 [*Parkes*]), and as there is no indication that bacterial populations decrease more rapidly with increasing depth (except in hydrothermal sediments), it is likely that bacterial populations exist even deeper within the sediment. These deep populations are highly significant as the sum of bacterial populations to only 500 m (average oceanic sediment depth; Chester 1990) represents 10% of the surface biosphere.

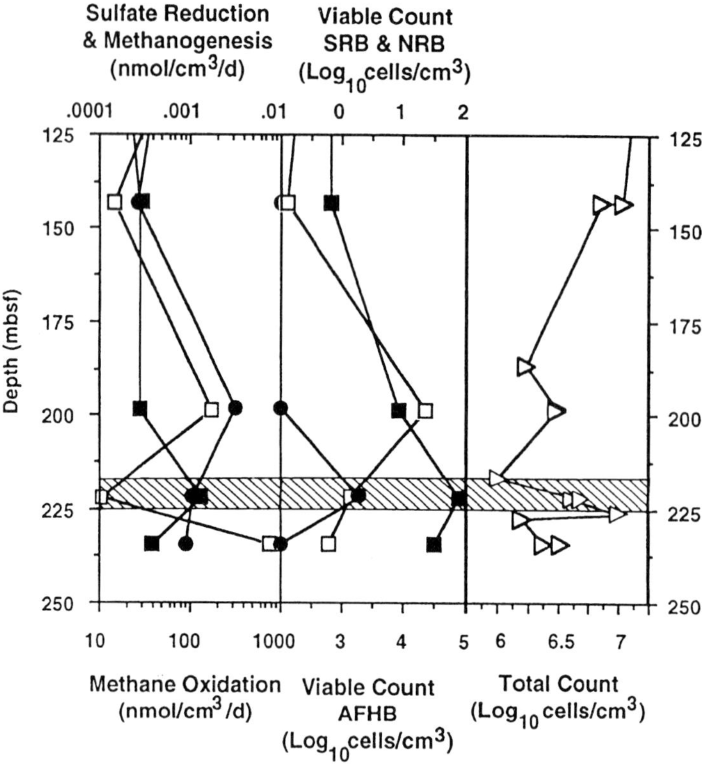

FIG. 2. *(Parkes)* Depth distributions of bacterial populations and activity in deep Gas-hydrate-containing sediments from the Cascadia Margin, Pacific Ocean. Bacterial activity: rates of sulfate reduction (●); methanogenesis (□); and methane oxidation (■). Culturable bacterial populations: sulfate-reducing bacteria (SRB) (●); nitrate-reducing bacteria (NRB) (■); and anaerobic fermentative heterotrophic bacteria (AFHB) (□). Triangles indicate direct bacterial count. The horizontal shaded area at a depth of approximately 225 m below sea floor (mbsf) indicates the gas hydrate zone. (From Cragg et al 1995.)

But bacterial populations don't simply decrease with sediment depth as they slowly run out of available organic matter, they can also be stimulated at depth. For example, in a subduction zone in the Pacific Ocean (Cascadia Margin, ODP Leg 146), where fluid and gas venting produce a discrete gas hydrate zone above a bottom stimulating reflector (BSR), bacterial populations and activity are stimulated. These populations are utilizing the elevated methane concentrations in the hydrate zone between 200 and 225 m below the surface (Fig. 2 [*Parkes*], Cragg et al 1995) and hence there is a deep bacterial biosphere which is independent of buried organic matter. Similar results have recently been found for bacterial populations in deep aquifers, but here the energy source was hydrogen (Stevens & McKinley 1995).

As temperature rises with sediment depth between about 30–100 °C/km and bacteria have been shown to grow up to about 115 °C, temperature would only be limiting at a depth of several kilometres, except in hydrothermal regions. Bacterial populations indeed decrease more rapidly with depth in hydrothermal regions due to the combined stress of organic carbon limitation and elevated temperatures. But as I have previously mentioned, at two sites only a few metres away from an active (276 °C) hydrothermal vent, in a zone associated with lateral hydrothermal fluid flow, bacterial populations actually increase (See Fig. 1, p 13; Cragg & Parkes 1994). This zone was approximately 17 m below the surface and was at 169 °C.

Therefore, there is the possibility of a deep 'hot' bacterial biosphere on Earth. In addition, some deep bacteria may even survive oil formation and migration, to be reactivated during oil recovery, causing problems such as oil souring due to bacterial sulfide formation (Parkes & Maxwell 1993). Hence, further studies on this deep bacterial biosphere on Earth would seem a necessary prerequisite for any similar studies on Mars.

Trewin: There is a great deal of work taking place on souring of oil reservoirs and introduction of bacteria by meteoric flow or contamination. There is also the possibility that viable organisms can be buried within the sedimentary pile. People tend to think organisms buried with sediment in the sedimentary basin are going to die, but very rapid rates of burial, do occur—within a million years or so burial depths of a couple of kilometres can be achieved in rapidly subsiding basins, and there is no reason why bacteria should all die under those conditions. There are many possibilities.

Parkes: I can't provide information about unique bacterial isolates from deep sediments, as the isolates we have so far obtained are genetically closely related to near-surface bacterial types and have probably adapted to increasing temperature and pressure during burial (Parkes et al 1994). However, 16S rRNA gene sequence analysis of DNA extracted directly from the sediment demonstrates the presence of some unique bacterial types (unpublished results). The challenge is now for us to isolate these.

Hence, it appears that there is a unique deep bacterial biosphere that is well adapted to this environment. Therefore it is not necessary for deep environments to have been inoculated with surface bacteria. It may well have been the other way round in some circumstances.

Nisbet: If you believe the US Geological Survey estimates, there is 10^{16}–10^{19} g carbon in the methane hydrates (Kvenvolden 1988). Was that the BSR layer you were showing from the Japan Sea?

Parkes: The site was the Cascadia Margin, and the highlighted zone was the gas hydrate, which happened to be just above BSR in that case.

Nisbet: This BSR layer is very widespread—you're talking about an amount of carbon equivalent to that lying around on the surface biosphere. There's plenty of carbon in sediment. The trouble in the mud biosphere is the energy source. On the surface you're dealing with a solar input to the top of the atmosphere at the equator of 343 W/m^2, whereas the energy you're getting geothermally is around 100 mW/m^2, unless you're close to a vent. In other words, the available energy flow from geothermal sources, globally, is much less.

Shock: It is all run on chemical energy.

Nisbet: Something has got to drive your chemical disequilibrium: that something, ultimately, is either the sun via photosynthesis, or geological processes.

Shock: Mixing?

Nisbet: Yes, but the mixing is done by the plate system, and something has got to make the plate system work. Basically it is a heat system, run by heat loss from the interior, and that is very small compared with solar input.

Shock: You can say that deep life is difficult because you don't have geophysical energy, but it doesn't rely on some physical form of energy, it relies on chemical energy. Compounds are out of equilibrium and the organisms take them closer to equilibrium.

Nisbet: The disequilibrium will come because you're depositing sediments or making volcanic rock or something like that, and that all comes back to the plate system, which is still very little energy compared with what you get from sunlight. It may be a very extensive part of the biosphere in its physical extent, but the subsurface biopshere is limited unless it has access to the products of sunlight.

Shock: I don't see that that is a problem.

Stetter: John Parkes mentioned his fascinating observation of microbes within deep ocean drilling cores with original temperatures of 169 °C. How were these samples recovered in order to avoid artefacts? I also took hot sediment core samples at the Guaymas deep sea vents with original temperature gradients of up to 200 °C and brought them up to the surface by the deep-diving submersible Alvin. In the laboratory, I found living organisms at sites which had original temperatures of 200 °C, but which we could cultivate only up to 100 °C. There was a simple explanation: during the ascent to the surface, because of decompression there was vigorous out-gassing which caused a lot of turbulence within the cores and mixed things together. How did you avoid this problem?

Parkes: All the samples I talked about were collected as part of the international ODP. The core technology is very well developed and you can tell if there are disturbance problems in terms of sampling because all the geochemistry is measured. In addition, we took our samples from the middle of the core to avoid any contamination from sediment being smeared along the core liner. Contamination problems are very much reduced as we were using direct microscopic analysis of bacteria and not growth of bacteria in enrichment media. As bacterial populations above and below this 169 °C zone were much lower than within the zone, coring disturbances can't explain these results, especially as identical results were obtained in two close but separate sites (Fig. 1, p 13).

References

Chester R 1990 Marine geochemistry. Unwin Hyman, London

Cragg BA, Parkes RJ 1994 Bacterial profiles in hydrothermally active deep sediment layers from Middle Valley (N. E. Pacific) Sites 857 and 858. Proc Ocean Drilling Prog, Sci Res 139:509–516

Cragg BA, Parkes RJ, Fry JC et al 1995 The impact of fluid and gas venting on bacterial populations and processes in sediments from the Cascadia Margin Accretionary System (Sites 888–892) and the geochemical consequences. Proc Ocean Drilling Prog, Sci Res 146:399–411

Huber R, Burggraf S, Mayer T, Barns SM, Rossnagel P, Stetter KO 1995 Isolation of a hyperthermophilic Archaeum predicted by *in situ* RNA analysis. Nature 376:57–58

Kvenvolden KA 1988 Methane hydrates and global climate. In: Global biogeochemical cycles, vol 2. Academic Press, London, p 221–229

Parkes RJ, Maxwell JR 1993 Some like it hot (and oily). Nature 365:694–695

Parkes RJ, Cragg BA, Bale SK et al 1994 Deep bacterial biosphere in Pacific Ocean sediments. Nature 371:410–413

Stevens TO, McKinley JP 1995 Lithoautotrophic microbial ecosystems in deep basalt aquifers. Science 270:450–454

Zillig W, Gropp F, Hienschen A et al 1986 Archaebacterial virus host systems. System Appl Microbiol 7:58–66

Hydrothermal systems as environments for the emergence of life

Everett L. Shock

Department of Earth and Planetary Sciences and McDonnell Center for the Space Sciences, Washington University, St Louis, MO 63130, USA

Abstract. Analysis of the chemical disequilibrium provided by the mixing of hydrothermal fluids and seawater in present-day systems indicates that organic synthesis from CO_2 or carbonic acid is thermodynamically favoured in the conditions in which hyperthermophilic microorganisms are known to live. These organisms lower the Gibbs free energy of the chemical mixture by synthesizing many of the components of their cells. Primary productivity is enormous in hydrothermal systems because it depends only on catalysis of thermodynamically favourable, exergonic reactions. It follows that hydrothermal systems may be the most favourable environments for life on Earth. This fact makes hydrothermal systems logical candidates for the location of the emergence of life, a speculation that is supported by genetic evidence that modern hyperthermophilic organisms are closer to a common ancestor than any other forms of life. The presence of hydrothermal systems on the early Earth would correspond to the presence of liquid water. Evidence that hydrothermal systems existed early in the history of Mars raises the possibility that life may have emerged on Mars as well. Redox reactions between water and rock establish the potential for organic synthesis in and around hydrothermal systems. Therefore, the single most important parameter for modelling the geochemical emergence of life on the early Earth or Mars is the composition of the rock which hosts the hydrothermal system.

1996 Evolution of hydrothermal ecosystems on Earth (and Mars?). Wiley, Chichester (Ciba Foundation Symposium 202) p 40–60

The close genetic proximity of hyperthermophilic bacteria and archaea to a common ancestor of all life suggests that such an ancestor was also a hyperthermophile (Woese et al 1990). The reasons for the success of modern hyperthermophiles may reveal why the common ancestry of life appears to point back to hydrothermal systems. As illustrated here, hydrothermal systems are extraordinary settings for organic synthesis from CO_2 or carbonic acid. Because carbon fixation is exergonic in these systems, there is no better-suited environment for chemolithoautotrophic organisms.

Energetically, hyperthermophilic anaerobes in present-day hydrothermal systems lead an extremely easy life. Geological and geochemical processes supply them with abundant energy in easily used chemical forms (McCollom & Shock 1994, 1996).

Tapping even a small amount of this energy leads to a rate of biological productivity which may be the highest on the planet (Lutz et al 1994). The key to this productivity is the biological dissipation of geochemical disequilibrium provided by the mixing of hydrothermal fluids and cold seawater. Efforts to extrapolate to the early Earth reveal that the quality and quantity of geochemically supplied energy depend on the composition of the crust and seawater. The dependence on crustal composition predominates. The following discussion touches on some of the salient features of hydrothermal systems, some reasons why these systems support thriving biological communities, and summarizes recent work on organic synthesis and cellular organization in analogous systems on the early Earth.

Hydrothermal systems, rock alteration and geochemical disequilibrium

Convective circulation of water through cooling rocks is an enormously effective way of dissipating heat. Shrinkage accompanies the solidification of newly formed volcanic rocks and provides pathways for water to enter the rock and remove heat by convection. As a result, hydrothermal systems are the inevitable consequence of the combination of water and volcanic activity on any planet. On the early Earth, hydrothermal systems would have existed after the surface of the planet had cooled to the point that liquid water could exist. If the abundance of H_2O on the early Earth was similar to today's, the upper temperature for liquid water would be its critical temperature of 374 °C. Lower surface temperatures would enhance the effectiveness of heat dissipation by circulating fluids. Regardless, the consequences of these high temperature reactions are that the rock will be altered and the fluid composition will be transformed.

At present, these transformations are dramatic for submarine hydrothermal systems. Mid-ocean ridge basalt solidifies at about 1170 °C (Usselman & Hodge 1978). It reacts with seawater at lower temperatures (2 to >400 °C) to yield new mineral assemblages indicative of the alteration conditions (sometimes referred to as greenschist metamorphism by petrologists). Addition of some chemical components from seawater to the rock (especially H_2O, Mg and carbonate), and varying amounts of leaching (of silica, Al, Ca, Fe, Cu, Mn, Zn and many other trace elements) accompany the alteration in hydrothermal systems. The leached elements are transferred to the reacting seawater as it is transformed into the hydrothermal fluid. Comparison of the compositions of seawater and a vent fluid in Table 1 illustrates the changes which occur during hydrothermal circulation of seawater in basalt in present-day submarine systems.

The fluids described in Table 1 are far from thermal and chemical equilibrium when they mix in the general environment of a submarine hydrothermal vent field. On the one hand, this disequilibrium is partially dissipated in the zone of fluid mixing by the precipitation of minerals and the growth of organisms. On the other hand, the disequilibrium state is maintained by the continued circulation of seawater through basalt driven by the volcanic heat. It is the disequilibrium provided by geological and geochemical processes that is the chemical energy source for life in and around submarine hydrothermal systems. The amount of geochemical energy available to organisms can

TABLE 1 **Composition of hydrothermal vent fluid and seawater used by McCollom & Shock (1996) in mixing calculations**

	Hydrothermal vent fluid (Von Damm et al 1985)	*Seawater (Von Damm 1990)*
T (°C)	350	2
log fO_2	−30.00	−1.240
O_2,aq	0	0.1
H_2,aq	1.3	0
SO_4^{2-}	0	27.9
H_2S,tot	7.5	0
pH	3.5	7.8
Na^+	439	464
Ca^{2+}	16.6	10.2
Mg^{2+}	0	5.27
Ba^{2+}	0.010	0.00014
Al^{3+}	0.0047	0.00002
Fe	0.750	0.0000015
Cl^-	496	541
HCO_3^-	5.72	2.3
SiO_2,aq	17.3	0.16
K^+	23.2	9.8
Mn^{2+}	0.699	0
Cu^{2+}	0.00970	0.000007
Pb^{2+}	0.000194	0
Zn^{2+}	0.089	0.00001

The hydrothermal vent fluid is represented by the fluid from the Southwest vent on the East Pacific rise at 21 °N. All concentrations millimolal.

be quantified by calculating the chemical affinity of reactions from which the organisms get energy.

Geochemical disequilibrium, chemical affinity and hyperthermophiles

Chemical affinity is a concept proposed by de Donder (1920), developed in considerable detail by Prigogine & Defay (1954) and applied to geochemistry by Helgeson (1968, 1979). It is a particularly useful thermodynamic property because it quantifies the extent to which a reaction or an entire chemical system is out of equilibrium. Because chemical affinity is defined in terms of the reaction progress variable, ξ, changes in chemical affinity can be tied explicitly to changes in the progress of chemical reactions that describe the state of a system. One definition of chemical affinity, **A**, is given by

$$\mathbf{A} = -(\partial \Delta_r G / \partial \xi) \tag{1}$$

where $\Delta_r G$ stands for the overall Gibbs free energy of reaction, which is in turn defined by

$$\Delta_r G = \Delta_r G^\circ + RT \ln Q_r \tag{2}$$

where $\Delta_r G^\circ$ represents the standard Gibbs free energy of reaction, R designates the gas constant, T indicates temperature in Kelvin, and Q_r stands for the activity product given by

$$Q_r = \Pi a_i^{\nu_{i,r}} \tag{3}$$

where a_i designates the activity of the ith species, and $\nu_{i,r}$ designates the stoichiometric reaction coefficient for the ith species in the rth reaction. At equilibrium, Q_r is equivalent to the equilibrium constant K_r, and since $\Delta_r G$ is zero by definition at equilibrium,

$$\Delta_r G^\circ = -RT \ln K_r \tag{4}$$

Combination of equations (1), (2) and (4) leads to the expression

$$\mathbf{A} = RT \ln (K_r / Q_r) \tag{5}$$

from which it is evident that the difference between the equilibrium constant for a reaction and the activity product for that reaction in a given situation yields a value for the chemical affinity. In this sense, chemical affinity can be used to convert concentrations into the amount of energy available in a system which is out of chemical equilibrium.

One application of chemical affinity to quantifying available energy during mixing of hydrothermal fluid and seawater is given by McCollom & Shock (1996), who evaluated the biological energy provided by geochemical disequilibrium between sulfate and H_2S. Hyperthermophiles are known to reduce sulfur and sulfate to sulfide (Karl 1995, Baross & Deming 1995). It is also known that organisms that live at lower temperatures can oxidize sulfur or sulfide to sulfate. In fact, H_2S oxidation is widely regarded as the basis of chemosynthesis in hydrothermal vent environments (Jannasch & Mottl 1985, Tunnicliffe 1991). The potential for sulfate reduction or H_2S oxidation in a given situation can be evaluated by comparing values of Q_r for the reaction

$$SO_4^{2-} + 2\,H^+ + 4\,H_2(aq) = H_2S(aq) + 4\,H_2O \tag{6}$$

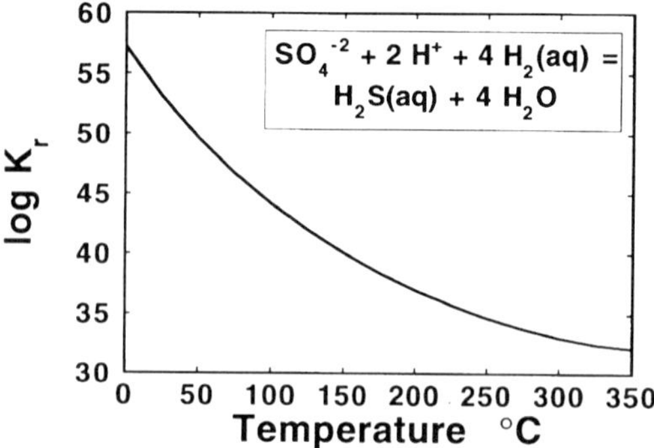

FIG. 1. Plot of the logarithm of the equilibrium constant calculated for reaction (6) as a function of temperature at pressures corresponding to vapour–liquid equilibrium for H_2O.

to values of K_r shown[1] in Fig. 1. Both Q_r and K_r depend on temperature. To evaluate the chemical energy available as seawater and hydrothermal fluids mix, McCollom & Shock (1996) performed mixing calculations starting with the two fluid compositions given in Table 1. They assumed that conductive cooling is minimal and evaluated the temperature of the mixture based on the mass of each of the two fluids. Some of the results of this study are summarized in Fig. 2.

The plot in Fig. 2A shows the temperature of the mixture as cold seawater is mixed into hot (350 °C) hydrothermal fluid. Note that at high temperatures the temperature drops dramatically with a small addition of seawater, but that at low temperatures increasingly large amounts of seawater must be added to affect a change in the temperature. For this type of mixing calculation, each temperature corresponds to a different bulk composition, thus explaining why Q_r is temperature dependent. These changes are reflected for the chemical species of interest in reaction (6) in Figs 2B and 2C. Note that sulfate and H_2S are not allowed to equilibrate in this calculation. This is meant to simulate the natural process, where the lack of equilibration of sulfate and H_2S allows organisms to gain energy from sulfate reduction or sulfide oxidation. Note also that the calculated H_2 content of the mixed fluid decreases with temperature and is effectively depleted at around 50 °C, at which point the calculated O_2 content increases with decreasing temperature to the value for seawater in Table 1. The resulting fluctuations in the overall oxidation state of the mixed fluid are recorded by changes in the fugacity of oxygen (fO_2), which goes through an enormous fluctuation around 50 °C as shown in Fig. 2C.

Values of the chemical affinity for reaction (6) are shown in Fig. 2D. The values plotted are per kilogram of H_2O in the mixed fluid. Positive values correspond to

[1]Values of K_r are obtained from standard state Gibbs free energies for reaction (6) which were evaluated with the revised Helgeson–Kirkham–Flowers (HKF) equation of state (Shock et al 1992) using data and parameters given by Shock & Helgeson (1988) for SO_4^{2-} and H^+, Shock et al (1989) for the dissolved gases H_2 (aq) and H_2S (aq), and Johnson & Norton (1991) for H_2O.

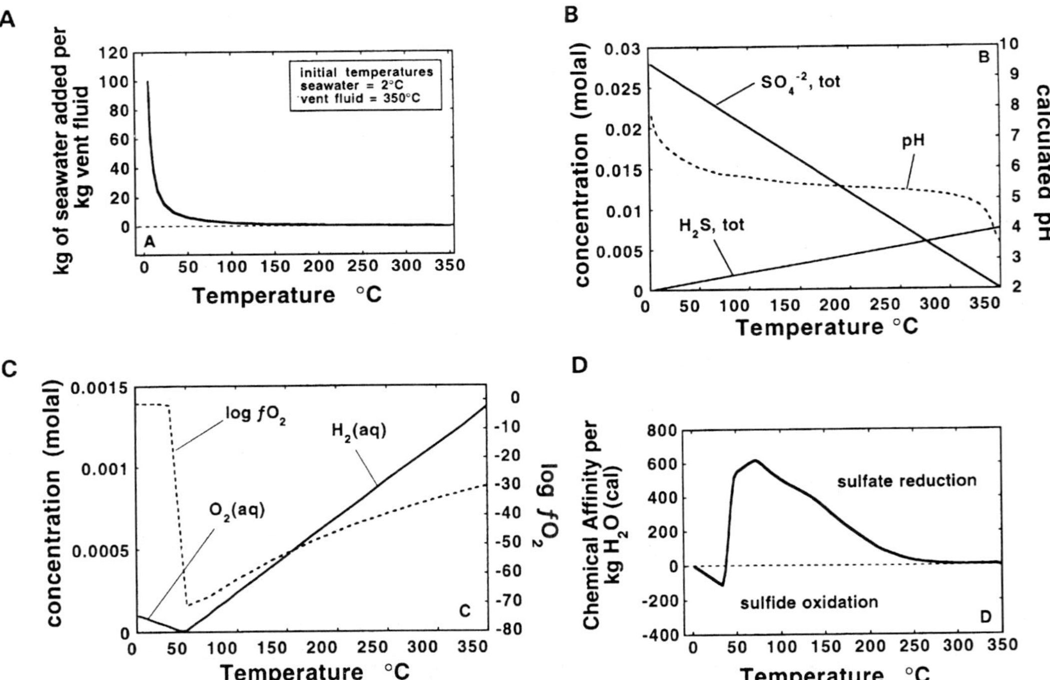

FIG. 2. Results of theoretical calculations of mixing between the hydrothermal vent fluid and seawater compositions listed in Table 1. (A) The temperature of the mixed fluid in terms of the number of kilograms of cold seawater mixed into a kilogram of vent fluid. (B) The variation in pH, and the total concentrations of SO_4^{2-} (including HSO_4^- and metal complexes) and H_2S (including HS^-) in the mixed fluid as functions of temperature. (C) The concentrations of dissolved gases H_2(aq) and O_2(aq) as well as the logarithm of the fugacity of O_2 in the mixed fluid. (D) The resulting chemical affinity for reaction (6) per kg H_2O in the mixed fluid, calculated with equation (5) using concentrations shown in (B) and (C), as well as the equilibrium constants shown in Fig. 1. Positive values indicate temperatures at which sulfate reduction is favourable and negative values indicate temperatures at which sulfide oxidation is favourable.

conditions at which reaction (6) is favoured to proceed from left to right as written, and negative values indicate that the reaction is favoured to proceed from right to left. Therefore, energy is available to H_2S oxidizers at temperatures below ~45 °C, but energy is available to sulfate reducers at all temperatures >45°C. This fluctuation mimics the fluctuation in fO_2 in Fig. 2C. Note that there is considerable energy available from sulfate reduction at the 100–150 °C range where hyperthermophiles are known or suspected to thrive (Daniel 1992). If a hyperthermophile is capable of reducing sulfate, then it can combine H_2 from the vent fluid with sulfate from seawater (both of which are present and out of equilibrium in the mixed fluid) to form H_2S and H_2O, capture some of the energy released by this reaction, and use it for its various energy-demanding metabolic processes.

The type of calculations summarized in Fig. 2 provide insight into the metabolic strategies of hyperthermophilic organisms, and explanations for the vast biological productivity of hydrothermal systems. They also provide a framework for quantifying available energy in hydrothermal systems that involve initial fluid compositions other than seawater, and different types of host rocks. Extrapolations to the early Earth (Shock 1990, 1992a,b, MacLeod et al 1994, Russell & Hall 1995, 1996, Shock et al 1995, Shock & Schulte 1995, 1996) and to Mars (Griffith & Shock 1995, Shock & Schulte 1995, 1996) have already provided insights into the geochemical environment from which hydrothermal life may have emerged.

Hydrothermal organic synthesis

Before extrapolating to the early Earth where there are so many unknowns, it is useful to return to the present-day hydrothermal systems and evaluate the energetic demands for fixing carbon into organic compounds. The surprising result is that there is a thermodynamic drive to form simple organic compounds from dissolved CO_2 or bicarbonate, and that organisms can actually gain energy by conducting simple organic synthesis because it lowers the Gibbs free energy of the mixture of seawater and hydrothermal fluid (Shock 1995, Shock & Schulte 1995, 1996). Some of these results are summarized in this section.

In much the same way that the chemical affinity of the sulfate reduction reaction becomes positive during mixing of seawater and hydrothermal fluids, the chemical affinities of organic synthesis reactions are also positive. The underlying reason is that bicarbonate present in seawater is out of equilibrium with the dissolved H_2 in the hydrothermal fluid. In addition, any dissolved CO_2 in the hydrothermal fluid will be out of equilibrium with the dissolved H_2 as the temperature is lowered (Shock 1990, 1992a,b). The organic synthesis reactions can be written as

$$a\,HCO_3^- + a\,H^+ + b\,H_2(aq) = c\,\text{organic(aq)} + d\,H_2O \qquad (7)$$

and

$$w\,CO_2(aq) + x\,H_2(aq) = y\,\text{organic}(aq) + z\,H_2O \tag{8}$$

where the reaction coefficients are indicated as variables that depend on the composition of the organic compound formed. Mass action expressions for reactions of this type can be combined with mass balance constraints obtained from geochemistry to evaluate the energetics of organic synthesis in hydrothermal systems.

Mixing calculations of the type conducted by McCollom & Shock (1996) can be modified to include organic synthesis reactions[2] (Shock 1995, Shock & Schulte 1995, 1996). These efforts demonstrate that the results are highly sensitive to the concentration of aqueous H_2 in the hydrothermal fluid. This makes sense because $H_2(aq)$ is the most abundant reductant in the fluid. Fluid–rock reactions control the concentration of $H_2(aq)$ that is generated from H_2O in response to the prevailing redox state of the system. In turn, the composition of the rock, and specifically the ratio of ferrous to ferric iron, controls the redox state. This is illustrated in Fig. 3, where the percentage of carbon present as inorganic forms at metastable equilibrium is plotted as a function of temperature. The two curves in this figure represent results for two mixing calculations where the only difference is the initial concentration of $H_2(aq)$[3]. Both of these concentrations of $H_2(aq)$ are plausible in present-day hydrothermal systems. The fayalite–magnetite–quartz (FMQ) assemblage is appropriate for reactions between fresh basalt and seawater, and the pyrrhotite–pyrite–magnetite (PPM) assemblage is closely consistent with conditions in the sulfide-rich mounds which are the surface expression of submarine hydrothermal systems.

In the mixing calculations involving carbon, inorganic carbon is allowed to be converted to organic compounds if the Gibbs free energy of the system is lowered. In other words, organic synthesis is permitted if it is thermodynamically favoured. The deviation of the curves in Fig. 3 from 100% inorganic carbon indicate that organic synthesis would be favoured at temperatures from ~250 to 50 °C in either case. This is shown in more detail in Fig. 4 for the calculations which begin with the $H_2(aq)$ concentration of the vent fluid set by FMQ. In this figure the curves indicate the percentage of carbon in the system that is present as each family of organic compounds. Note the shift in the group that predominates as the overall redox state changes during the mixing of seawater and hydrothermal fluid. The oxidation state of the system is not

[2]Organic compounds considered in these calculations include carboxylic acids, alcohols, aldehydes and ketones. Hydrocarbons are excluded from the calculations to remain consistent with observations about metastable equilibrium in natural ecosystems (Shock 1989, 1994, Helgeson et al 1993). Thermodynamic data and revised HKF equation-of-state parameters are taken from Shock & Helgeson (1990) and Schulte & Shock (1993) who provide values for straight chain compounds with numbers up to eight.

[3]Note that the initial concentrations of $H_2(aq)$ used in the calculations are referenced to mineral assemblages which are capable of buffering the activity of $H_2(aq)$ in reactions involving H_2O. These assemblages and the buffering reactions are pyrrhotite–pyrite–magnetite (PPM) given by $2FeS+4/3\,H_2O = FeS_2+1/3Fe_3O_4+4/3H_2(aq)$ and fayalite–magnetite–quartz (FMQ) given by $3/2Fe_2SiO_4+H_2O = Fe_3O_4+3/2SiO_2+H_2(aq)$.

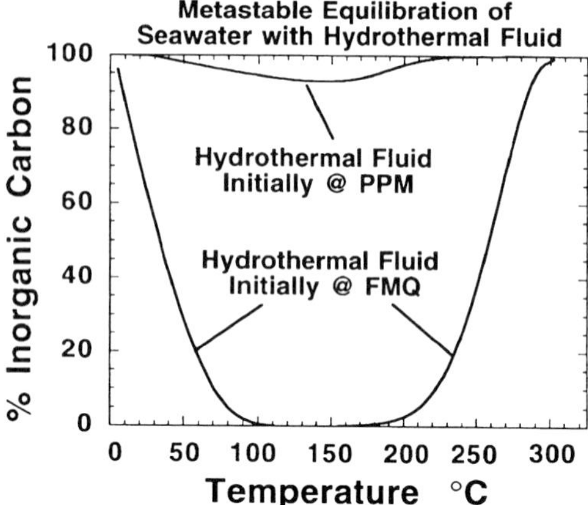

FIG. 3. Plots of the calculated percentage of the carbon which is inorganic at metastable equilibrium between CO_2 and aqueous organic compounds during mixing of hydrothermal fluids and seawater. The differences in the curves are a consequence of the initial H_2 content of the hydrothermal fluid, which is taken to be set by equilibrium with two different mineral assemblages (see text). There is a larger potential for organic synthesis from a fluid with a content of H_2 initially set by FMQ than one set by PPM.

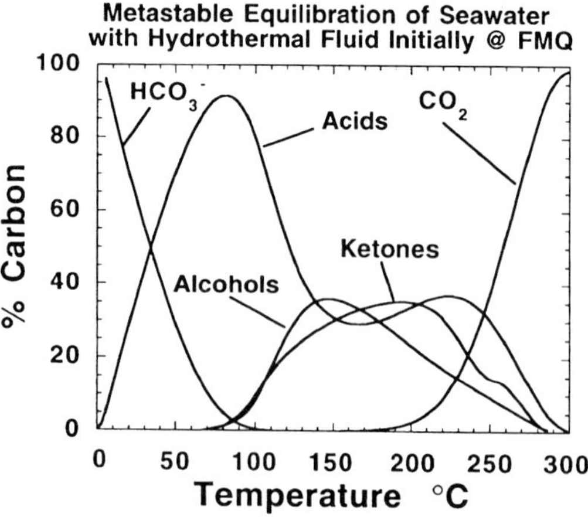

FIG. 4. Calculated distribution of carbon among organic and inorganic forms at metastable equilibrium during the mixing of seawater with a hydrothermal fluid with a concentration of H_2 set by the FMQ assemblage (corresponding to the lower curve in Fig. 3).

buffered in these calculations, but is controlled by the initial composition of the two fluids and the relative masses of the fluids at any given temperature. The predominance of carboxylic acids at lower temperatures is also influenced by changes in pH which help to stabilize the acids and their anions relative to other compounds.

The results shown in Figs 3 and 4 indicate that there is a thermodynamic drive in modern systems to form organic compounds as seawater and hydrothermal fluids, which are far from equilibrium, mix and move towards a more stable state. Hyperthermophiles must take advantage of this situation, because they can fix carbon at no energetic cost. In fact, they could even gain energy from the organic synthesis process. In addition, minimizing the Gibbs free energy of the system favours organic compounds that are used to make membranes, as shown in Fig. 5 where the distribution of carboxylic acids is plotted as a function of temperature for the mixing calculation summarized in Fig. 4. The numbers labelling the curves refer to the number of carbon atoms in a molecule of the acid indicated. Note that larger compounds are favoured as temperature decreases. Similar results are found for the alcohols and ketones as well. In the range around 100 °C where hyperthermophiles are known to thrive, octanoic acid and octanoate predominate in these calculations over the other carboxylic acids. This suggests that the monomeric constituents of cell membranes and other biomolecules are synthesized by hyperthermophiles at no energetic cost. Polymerization requires energy just as it does at room temperature, but in the case of peptide bonds, increasing temperature lowers the energy required to make the bond (Shock 1992a,b).

Model for a hydrothermal cell

The remarkable potential for organic synthesis in hydrothermal systems has led other investigators to propose models for early cells that are composed of materials

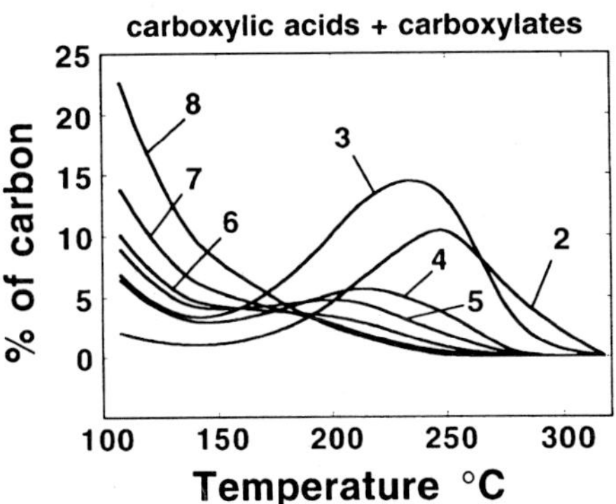

FIG. 5. Calculated distribution of carboxylic acids (the sum of associated molecules and anions) as functions of temperature in the fluid-mixing calculation shown in Fig. 4. Curves are identified by the number of carbons in the acid molecule (2 = acetic acid + acetate, etc.).

indigenous to hydrothermal systems. Russell & Hall (1996), building on their earlier results (Russell et al 1988, 1989, 1993, 1994, MacLeod et al 1994, Russell & Hall 1995) envision the following scenario for a hydrothermal cell. They consider the mixing of medium-temperature hydrothermal fluids (100–200 °C) that have formed through reactions with komatiite, a rock type that is more common in Archaean rocks than more recent formations. The resulting fluid has a basic pH, and is allowed to mix with seawater which is taken to be more acidic than modern seawater owing to a higher concentration of CO_2 in the atmosphere of the early Earth. Russell et al (1989, 1994) demonstrate through experiments that ferrous sulfide will grow at the mixing front between these two fluids, and that the precipitate grows in hollow, bubble-like forms. The growth of these sulfide membranes serves to separate the two fluids of differing pH and redox state. In this model, the separation of these fluids permits a primitive sort of proton-motive force, if there is a mechanism for exchange across the membrane. This type of model is highly complementary to the theoretical demonstration of the great potential for organic synthesis described above. Somehow the confluence of these ideas could be leading to an altogether new paradigm for the emergence of life. This paradigm is strongly rooted in realistic geochemistry of one of the most ordinary and inescapable processes on any planet with water and volcanoes, including Mars.

Concluding remarks

Others have argued that the hyperthermophilic ancestry of life is a consequence of a bottleneck in evolution engendered by the heavy cosmic bombardment of the early Earth. Sleep et al (1989) argue that life in submarine hydrothermal systems would be likely to survive the type of large-sized impacts which would have been common at the stage of Earth's history when life is thought to have appeared. If so, life could have begun in another setting through different and unknown processes, but only life in hydrothermal systems would have survived to produce the variety of life presently on the planet (Lazcano 1993). This argument is difficult to disprove, and I see no need to try. On the other hand, it is simpler, and I believe more profitable, to envision life emerging from the environment that is so deeply recorded in the genetic code.

Acknowledgements

This report summarizes ongoing research at Washington University by Tom McCollom, Mitch Schulte, Laura Wetzel, Laura Griffith and myself, as well as work by Mike Russell, Allan Hall and their colleagues at the University of Glasgow. I thank all of them for many helpful discussions. The research from Washington University is supported in part by NSF grant OCE-9220337 and NASA grant NAGW-2818. This is GEOPIG Contribution #106.

References

Baross JA, Deming JW 1995 Growth at high temperatures: isolation and taxonomy, physiology, and ecology. In: Karl DM (ed) The microbiology of deep-sea hydrothermal vents. CRC Press, Boca Raton, FL, p 169–217

Daniel RM 1992 Modern life at high temperatures. Orig Life Evol Biosphere (suppl) 22:33–42

de Donder T 1920 Lecons de thermodynamique et de chimie-physique. Gauthier-Villars, Paris

Griffith LL, Shock EL 1995 A geochemical model for the formation of hydrothermal carbonates on Mars. Nature 377:406–408

Helgeson HC 1968 Evaluation of irreversible reactions in geochemical processes involving minerals and aqueous solutions. I. Thermodynamic relations. Geochim Cosmochim Acta 32:853–877

Helgeson HC 1979 Mass transfer among minerals and hydrothermal solutions. In: Barnes HL (ed) Geochemistry of hydrothermal ore deposits, 2nd edn. Wiley, New York, p 568–610

Helgeson HC, Knox AM, Owens CE, Shock EL 1993 Petroleum, oil field waters and authigenic mineral assemblages: are they in metastable equilibrium in hydrocarbon reservoirs? Geochim Cosmochim Acta 57:3295–3339

Jannasch HW, Mottl MJ 1985 Geomicrobiology of deep-sea hydrothermal vents. Science 229:717–725

Johnson JW, Norton D 1991 Critical phenomena in hydrothermal systems: state, thermodynamic, electrostatic, and transport properties of H_2O in the critical region. Am J Sci 291:541–648

Karl DM 1995 Ecology of free-living, hydrothermal vent microbial communities. In: Karl DM (ed) The microbiology of deep-sea hydrothermal vents. CRC Press, Boca Raton, FL, p 35–124

Lazcano A 1993 Biogenesis: some like it hot. Science 260:1154–1155

Lutz RA, Shank TM, Fornari DJ et al 1994 Rapid growth at deep-sea vents. Nature 371:663–664

MacLeod G, McKeown C, Hall AJ, Russell MJ 1994 Hydrothermal and oceanic pH conditions of possible relevance to the origin of life. Orig Life Evol Biosphere 23:19–41

McCollom TM, Shock EL 1994 Energetics of biological sulfate reduction within chimney walls at mid-ocean ridges: proceedings of the American Geophysical Union Fall Meeting, San Francisco, Dec 1994. Eos 75:707

McCollom TM, Shock EL 1996 Geochemical constraints on chemolithoautotrophic metabolism by microorganisms at seafloor hydrothermal systems. Geochim Cosmochim Acta, in press

Prigogine I, Defay R 1954 Chemical thermodynamics. Jarrold & Sons, London

Russell MJ, Hall AJ 1995 The emergence of life at hot springs: a basis for understanding the relationships between organics and mineral deposits. In: Pasava J, Kribek B, Zak K (eds) Mineral deposits: from their origin to their environmental impacts. Proceedings of the third biennial Society for Geology Applied to Mineral Deposits Meeting, Prague, Aug 1995. AA Balkema, Rotterdam, p 793–795

Russell MJ, Hall AJ 1996 The emergence of life from iron monosulphide bubbles at a hydrothermal redox front. J Geol Soc, in press

Russell MJ, Hall AJ, Cairns-Smith AG, Braterman PS 1988 Submarine hot springs and the origin of life. Nature 336:117

Russell MJ, Hall AJ, Turner D 1989 *In vitro* growth of iron sulphide chimneys: possible culture chambers for origin-of-life experiments. Terra Nova 1:238–241

Russell MJ, Daniel RM, Hall AJ 1993 On the emergence of life via catalytic iron sulphide membranes. Terra Nova 5:343–347

Russell MJ, Daniel RM, Hall AJ, Sherringham J 1994 A hydrothermally precipitated catalytic iron sulphide membrane as a first step toward life. J Mol Evol 39:231–243

Schulte MD, Shock EL 1993 Aldehydes in hydrothermal solution: standard partial molal thermodynamic properties and relative stabilities at high temperatures and pressures. Geochim Cosmochim Acta 57:3835–3846

Shock EL 1989 Corrections to 'Organic acid metastability in sedimentary basins.' Geology 17:572–573

Shock EL 1990 Geochemical constraints on the origin of organic compounds in hydrothermal systems. Orig Life Evol Biosphere 20:331–367

Shock EL 1992a Chemical environments in submarine hydrothermal systems. Orig Life Evol Biosphere (suppl) 22:67–107

Shock EL 1992b Stability of peptides in high temperature aqueous solutions. Geochim Cosmochim Acta 56:3481–3491

Shock EL 1994 Application of thermodynamic calculations to geochemical processes involving organic acids. In: Lewan MD, Pittman ED (eds) Organic acids in geological processes. Springer-Verlag, New York, p 270–318

Shock EL 1995 Fluid mixing, organic synthesis and life at high temperatures: proceedings of the Geological Society of America annual meeting, November 1995, New Orleans, LA. Abstr Prog 27:312

Shock EL, Helgeson HC 1988 Calculation of the thermodynamic and transport properties of aqueous species at high pressures and temperatures: correlation algorithms for ionic species and equation of state predictions to 5 kb and 1000 °C. Geochim Cosmochim Acta 52:2009–2036

Shock EL, Helgeson HC 1990 Calculation of the thermodynamic and transport properties of aqueous species at high pressures and temperatures: standard partial molal properties of organic species. Geochim Cosmochim Acta 54:915–945

Shock EL, Schulte MD 1995 Hydrothermal systems as locations of organic synthesis on the early Earth and Mars: proceedings of the American Geophysical Union Fall Meeting, San Francisco, 1995. American Geophysical Union, Washington DC

Shock EL, Schulte MD 1996 Hydrothermal organic synthesis on the early Earth and Mars. Orig Life Evol Biosphere, in press

Shock EL, Helgeson HC, Sverjensky DA 1989 Calculation of the thermodynamic and transport properties of aqueous species at high pressures and temperatures: standard partial molal properties of inorganic neutral species. Geochim Cosmochim Acta 53:2157–2183

Shock EL, Oelkers EH, Johnson JW, Sverjensky DA, Helgeson HC 1992 Calculation of the thermodynamic properties of aqueous species at high pressures and temperatures: effective electrostatic radii, dissociation constants, and standard partial molal properties to 1000 °C and 5 kb. J Chem Soc Faraday Trans 88:803–826

Shock EL, McCollom T, Schulte MD 1995 Geochemical constraints on chemolithoautotrophic reactions in hydrothermal systems. Orig Life Evol Biosphere 25:141–159

Sleep NH, Zahnle LJ, Kasting JF, Morowitz HJ 1989 Annihilation of ecosystems by large asteroid impacts on the early Earth. Nature 342:139–142

Tunnicliffe V 1991 The biology of hydrothermal vents: ecology and evolution. Oceanogr Mar Biol Ann Rev 29:319–407

Usselman TM, Hodge DS 1978 Thermal control of low-pressure fractionation processes. J Volc Geotherm Res 4:265–281

Von Damm KL 1990 Seafloor hydrothermal activity: black smoker chemistry and chimneys. Ann Rev Earth Planet Sci 18:173–204

Von Damm KL, Edmond JM, Measures CI et al 1985 Chemistry of submarine hydrothermal solutions at 21° N, East Pacific Rise. Geochim Cosmochim Acta 49:2197–2220

Woese CR, Kandler O, Wheelis ML 1990 Towards a natural system of organisms: proposal for the domains Archaea, Bacteria, and Eucarya. Proc Natl Acad Sci USA 87:4576–4579

DISCUSSION

Russell: What is the composition of Archaean seawater?

Shock: In our calculations we took present-day seawater and removed the dissolved oxygen. There are many more things you can do to perhaps make this more realistic.

Nisbet: If the mid-ocean ridges were komatiitic in the late Hadean, and if the continents were fairly restricted, then the hydrothermal exchange reactions with komatiites would have controlled the seawater composition. What do you think the late Hadean or early Archaean seawater was like, say 4–4.2 Ga ago? Can you make some guesses on sodium content, for instance?

Shock: Not at this point.

Nisbet: The systems were presumably dominated by the komatiites.

Shock: It is true that if there weren't any continents, it would be rather hard to have continental-style weathering. I haven't thought through the models that you would have to use to show this.

Nisbet: I should explain, because this is a fairly abstruse geological argument. We're trying to measure the temperature history of the Earth's mantle, and when you get back beyond about 2.5 Ga, i.e. into the Archaean, there are these rather strange rocks which are high magnesium in composition and which flow very far, called komatiites (from the Komati River in South Africa). These are formed by extremely hot lavas and rarely occur in recent geological history. But some of the komatiite islands would have been wide and may have had an extensive land, so there would be subaerial weathering, even if the continental crust was limited.

Shock: Again, it comes down to the composition of the atmosphere. There are many unknowns.

Des Marais: Recently, I heard it proposed that komatiites were produced only at the locations of hotspots and that there was much more basalt than komatiites during the Archaean.

Nisbet: In the late Archaean, the hot areas in the mantle would have produced komatiite islands, like giant flat Hawaiis, over hotspots. But in the very early Archaean, 4–4.2 Ga, it's quite likely that the mid-ocean ridges themselves were komatiitic. As time went on, the ridges changed to erupting basalt, as they do today, but the hotspots continued to produce komatiites until the end of the Archaean.

Davies: Do we know anything at all about the depth of the ocean of 4.2 Ga? Would there have been any dry land at that time?

Nisbet: The komatiite volcanoes would have gone to build edifices on the surface, so there would have been islands. There's something called the 'Hess relationship' between the depth of the ocean and the continents (Hess 1962; see also Nisbet 1987, p 146–151). If you get a deep ocean, you get continental volume that's considerable but the area is small. We know from the Acasta gneiss that at 4 Ga there is continental material around. Even as early as 4.2 Ga ago there are continental minerals in Western Australia. So there were continents, but they may not have been as extensive as they are today.

Davies: And by continents, you mean necessarily something sticking above the water?

Shock: Do you mean something that is compositionally distinct?

Nisbet: Compositionally, yes, but the Hess isostasy rules mean that something must stick out of the water somewhere.

Des Marais: There could have been a lot of continental material at that time, but we don't know how emergent it was. There was a steeper geothermal gradient, but it is unclear how deep the continental roots were. It could be that there was less emergent land simply because the continents were generally thinner and laterally more extensive during the Archaean than they were later on during the Proterozoic and Phanerozoic.

Walter: We know that by 3.5 Ga there were numerous emergent volcanic edifices.

Russell: I would like to flag the fact that from these systems Everett Shock is giving us octanoate and perhaps longer-chained carboxylic acids 'for free'. Luigi Luisi's group can get enzymic reactions and even polymerase chain reactions to occupy liposomes or micelles composed of octanoate and similar material (Walde et al 1994, Oberholzer et al 1995). This is an astonishing realization for us all: here we've got the potential for abiotic organic synthesis on a plate.

Des Marais: Of course, the essence of this argument concerns the kinetic barriers. One of the lines of evidence for methane being abiogenic in mid-ocean ridge systems is that the temperature for equilibration of CO_2 is 600–700 °C, which is consistent with the fracture front near to where the fluids perhaps most closely approach the magmas. The implication is that if you're able to read those kinds of isotopic temperatures from the methane which has ascended to the seafloor and cooled to ambient temperatures, then equilibration must become difficult below 600 °C. As CO_2 enters a reaction sequence which leads to methane, and a kinetic barrier exists somewhere in the sequence thus allowing intermediate organic species to accumulate, where would this barrier be located in the sequence?

Shock: I'm not sure. It appears from the composition of brines in sedimentary basins that the organic acids (e.g. acetic acid, propanoic acid) in the brines are in redox equilibrium with CO_2 but not methane. Since it is impossible for these fluids to be getting to 600 °C in the sea floor hydrothermal system, the lack of equilibration of the isotopes just means that the CO_2 and methane are not equilibrating in the system isotopically. Thus it is hard to argue that they're equilibrating thermodynamically. CO_2 and methane are out of equilibrium in sedimentary basins and certainly in the hydrothermal vent fluid. Using the reported concentrations of CO_2 and H_2 in an equilibrium calculation, you can argue that there is 'too much' methane. So a kinetic barrier is definitely there between CO_2 and methane: obviously, there would be no methanogens if there wasn't one. The evidence from lower temperatures is that CO_2 can be in redox equilibrium with simple organic compounds. Therefore, at some more-or-less final step of generating the methane, that is the kinetic barrier, and other reactions between CO_2 and organic compounds are plausible. That's why these results look the way they do, because CO_2 is allowed to go to organic compounds, but not to methane.

Des Marais: An important aspect of brines in sedimentary rocks is that the system is initially endowed with an inventory of biologically produced organic matter. Thus, as long as redox conditions permit organics to survive when the sediments are heated, this organic inventory provides a convenient supply of organic compounds without confronting the challenge of achieving a net synthesis of organic compounds from inorganic precursors. Thus one can easily produce organic acids, and the equilibrium mixture might even include CO_2. In a prebiotic hydrothermal system, perhaps the redox and the concentrations of various constituents favoured net organic synthesis, but we are still searching for the mechanism of this synthesis.

Shock: Clearly, the reaction has to be going the other way in a sedimentary basin where you may have isotopic evidence that the carbonate cement in the rocks started out as organic carbon. There are several ways to identify a pathway in a natural system between organic compounds and inorganic compounds. You can trace it isotopically, which tells you that at least something is going on. You can also show that open pathways exist by doing free energy minimization and identifying the compounds present in equilibrium with one another in this metastable sense. If there's a pathway present and it can go one way, why can't it go the other way? Meanwhile, while we are trying to figure out how to do the experiment, there are all these methanogens down there taking advantage of it.

Des Marais: But how was organic matter made on earth before life arose? Methanogens have these wonderful enzymes with which to make their cellular constituents.

Russell: They're [Fe_3NiS_4] enzymes. Awaruite (Ni_3Fe) would have been present in the ancient serpentinized crust (Krishnarao 1964) and it could have played a somewhat similar catalytic role, at least for hydrogenation reactions.

Shock: Mechanistically, it is unclear. All you can do with this thermodynamic approach is to find out where the potential is; where it is worth looking.

Russell: Can I make an appeal? We are becoming besotted with these visually extraordinary black smokers, and are not considering other submarine springs and seepages. For example, today, 20 times as much water circulates through off-ridge systems and deep ocean springs and seepages than through those on-ridge (COSOD II 1987). I think we should be taking the off-ridge systems seriously, especially on a hot early earth (Fehn & Cathles 1986). I don't know of anyone who has managed to get money to study off-ridge oceanic springs and/or seepages. This is what we should be looking for now; we don't need to find yet another black smoker.

Giggenbach: How seriously should one take your finding that carboxylic acids and ketones are stable above 200 °C? There is just the potential that they could form; there's no indication that they are in any way stable. For instance, one can take a geothermal discharge as a model of what is stable and what is not. If you go to a natural gas deposit up to 180 °C, hydrocarbons are stable but all the oxygen-containing compounds have virtually decomposed. Above 180 °C, geothermal systems show that the hydrocarbons and especially methane are increasingly in close equilibrium. If one analyses, for instance, hydrocarbons in a geothermal system like Wairakei, then one still gets approximately the same composition as in a normal natural gas (methane, ethane, propane, butane, etc.). If one goes to 280 °C, they're completely gone and only methane and benzene predominate. So in a hydrothermal system with a lifetime of about 100 000 years, all organics are gone. At 260 °C they are in equilibrium.

Shock: That has never been demonstrated with a metamorphic rock. There are many metamorphic rocks that have considerable carbonaceous material in them that is not graphite. Equilibrium in the C–H–O system is not reached below 400–500 °C during metamorphism. You would have to predict, at 200 °C say, exactly what the equilibrium concentration of the particular organic compound you're looking for would be. If you do that, sometimes you find that the equilibrium concentrations are extraordinarily low and perhaps below the detection limit of whatever equipment you're using to identify them. It depends on the overall oxidation state of the system.

Des Marais: One can gain some insight by an isotope measurement, because you can predict what the isotope distribution should be in equilibrium. You don't have to know what the conditions were, necessarily, to establish whether or not equilibrium occurred. From every set of gases that we've looked at from geothermal systems at temperatures below 250 °C, it looks like it's a sedimentary organic source that's been cracked to form hydrocarbons. This is consistent with disequilibrium, as opposed to complete stable equilibrium. Even where the relative concentrations of the gases can mimic equilibrium in continental geothermal systems, the isotopes in these gases are out of equilibrium (Des Marais et al 1988).

Giggenbach: In present day environments, the hydrocarbons in hydrothermal systems come from decomposition of organic matter. But they approach equilibrium. The destruction of organic matter is fast enough to reduce concentrations to those expected for equilibrium.

Shock: I think you need to identify exactly what the equilibrium concentrations are so that you know exactly what you're looking for.

Russell: Everett Shock is starting from what we know, but as his is such a successful approach, we can look at the other possibility: that is, off-ridge springs. These will have been about 150 °C, and have operated for much longer in a seawater that may have been supporting 10 bars of CO_2 4.2 Ga ago. So we can imagine replacing present day seawater and black smoker chemistry with an acid Hadean seawater providing CO_2 or H_2CO_3 to whatever the vesicle or container might have been. And in this container, or culture chamber, the hydrogen is being produced from the hydrothermal system. They mix and perhaps react within the vesicles on the hot spring mound.

Shock: It is important to keep in mind that the calculations I was showing relate to what happens when you mix two fluids which are far from equilibrium with one another, like seawater and hydrothermal vent fluid. The vent fluid perhaps has equilibrated with the rock to a large extent, but when the two fluids mix rapidly and dynamically, there's an enormous potential for kinetic barriers to be in place. Now if you tap into a hydrothermal system that's been bubbling away on the continent and is in a less dynamic setting, perhaps it has had longer to reach an equilibrium state. I am unaware of any calculations that have shown convincingly that the distribution of organic compounds in any hydrothermal fluid from anywhere corresponds to an equilibrium assemblage — it would be worth seeing.

Davies: How extensive would these off-ridge spring systems have been?

Russell: At present they occur throughout the sea floor at distances of 7–8 km. They were discovered by Anderson et al (1977), and they're found as thermal cusps. There's room for about a million of them on the sea floor (Macleod et al 1994).

Davies: What does that translate to as a proportion of the sea floor?

Shock: It depends on how you want to characterize the location of these systems. One way is to look for areas where the heat flow that you measure is not conductive and has been altered by convective transport of water. This is true around all of the ridges out to rocks that are about 65 Ma old on either side of the ridge. Ridges spread at different rates, but rocks younger than 65 Ma show evidence of hydrothermal circulation from heat flow measurements. If you add that up, it's equivalent to the area of the continents on Earth at present. In that entire area there could be more-or-less diffuse hydrothermal systems (in the sense of temperatures that are higher than ambient sea water, not necessarily 350 °C).

Russell: One other thing about these hydrothermal circulation systems is that, unlike the acid black smokers, low temperature seepages on the ocean floor will be alkaline. The chemistry of the springs is quite different to that of the black smokers and they would react differently with ocean water. Moreover, there would have been many more of them 4.2 Ga ago because the oceanic crust would have been subject to a much higher heat flow than today.

Everett Shock demonstrates that there is a thermodynamic tendency for organic molecules to be generated when highly reduced hydrothermal fluid mixes with ocean waters containing carbonate. Yet there must be kinetic barriers, otherwise such important reactions would be a part of everyday chemistry. I would like to address the question, 'what were the natural catalytic mixing chambers in which hydrogenation of carbonate took place about 4.2 Ga ago?' Firstly I would like to describe one possible container suggested from our studies of ore deposits. These are hollow pyrite botryoids about 0.1 mm in diameter from the 350 Ma old Tynagh base metal sulfide deposit in Ireland (Russell et al 1994, Russell 1996a). I think they represent fossilized bubbles of iron monosulfide. We have managed to grow similar structures in the laboratory which are comprised of iron monosulfide membrane. The membrane is about a micrometre or so thick. The compartments so enclosed are between 10 and 20 μm across. To do this we simulate the medium temperature off-ridge spring waters (pH~9 according to Macleod et al 1994) with a 0.5 M solution of sodium sulfide which is introduced through a fine hole in the base of a visijar containing a 0.5 M aqueous solution of ferrous chloride. The latter solution represents the Hadean ocean, made acid (pH~5.5) and hot (~90 °C) by the high partial pressure of CO_2 in the early atmosphere (Kasting & Ackerman 1987). You can watch (Fig. 1 *[Russell]*) as the iron monosulfide bubbles are distended and weaken and fresh bubbles are formed at the surface of the growing sulfide mound.

The spontaneous precipitation of the membrane serves to hold the two solutions apart and far from equilibrium. The electrical and electrochemical potential either side of such a membrane developing on the floor of the Hadean ocean would have amounted to about 0.5 V (Russell et al 1994). Moreover, the hydrothermal solution, having equilibrated with the magnesium iron silicates, magnetite and pyrrhotite comprising the basaltic or komatiitic ocean crust, would have delivered hydrogen, ammonia and, according to Everett (Shock 1992), minor quantities of acetate and possibly some organic sulfides (Kaschke et al 1994). Obviously, earliest life could not have been

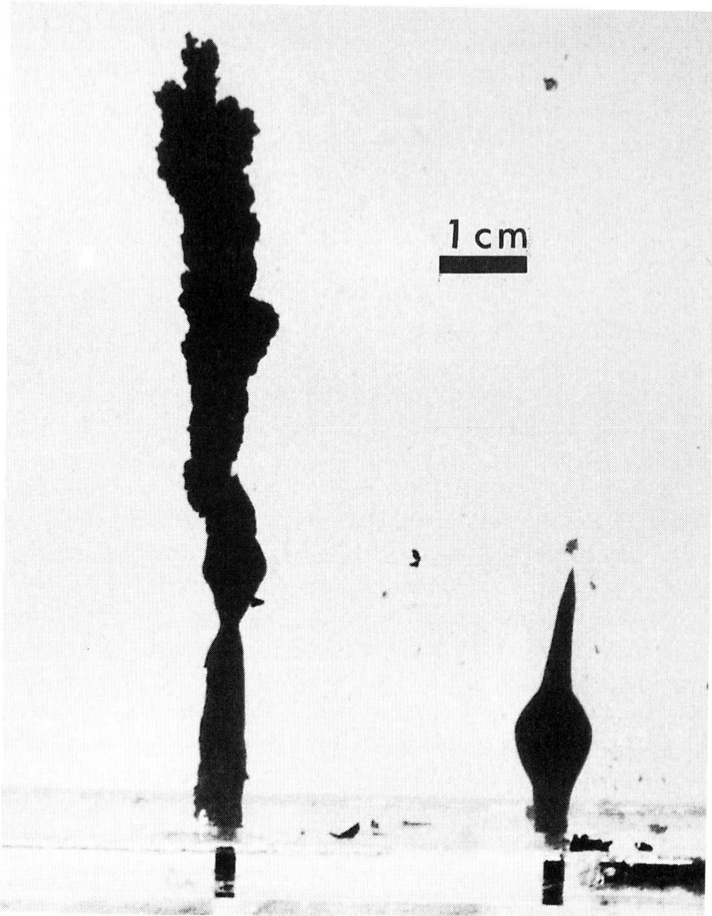

FIG. 1. *(Russell)* Structure formed when 0.5 M sodium sulfide solution is injected into a 0.5 M solution of ferrous chloride (field of view 8 cm).

inventive. Yet all known living cells use a common energy transduction system, the so-called protonmotive force. The cell pumps hydrogen ions to the outer surface of the membrane, so that on their return journey they generate ATP from ADP and inorganic monophosphate. ATP is the energy currency of the cell. In our model of the emergence of life, where an alkaline hydrothermal solution interfaces with an acid ocean across an iron monosulfide membrane, a potential protonmotive force already exists as an aspect of the initial geochemical conditions on the early Earth. In fact the 0.5 V potential that I have alluded to is about twice what is required to run modern cellular metabolism, although we must admit that we don't know how this energy potential was originally harnessed. Yet the iron monosulfide comprising the membrane is likely to have had catalytic properties, augmented by micro-porosity (cf. Bedard et al 1989), protoferredoxin formation (e.g. Bonomi et al 1985) and the presence of some nickel. This is the milieu in which we imagine Everett's reactions to have been catalysed by the hydrogenation of the oceanic carbonate. Let us remind ourselves that the hydrogen in the Kansas gas fields has been generated by the reduction of water as it reacted with mafic intrusives (Coveney et al 1987, and see Neal & Stanger 1983,1984). Perhaps hydrothermal acetate and organic sulfides were the initiators and primers for such organic synthesis.

The generation of the small quantities of the long-chained thiols as well as the carboxylic acids and other organic molecules suggested by Everett, would have given the predominantly inorganic membranes durability, flexibility and an enhanced catalytic activity (Russell et al 1994). But as the syntheses became more successful — and there would have been a positive feedback here — then suddenly these reactive thiols and carboxylic acids would have threatened the integrity of the iron monosulfide membrane, eventually destroying it. They could then organize themselves for an organic take-over. The newly successful organic membrane would have enclosed protocells an order of magnitude smaller than their inorganic parent. And they would have been much more flexible, and could have retained the best of the old order, that is the protoferredoxins. These iron sulfur peptides, with their propensity to transfer electrons, would have catalysed both the synthesis and the cleavage of organic molecules, much as they do today.

At first, these reproducing organic cells would also have remained coupled to the off-ridge hydrothermal convection systems. Such coupling is significant to the exploration for fossil evidence of life on Mars. Sulfide microbialites generated at palaeo hot springs or seepages might be expected to occur over faults, whereas petrified hydro-magnesite microbialites could have developed over buoyant meteoric plumes emanating from deltas in marine or lacustrine sediments (Russell et al 1993, 1996b, cf. Ilich 1952, Lur'ye & Gablina 1972).

References

Anderson RB, Langseth MG, Sclater JG 1977 The mechanisms of heat transfer through the floor of the Indian Ocean. J Geophysical Res 82:3391–3409

Bedard RL, Wilson ST, Vail LD, Bennett JM, Flanigen EM 1989 The next generation: synthesis, characterization and structure of metal-sulfide-based microporous solids. In: Jacobs PA, Van Santen RA (eds) Zeolites: facts figures and future. Elsevier, Amsterdam, p 375–387

Bonomi F, Werth MT, Kurtz DM 1985 Assembly of $Fe_nS_n(SR)^{2-}$ (n = 2,4) in aqueous media from iron salts, thiols and sulfur, sulfide, thiosulfide plus rhodonase. Inorganic Chem 24:4331–4335

COSOD II 1987 Report of the Second Conference on Scientific Ocean Drilling. Joint Oceanic Institutions for Deep Earth Sampling/European Science Foundation, Washington/Strasbourg

Coveney RM, Goebel ED, Zeller EJ, Dreschhoff GAM, Angino EE 1987 Serpentenization and the origin of hydrogen gas in Kansas. Bull Am Assoc Petroleum Geologists 71:39–48

Des Marais DJ, Stallard ML, Nehring NL, Truesdell AH 1988 Carbon isotope geochemistry of hydrocarbons in the Cerro Prieto Geothermal Field, Baja, California Norte, Mexico. Chem Geol 71:159–167

Fehn U, Cathles LM 1986 The influence of plate movement on the evolution of hydrothermal convection cells in the oceanic crust. Tectonophysics 125:289–312

Hess HH 1962 History of ocean basins. In: Engel AEJ, James HL, Leonard BF (eds) Petrological studies: a volume in honor of A. F. Buddington. Geol Soc Am, Boulder, CA, p 599–620

Ilich M 1952 Magnezitso leziste 'Bela Stena'. Zbornik radova Geoloskog I Rudarskogfaculteta TVS. Sv 1. Beograd

Kaschke M, Russell MJ, Cole JW 1994 [FeS/FeS_2]. A redox system for the origin of life. Orig Life Evol Biosphere 24:43–56

Kasting JF, Ackerman TP 1987 Climatic consequences of very high carbon dioxide levels in the Earth's early atmosphere. Science 234:1383–1385

Krishnarao JSR 1964 Native nickel-iron alloy, its mode of occurrence, distribution and origin. Econ Geol 59:443–448

Macleod G, McKeown C, Hall AJ, Russell MJ 1994 Hydrothermal and oceanic pH conditions of possible relevance to the origin of life. Orig Life Evol Biosphere 23:19–41

Lur'ye AM, Gablina IF 1972 The copper source in production of Mansfeld-type deposits in the West Ural foreland. Geochem Int 9:56–67

Neal C, Stanger G 1983 Hydrogen generation from mantle source rocks in Oman. Earth Planet Sci Lett 66:315–320

Neal C, Stanger G 1984 Calcium and magnesium hydroxide precipitation from alkaline groundwater in Oman, and their significance to the process of serpentinization. Mineralogical Mag 48:237–241

Nisbet EG 1987 The young earth. Allen & Unwin, London

Oberholzer T, Albizio M, Luisi PL 1995 Polymerase chain reaction in liposomes. Chem Biol 2:677–682

Russell MJ 1996a Life from the depths: the generation of ore deposits and life at submarine hot springs. Science Spectra 4:27–31

Russell MJ 1996b The generation at hot springs of ores, microbialites and life. Ore Geol Rev, in press

Russell MJ, Daniel RM, Hall AJ 1993 On the emergence of life via catalytic iron sulphide membranes. Terra Nova 5:343–347

Russell MJ, Daniel RM, Hall AJ, Sherringham J 1994 A hydrothermally precipitated catalytic iron sulphide membrane as a first step toward life. J Mol Evol 39:231–243

Shock EL 1992 Chemical environments of submarine hydrothermal systems. Orig Life Evol Biosphere (Suppl) 22:67–107

Walde P, Goto A, Monnard P-A, Wessicken M, Luisi PL 1994 Oparin's reactions revisited: enzymatic synthesis of poly(adenylic acid) in micelles and self-reproducing vesicles. J Am Chem Soc 116:7541–7547

Chemical and physical context for life in terrestrial hydrothermal systems: chemical reactors for the early development of life and hydrothermal ecosystems

Richard W. Henley

Etheridge Henley Williams, PO Box 250, Deakin West, ACT 2600, Australia

Abstract. The diversity of terrestrial hot spring systems, resulting from the large scale coupled transfer of heat and mass in the Earth's crust, maximizes opportunities for evolving ecosystems by the continuous supply of nutrients (P, N, C, S) together with the metals (e.g. K, Mg, Mo, Zn) essential to biogenesis. Cyclic, evaporative micro-environments are common, and potentially catalytic mineral surfaces are also continually created through rock alteration and mineral deposition in and around hot springs. These dynamical systems constitute highly interactive, open, chemical environments capable of establishing complex biochemical microreactors. Volcanic collapse settings on oceanic islands, provide a highly dynamic scenario for the initiation of life and development of diverse ecosystems at the earliest stages of development of the Earth's crust.

'Diversity is a survival factor for the community itself' Daniel Quinn—Ishmael

1996 Evolution of hydrothermal ecosystems on Earth (and Mars?). Wiley, Chichester (Ciba Foundation Symposium 202) p 61–82

What chemical stages were involved in the early steps of biogenesis (Joyce 1989) and the evolution of ecosystems? Were inorganic substrates essential to the catalytic synthesis of proteinoids and protobiont molecular aggregates (Ferris 1994)? Complementary questions focus on energy conversion as a prime requirement of life, and channel debate onto the availability of key molecular components including carbon, nitrogen, sulfur and phosphorus compounds. Where could the combinations of appropriate chemical and physical processes commonly occur which might establish a high probability for these steps to have been taken to initiate life?

In this paper, the microenvironments generated within terrestrial[1] hydrothermal systems will be reviewed in the context of their diversity and, therefore, their potential to develop not one but several early biogenic pathways and ecosystems. The hot spring environments under consideration are all characterized by extreme complexity and periodicity. All represent the natural equivalent of stirred, open-system chemical reactors replenished by both reactive carbon, nitrogen and sulfur species, and trace elements. All incorporate precipitates and alteration minerals suitable for the templating of biogenic reactions. Moreover, hydrothermal system discharges *focus* chemical processes in space and time whereas many of the other common physicochemical processes in the hydrosphere are *dispersive* and therefore potentially lower the probability of organic evolution. The emphasis here is on the common factors from which the diversity of terrestrial hot spring and vapour discharges arises, tracking in turn through to the wide variety of physical and chemical conditions available for the initiation of life and the development of early terrestrial ecosystems.[2]

Dynamics of hydrothermal systems

Hot springs in volcanic settings represent only part of the discharge of large scale, high power hydrothermal systems. Drilling to depths of 3 km or so into these systems for exploitable energy (Fig. 1) has encountered temperatures up to and beyond the critical point of pure water (374.1 °C). The chemical and physical systematics of a large number of such active terrestrial systems and their fossil counterparts have been reviewed extensively elsewhere (e.g. White 1981, Henley & Ellis 1983). The geological character of these systems is also well known through the exploration of their fossil equivalents for metal deposits. Indeed, the distribution of major gold deposits is itself a record of the time and space distribution of strong crust and mantle lithosphere interactions.

Hydrothermal systems occur on the scale of tens of kilometres in volcanic settings. They are manifest subaerially as hot springs and vapour discharges whose compositions and temperatures reflect the extent of deep hydrothermal fluid and shallow ground water interactions (Fig. 1). They are commonly associated with strongly altered rocks (Browne 1978). Fluid flow within these systems occurs through a complex interplay of high permeability fracture networks and lower-permeability, partially to completely altered, host rocks. The former are most commonly developed as a consequence of co-active fault movement so that hydrothermal discharges are localized by crustal scale structures. Individual hydrothermal systems appear to operate on time scales from tens of thousands to a few million years dependent on the intrusive and cooling history of the magma systems which power them (Cathles 1981). However,

[1]Terrestrial hydrothermal ecosystems, which discharge to the land–atmosphere interface are distinguished in this paper from submarine systems which discharge to the ocean–sediment interface. The characteristics of these have been extensively reviewed elsewhere.

[2]I here use the term ecosystem to refer to any simple combination of 'individual' living entities. As such it applies immediately that life is initiated, however transiently.

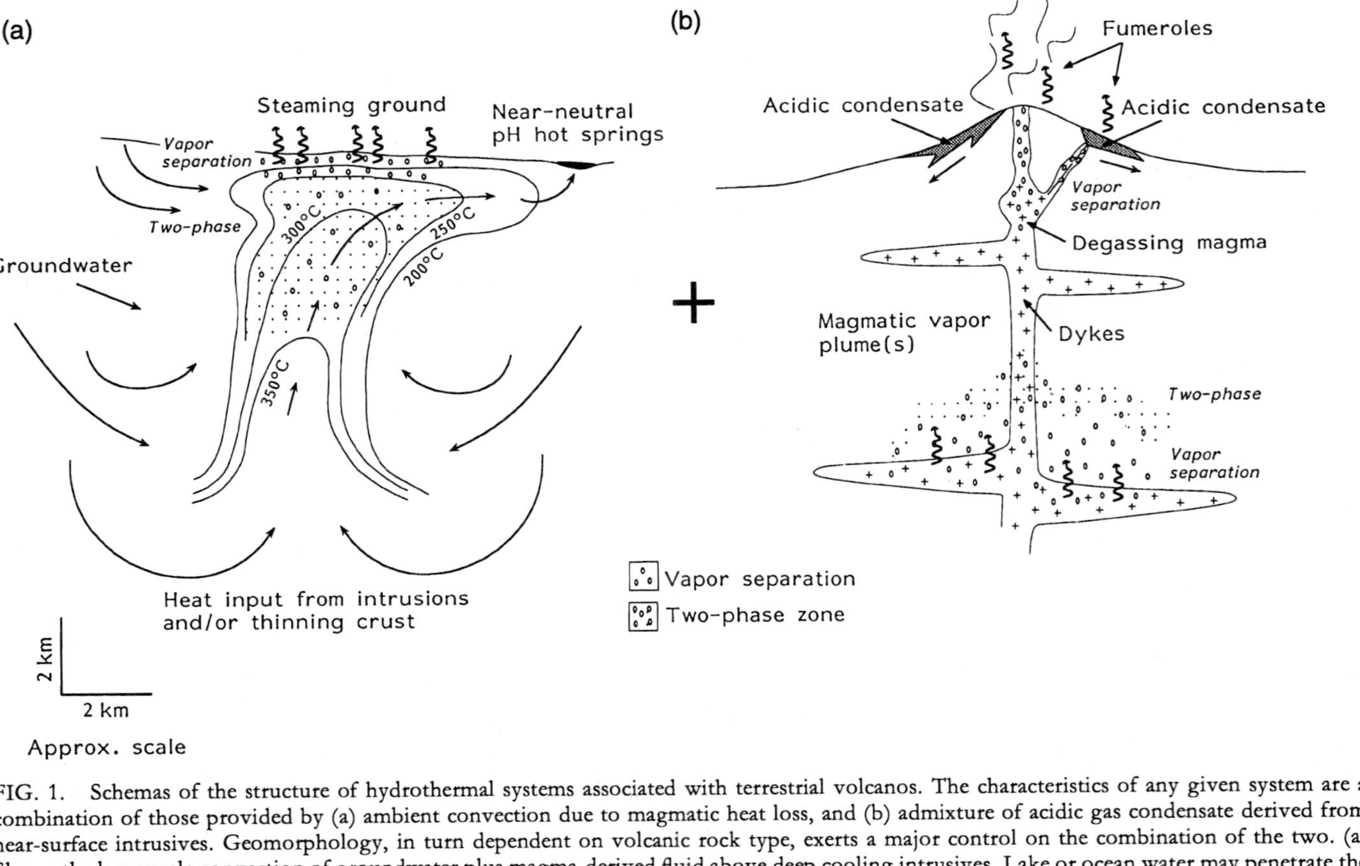

FIG. 1. Schemas of the structure of hydrothermal systems associated with terrestrial volcanos. The characteristics of any given system are a combination of those provided by (a) ambient convection due to magmatic heat loss, and (b) admixture of acidic gas condensate derived from near-surface intrusives. Geomorphology, in turn dependent on volcanic rock type, exerts a major control on the combination of the two. (a) Shows the large scale convection of groundwater plus magma-derived fluid above deep cooling intrusives. Lake or ocean water may penetrate the systems as part of the shallow recharge shown here to be dominated by rainwater. (b) Shows the loss of HCl, SO_2, CO_2, etc., from shallow intrusives in volcanos and the development of downward-flowing condensate which is enhanced in trace metal content by rock leaching.

they wax and wane over longer time scales ($>$15 Ma) where there is a protracted, belt-scale history of magmatic activity. As well as these long term system-scale changes, the outflow rates and chemistry of individual springs undergo transient changes in response to seismicity and progressive changes due to the effects of rapid burial and erosion in active volcanic settings.

Hot springs, fumaroles and solfatara

The characteristics of any given hydrothermal system are determined by the interplay of two sets of processes as shown schematically in Fig. 1. Near-surface magma degassing (Fig. 1) causes the development of high temperature fumaroles ($>$600 °C)[3] and vigorously boiling, highly acidic hot springs. Fluids from these and their sub-surface reservoirs react rapidly with consanguineous volcanics to produce mineral assemblages dominated by alunite, kaolinite and sulfur, and polymorphs of silica. Pyrite precipitates are also common in acid springs, and warm, iron-rich springs may occur in the distal outflows on the flanks of volcanoes. In systems developed from much deeper degassing magmas, acidic gaseous components (HCl, SO_2) are neutralized deep in the system by rock reaction so that fluid mixtures discharge as near neutral pH to alkaline hot springs dominated by recycled surface and groundwaters rather than by acidic magmatic gas condensates.

Essentially all of the naturally occurring elements are present within hydrothermal system discharges to some degree. The major element compositions of deep system fluids are determined by aluminosilicate, sulfide and carbonate mineral reactions (Ellis 1970, Giggenbach 1980, 1981, Henley & Ellis 1983, Henley et al 1984). Gas concentrations are controlled by magma degassing and their ratios by temperature-dependent homogenous and heterogeneous reactions (Giggenbach 1980). For example, ferrous iron released by alteration of common ferrous aluminosilicates (pyroxenes, amphiboles, etc.) reacts with H_2S to produce hydrogen and pyrite (FeS_2) via an FeS* intermediate, thus controlling the redox state of the system fluid. Carbon dioxide, commonly up to a few weight percent in the fluid, and nitrogen, react with hydrogen to control methane and ammonia contents. The abundances of trace metals are related to the temperature, salinity and gas content of the fluid (Henley 1985).

Figure 2, based on well-studied systems, provides a summary of the hot-spring environment, emphasizing: (a) the diversity of environments developed by mixing processes; (b) the availability of key nutrients for biosynthesis and the support of ecosystems; and (c) the extent of sub-surface regimes which may support microbial life (below 150 °C) in hydrothermal systems.

---

[3]Werner Giggenbach (unpublished ms.) has shown that low concentrations of aliphatic hydrocarbons, and simple aromatic and substituted aromatic compounds are also present in hot springs and in high temperature volcanic fumaroles. Based on their redox potential and CO_2 partial pressures, these data allow an order of magnitude estimate of 0.2 to 20 tonnes of benzene per year for the quiescent stage of volcanism in a single volcano.

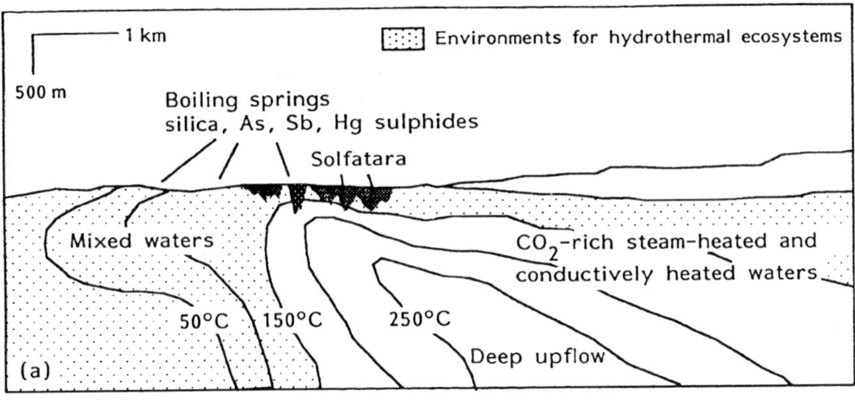

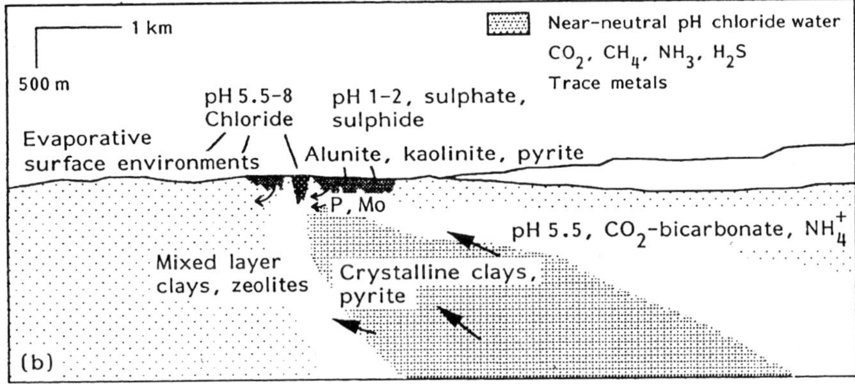

FIG. 2. Schematic cross section of the near-surface (a) groundwater interactions and thermal structure and relative locations of principle discharges, and (b) characteristics of water types and alteration minerals. Based on well-studied and drilled geothermal fields in New Zealand, Yellowstone National Park (USA) and elsewhere.

The thermal and chemical characteristics of discharges are determined by the extent of gas loss and steam condensation processes and their interface with either, or both, the atmosphere and the hydrosphere. These lead to a wide range of chemical compositions as well as physical characteristics. For example, hot springs range in temperature and composition from boiling, near-neutral-pH pools derived directly from deep system fluids (Table 1) through to warm springs derived by mixing of deep fluid with shallow groundwater. Simple mixing relations occur between temperature and the concentrations of non-volatile components.

The discharge of steam from shallow underground boiling occurs directly, as fumaroles with temperatures up to 180 °C, and indirectly by condensation into shallow perched groundwater flows to develop, through oxidation of H_2S, steam-heated

TABLE 1 A selection of analyses of hot waters from active geothermal fields in New Zealand

		Temperature (°C)	*pH*	*Li*	*Na*	*K*	*Rb*	*Cs*	*Mg*	*Ca*	SiO_2	*B*	NH_3	*Cl*	SO_4	*P*	HCO_3
Champagne Pool	Waiotapu	75	5.7	—	1130	161	—	—	0.04	33	385	—	—	1900	100	—	400
Frying Pan Lake	Waimangu	52	4.5	4.1	514	45	0.24	0.26	1.7	4.3	389	7	2.7	674	242	100–300	—
Fumarole Condensate	Tauhara	98	2.5	—	30	2	—	—	<3.0	16	—	—	—	—	—	—	—

All data in mg/kg. For sampling information and further data see Henley et al (1986).

waters and solfatara (Henley & Stewart 1983). The latter are characterized by the intense alteration of rock to kaolinite, alunite, amorphous iron oxide-hydroxides, sulfur and a variety of readily soluble alums. Mixed-layer smectite and smectite–illite–chlorite clays dominate kaolinite away from these acidic flow regimes. Pyrite deposition occurs in acidic springs where iron derived from local rock alteration reacts with H_2S under sulfur saturation conditions and black sulfur vacuoles are commonly present. Such settings tap vapour exsolved by deep fluid boiling and therefore carry relatively high ammonium concentrations. Steam derived by deep boiling may be condensed into overlying groundwater flows to develop CO_2-rich fluids (Fig. 2a) which discharge as warm bicarbonate-rich springs often associated with travertine deposits (Hedenquist & Stewart 1985).

The alteration minerals associated with lower-temperature, near-neutral-pH fluid discharges (Fig. 2b) are quite different and are dominated by only a few common minerals (Browne 1978). Of particular interest in the context of biogenesis are the zeolite and clay minerals, common sulfides such as pyrite and iron and manganese oxides and hydroxides. The common occurrence of ammonium-substituted alunite, clays and feldspar minerals (Altaner et al 1988, Krohn et al 1987) in hot spring environments is related to the transient invasion of hot water discharge sites by steam or derivative steam-heated waters. As noted below, such mixing scenarios involving acidic fluids maximize the availability of phosphate, particularly if evaporative processes also occur.

Boiling temperatures are common and accompanied by sinter deposition as amorphous to low crystallinity silica. Particularly interesting in such settings are the occurrences of periodic phenomena such as geysers which commonly produce evaporative splash zones tens of metres in extent. Less spectacular, but much more abundant, are a variety of exotic growth forms of amorphous silica (Fig. 3) in the surface outflows from hot springs. These develop by evaporation of capillary membranes in the run-off and splash from boiling springs and geysers. Cady & Farmer (1996, this volume) have shown that these contain biofilms and filamentous bacteria.

The delivery of dissolved silica and suspended alteration clays (largely mixed layer clays) to ephemeral catchment basins in the outflow, including lakes, leads to the accumulation of amorphous silica and clay sediments. In turn the flora and fauna-enriched ecosystems which evolve here relate to the warm microenvironments and to the abundances of major and minor elements.

Complex chemical systems

For any chemical reaction or ecosystem to be sustainable it must discharge its 'waste' as much as it adsorbs reactants or nutrients. Hot springs are always open systems receiving inputs from their environments and continuously or cyclically discharging products. Hydrothermal systems themselves are complex networks of such reactive settings as schematically illustrated in Fig. 4.

All of the microenvironments of hydrothermal systems are characterized by imperfect mixing and boiling or rapid evaporative cooling, leading to far-from-equilibrium

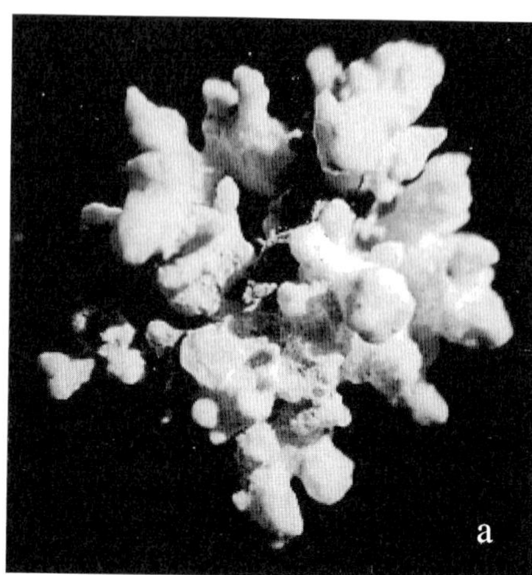

FIG. 3. Silica 'coccolithosilicii' developed by cyclic evaporation of a capillaric film drawn from the warm sub-horizontal outwash from hot springs. Such thin-film evaporative layers will have enhanced contents of major and minor elements and may provide an opportunity for simple membrane formation. (a) A single coccolithosilicus from the outwash of a hot spring (Waiotapu, New Zealand). (b) In this example the space between delicate silica spires becomes infilled by fine volcanic ash and reworked silica and clays (Rotokawa, New Zealand).

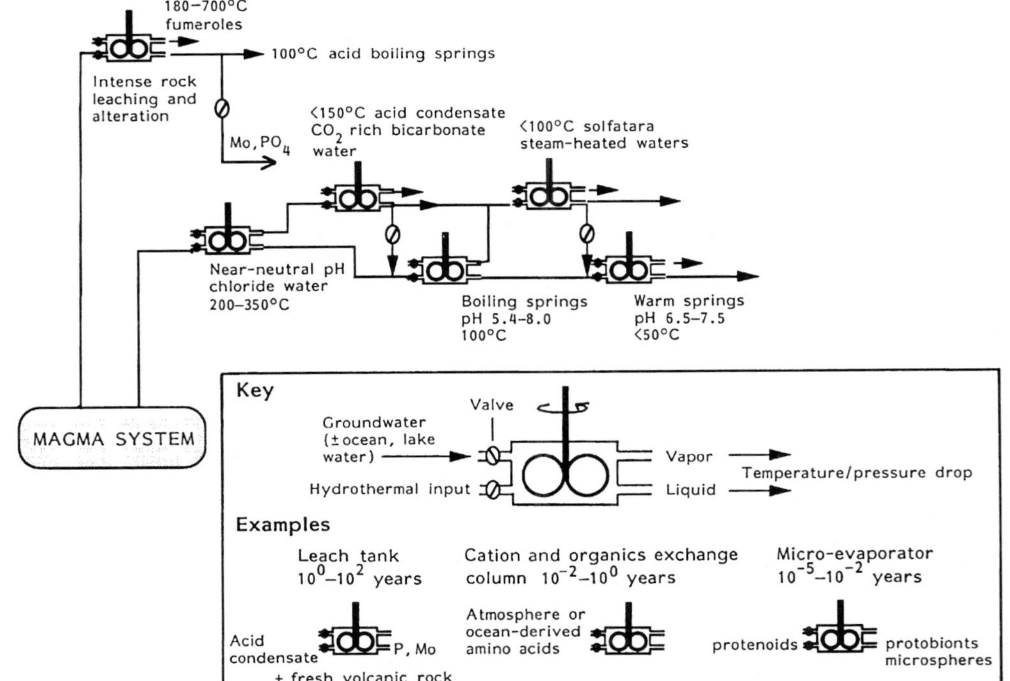

FIG. 4. Schema for the interactive chemical reactor systems which constitute a hydrothermal system. Each 'stirred' reactor has a variable input of hydrothermal fluid and groundwater, and outputs of vapour and liquid, both of which undergo downstream evaporation, cooling and mineral reaction (e.g. to clays and zeolites) and deposition. Note the feedbacks developed between derivative acidic and near-neutral-pH hot waters. Such complex interactivity may lead to conditions which can support biosynthesis and the evolution of diverse ecosystems.

conditions and the potential for autocatalytic chemical reactions (Prigogine 1980, Epstein 1995). Each environment can therefore be regarded as a stirred chemical reactor with variable inputs of groundwater, various derivatives of volcanic gas condensate and hydrothermal fluid (n.b. flux of any component may be zero), and controlled outputs of vapour and liquid. In each reactor different reaction rates determine the sequence of stable and metastable products which develop in response to changing chemical and thermal conditions. Together they constitute a dynamical system which, with the range of nutrients supplied from different sub-systems (e.g. Mo and P from acid leaching), may be capable of providing transient niches for biogenesis, and the subsequent development of diverse ecosystems. Ferris (1992), Shock (1990, 1993) and Simoneit (1993), for example, have highlighted the capacity for organic synthesis by dehydration–condensation reactions in hydrothermal settings.

Much of the debate on the origin of life or of its molecular building blocks, appears to have been oversimplified in that it seeks simple single or sequential reactions to produce self-replicating molecules. The contrary view is that extreme complexity may itself trigger self-replication through the autocatalysis of groups of polymers (Kauffman 1993). The variety of thermal and chemical conditions maintained by hydrothermal systems over periods of 10^5 years or more may have provided the settings within which such complexity and a range of autocatalytic chemical pathways may have been initiated. Their shallow flow regimes also recycle meteoric and oceanic waters which may already contain low concentrations of amino acids derived by electrical discharge in the atmosphere or via comet impact.

The specific ability of mineral surfaces to provide templates for organic synthesis at low concentrations of organic reactants is commonly cited but barely understood. Gedulin & Arrhenius (1994) have drawn attention to the favourability of double-layer hydroxide minerals over clays and zeolites for nucleating organic synthesis. Clays are known to concentrate amino acids and other monomers from dilute solution. Hot spring systems occur within intensely altered rocks dominated by clays, zeolites, sulfides and oxide–hydroxides.

Energy Conversion. A key characteristic of life is its ability to control the capture of energy from, and its release to, its environment, most commonly through ATP. It is open to speculation whether this characteristic evolved from protobiological compounds, or was itself an initiator of organic synthesis through, for example, phosphate adsorption onto mineral surfaces followed by bonding to organic molecules (Gedulin & Arrhenius 1994). The altered rocks around acidic and near-neutral-pH hot springs commonly contain phosphate (Stoffregen & Alpers 1987) and ammonium minerals. The phosphate concentrations of hot spring systems are poorly known. Stauffer & Thompson (1978) showed that the near-neutral-pH hot springs of Yellowstone contained less phosphorus (<2 mg/kg) than more acidic mixed water discharges (up to 21 mg/kg). These concentrations are likely to be controlled by the solubility of fluor- or hydroxy-apatite a common accessory mineral in intrusive rocks and volcanics. At Waimangu, New Zealand, H_2S oxidation at the surface of the large discharge lake, within a hydrothermal eruption crater, leads to acidic (pH 4.5) conditions, high phosphorus concentration (100 to 300 mg/kg) and the deposition of some unusual metal-enriched ferrihydrite precipitates distinguished by high tungsten and phosphorus content (Seward & Sheppard 1986).

An alternative to phosphate-based energy conversion is through reactions such as the deposition of pyrite in sulfur-rich environments. Russell et al (1994) and Keller et al (1994), for example, have discussed possible sea floor environments where such chemo-autotrophic processes may have occurred. The availability of pyrite in acidic hot spring systems is no less an opportunity for experimentation with biogenesis.

Membranes. A second basic characteristic of life is the separation by a membrane of the molecularly highly-organized 'organism' from its far less ordered, less information-enriched environment. These membranes have two key constituents; double layers of phospholipid molecules with minimum 10 C chain lengths (Deamer et al 1994) with polar phosphate groups on the exterior and interior surfaces, and reinforcing molecules such as cholesterol in eukaryotes and hopanoids in bacteria and archaea (Ourisson & Nakatani 1994). Archaea do not utilize lipids but appear to have phosphate groups anchored onto chain hydrocarbons through ether or glycerol, and structurally require no reinforcement. Ourisson & Nakatani (1994) therefore suggest that simple early membranes may have originated through binding of organic molecules (isopentanol) to a solid phosphate surface which fulfilled the role of catalyst. Miyamoto & Stockinius (1971) have shown that once concentrated and re-wetted, such amphiphilic molecules assemble into bilayered vesicles. Terrestrial hot spring systems naturally generate cyclic evaporative environments suitable for such membrane formation. The availability of high phosphate and ammonium contents in evaporative hot springs, the opportunity for osmotic swelling and shrinking of peptide-containing liposomes (Schopf 1983) as salt concentrations varied, and the abundance of new mineral surfaces may therefore have been a prime opportunity for membrane formation processes to occur. Coupled with access to low concentrations of amino acids in rain or seawater, the growth of silica 'coccolithosilicii' (Fig. 3) in and around hot springs has a striking potency for initiating biogenesis!

Role of metals. The varied microenvironments of subaerial hot springs provide essential magnesium by acid leaching of country rock and, in contrast to deep sea systems, are characterized by high potassium : sodium ratios, a characteristic also of living cells. Strongly polar transition metal ions are used both to catalyse organic synthesis and to tailor products through distortion of the electron field of reactant molecules. Living systems require metals (e.g. Zn, Cu, Fe and Mn in nuclei) to control redox processes and catalyse the growth of complex molecules (e.g. Anastassopoulou & Theophanides 1995); trace metals are important in, for example, the stabilization of DNA (Katsaros 1995) and the non-enzymic synthesis of oligothymidic acid (Oro 1994). The laboratory growth media for the yeast *Schizosaccharomyces* and the alga *Chlamydomonas*, for example, require manganese, zinc, molybdenum and copper (E. Williams, personal communication 1995). Hot spring systems discharge high concentrations of metals derived by rock leaching during alteration and directly from degassing magmas (Henley & Ellis 1983, Hedenquist 1995).

Cradles for life?

Given that hot springs and related hydrothermal discharges have the capacities to supply nutrients, remove waste and establish many varied interactive reactive settings, can we be more definitive about the most favourable loci on a primitive planet for the initiation of life?

In this paper, emphasis is placed on sub-aerial hot spring systems as highly favourable biogenic environments. In addition to their diversity, their chemical and physical states — with the possible exception of steam-heated discharges — are essentially independent of the nature of the early atmosphere.

Speculations on specific environments favourable to the initiation of life have frequently invoked oceanic volcanic island settings, recognizing the importance of the interface between volcanic discharges, the primordial ocean and the ambient atmosphere to establish a variety of prospective microenvironments for ecosystem evolution. The favourability of such environments is actually more profound than this because hot spring discharges are highly localized, along with their associated alteration products and mineral precipitates by the dynamic Ghyben-Herzberg buoyancy effect between freshwater and seawater which is characteristic of oceanic island hydrology (Fig. 5). The dynamics of this interface are enhanced by tidal effects. For example, on Lihir Island (Papua New Guinea) mixtures of seawater and thermal water discharge at and below the shore line while boiling steam-heated acidic pools and steam discharges form extensively acid clay-sulfate altered host rocks inland within the Luise caldera (Fig. 5) (Henley & Etheridge 1995).

The 'focusing' effect is amplified by the abundance of highly permeable brecciated rock resulting from volcano collapse and chemically developed secondary permeability due to dissolution of the anhydrite which cements the volcanic rocks and breccias. This powerful chemical and hydrodynamic feedback results in the restriction of almost the entire outflow of the system to a narrowly confined sub-horizontal layer near sea-level. Low concentrations of the phosphates englishite and leucophosphite [$KNaCaAl(PO_4)(OH),H_2O$ and $KFe(PO_4)(OH),H_2O$] have already been identified in kaolinite–alunite altered rocks (Carmen 1994). Other complex sulfate–phosphate minerals are to be expected on the basis of the few detailed studies of acid altered rocks in active and fossil systems (e.g. Stoffregen & Alpers 1987).

On Lihir, these coupled and focused processes precipitate a very large proportion of the trace gold ($<10\,\mu g/kg$) from the upflow system producing the world's largest known gold deposit (>42m ounces of gold) in less than 4000 years (Carmen 1994, Henley & Etheridge 1995). The process continues to this day! The same feedbacks

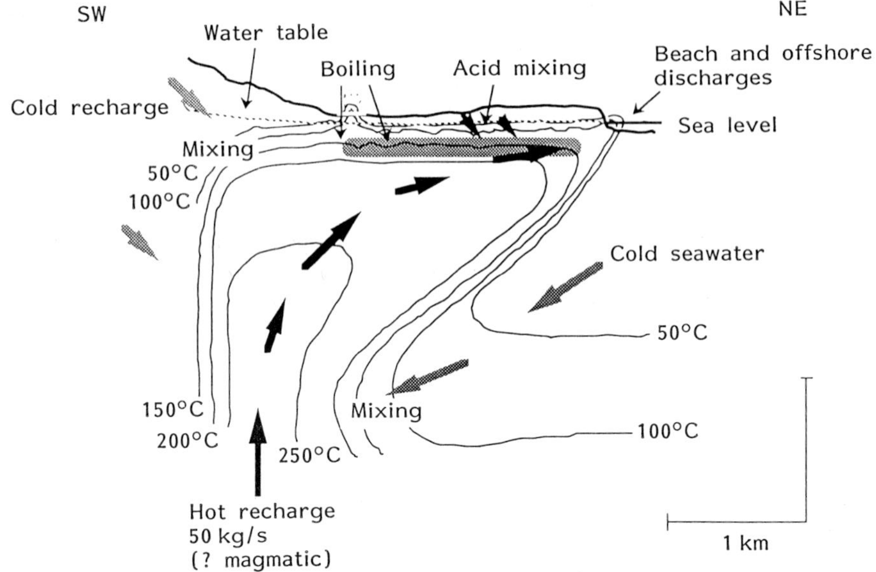

FIG. 5. Outflow systematics for the hydrothermal system in the collapsed Luise caldera, Lihir Island, Papua New Guinea. Note the focusing effect of the Ghyben-Herzberg lens developed between seawater and buoyant low salinity hydrothermal fluid. Alunite–kaolinite–silica alteration constitutes a blanket zone above the chloride water outflow where highly efficient gold deposition occurs. This volcanic hydrothermal scenario is common to many modern volcanoes and is recognized right back to Archaean volcanic settings. The focusing of flow within the Ghyben-Herzberg lens is enhanced by the concavity of the collapsed caldera.

may have been instrumental in maximizing the probability of biogenesis in analogous early Archaean volcanic island systems.[4]

Summary

The familiar hot springs of volcanic terrains each uniquely carry chemical information about the deeper, magmatically developed, hydrothermal system and its interaction with the hydrosphere and atmosphere. Each hot spring represents an open, far-from-equilibrium set of microreactor environments accessing the full range of major and minor elements required to sustain microbiological activity. As complex dynamical systems they may have had the capacity to trigger membrane formation and autocatalytic reaction pathways as the first steps in initiating life. Cyclic evaporative microenvironments could be a key to this process through access to reactants from hot spring,

[4]Lihir itself has no special ecological significance, but its hydrology is well known through the development of the gold resource.

rain and sea waters and an ability to increase their concentrations in films on large surface area solid substrates.

It would be inappropriate to suggest that only one highly specific circumstance initiated Archaean biogenesis. It is more likely that several reactor sequences were established, some of which became sustainable whereas others did not. For example, sulfur-based biogenetic pathways are one such parallel pathway; the availability of reactive mineral surfaces and the abundance of reduced sulfur for the development of sulfur-metabolic hyperthermophiles being common to submarine and some subaerial hot springs. Other pathways, initiated through the availability of a wider range of metals and of phosphorus in terrestrial hot springs, may have developed independently.

Acknowledgements

The invitation from the Ciba Foundation provided an exciting personal challenge to tackle another new interdisciplinary problem. Anthea Munroe helped in digging out technical information and Malcolm Walter was always available to discuss some of the wilder notions that evolved with this paper. Critique and comment from Jeff Hedenquist, Roger Summons, Bill Fyfe and especially Elizabeth Williams were extremely valuable.

References

Altaner SP, Fitzpatrick JJ, Krohn MD et al 1988 Ammonium in alunites. Am Mineralogist 73:145–152

Anastassopoulou J, Theophanides T 1995 The role of metal ions in biological systems and medicine. In: Kessissoglou DP (ed) Bioinorganic chemistry: an inorganic perspective of life. Proceedings of the NATO advanced study institute on bioinorganic chemistry, Rhodes Island, Greece, June 5–17, 1994. Kluwer, Norwell, MA, p 209–218

Browne PRL 1978 Hydrothermal alteration in active goethermal fields. Ann Rev Earth Plant Sci 6:229–250

Cady SL, Farmer JD 1996 Fossilization processes in siliceous thermal springs: trends in preservation along the thermal gradient. In: Evolution of hydrothermal ecosystems on Earth (and Mars?). Wiley, Chichester (Ciba Found Symp 202) p 150–173

Carmen GD 1994 Genesis of the Ladolam Gold Deposit, Lihir Island, Papua New Guinea. PhD thesis, Monash University, Australia

Cathles LM 1981 Fluid flow and genesis of hydrothermal ore deposits. Econ Geol (75th Anniversary vol) p 424–457

Deamer DW, Mahon EH, Bosco G 1994 Self-assembly and function of primitive membrane structures. In: Bengtson S (ed) Early life on Earth. Columbia, New York (Nobel Symp 84) p 107–123

Ellis AJ 1970 Quantitative interpretation of chemical characteristics of hydrothermal systems. Geothermics (suppl) 2:516–528

Epstein IR 1995 The consequences of imperfect mixing in autocatalytic chemical and biological systems. Nature 374:321–327

Ferris JP 1992 Chemical markers of prebiotic chemistry in hydrothermal systems. Orig Life Evol Biosphere 22:109–134

Ferris JP 1994 Chemical replication. Nature 369:184–185

Gedulin B, Arrhenius G 1994 Sources and geochemical evolution of RNA precursor molecules: the role of phosphate. In: Bengtson S (ed) Early life on Earth. Columbia, New York (Nobel Symp 84) p 91–106

Giggenbach WF 1980 Geothermal gas equilibria. Geochim Cosmochim Acta 44:2021–2032

Giggenbach WF 1981 Geothermal mineral equilibria. Geochim Cosmochim Acta 45:393–410

Hedenquist JW 1995 The ascent of magmatic fluid: discharge versus mineralization. In: Thompson JFH (ed) Magmas, fluids and ore deposits. Mineralogical Association of Canada, Nepean, Ontario, Short Course 23, p 263–289

Hedenquist JW, Stewart MK 1985 Natural steam-heated waters in the Broadlands-Ohaaki geothermal system, New Zealand: their chemistry, distribution and corrosive nature. Trans Geotherm Res Council 9:6P

Henley RW 1985 Ore transport and deposition in epithermal environments. In Herbert HK, Ho SE (eds) Stable isotopes and fluid processes in mineralization. Geology Department and University Extension, University of Western Australia, Perth, Publ No 23, p 51–69

Henley RW, Ellis AJ 1983 Geothermal systems ancient and modern: a geochemical review. Earth-Sci Rev 19:1–50

Henley RW, Etheridge MA 1995 The structural and hydrodynamic framework for epithermal exploration. Australasian Institute of Mining and Metallurgy, PACRIM Congress, Auckland. Australasian Institute of Mining and Metallurgy, Melbourne, Victoria, Australia, p 269–278

Henley RW, Stewart MK 1983 Chemical and isotopic changes in the hydrology of the Tauhara Geothermal Field due to exploitation at Wairakei. J Volc Geotherm Res 15:285–314

Henley RW, Truesdell AF, Barton P 1984 Fluid-mineral equilibria in hydrothermal systems. Society of Economic Geologists, Littleton, CO (Reviews in Economic Geology Series 1)

Henley RW, Hedenquist JW, Roberts PJ (eds) 1986 Guide to the active epithermal (geothermal) systems and precious metal deposits of New Zealand. Gebruder Borntraeger, Berlin

Joyce GF 1989 RNA evolution and the origins of life. Nature 338:217–224

Katsaros N 1995 The interaction of heavy metal ions with DNA. In: Kessissoglou DP (ed) Bioinorganic chemistry: an inorganic perspective of life. Proceedings of the NATO advanced study institute on bioinorganic chemistry, Rhodes Island, Greece, June 5–17, 1994. Kluwer, Norwell, MA, p 219–236

Kauffman SA 1993 The origins of order. Self-organization and selection in evolution. Oxford University Press, Oxford

Keller M, Blochl E, Wachtershauser G, Stetter KO 1994 Formation of amide bonds without a condensation agent and implications for the origin of life. Nature 368:836–838

Krohn MD, Altaner SP, Hayba DO 1987 Distribution of ammonium minerals at Hg/Au-bearing hot springs deposits: initial evidence from near-infrared spectral properties. In: Schafer RW, Cooper JJ, Vikre PG (eds) Mineable precious metal deposits of the Western United States. Geology Society of Nevada, Reno, NV, p 661–679

Miyamoto VK, Stockinius W 1971 Preparation and characteristics of lipid vesicles. J Memb Biol 4:252–269

Oro J 1994 Early chemical stages in the origin of life In: Bengtson S (ed) Early life on Earth. Columbia, New York (Nobel Symp 84) p 48–59

Ourisson G, Nakatani Y 1994 The terpenoid theory of the origin of cellular life: the evolution of terpenoids to cholesterol. Chem Biol 1:11–23 (erratum: 1995 Chem Biol 2:631)

Prigogine I 1980 From being to becoming. W. H. Freeman, New York

Russell MJ, Daniel RM, Hall AJ, Sherringham JA 1994 A hydrothermally precipitated catalytic iron sulphide membrane as a first step toward life. J Mol Evol 39:231–243

Schopf JW 1983 The Earth's earliest biosphere: its origin and evolution. Princeton University Press, Princeton, NJ

Seward TM, Sheppard DS 1986 Waimangu geothermal field. In: Henley RW, Hedenquist JW, Roberts PJ Guide to the active epithermal (geothermal) systems and precious metal deposits of New Zealand. Lubrecht & Cramer, Forestburgh, NY (Monograph Series on Mineral Deposits 26) p 81–92

Shock EL 1990 Geochemical constraints on the origin of organic compounds in hydrothermal systems. Orig Life Evol Biosphere 20:331–367

Shock EL 1993 Hydrothermal dehydration of aqueous organic compounds. Geochim Cosmochim Acta 57:3341–3349

Simoneit BRT 1993 Aqueous high-temperature and high-pressure organic geochemistry of hydrothermal vent systems. Geochim Cosmochim Acta 57:3231–3243

Stauffer RE, Thompson JM 1978 Phosphorus in hydrothermal waters of Yellowstone National Park, Wyoming. J Res US Geol Survey 6:755–763

Stoffregen RE, Alpers CN 1987 Woodhouseite and svanbergite in hydrothermal ore deposits: products of apatite destruction during advanced argillic alteration. Can Mineralogist 25:201–211

White DE 1981 Active geothermal systems and hydrothermal ore deposits. Econ Geol (75th Anniversary vol) p 392–423

DISCUSSION

Walter: I would like to return to one of the points you made about the springs focusing information from the broader environment. Maybe I missed part of that point, but you said that with a few rare exceptions, atmospheric information is not preserved within the spring systems. What were those exceptions? I guess you were saying that the broader environmental information, such as the composition of the ocean, can be recorded in some way within fossil mineralogical systems. This is important from the point of view of anybody trying to understand what the early Earth was like. Could you develop this idea?

Henley: There are lots of ways one can decode the information that systems leave behind. First of all, by sampling hot springs and fumaroles, one can track back from the chemical composition via either an empirical approach or a thermodynamic approach to figure out the composition of the deeper system fluids. This can be done for reactive species such as nitrogen, hydrogen and ammonia, which can be analysed directly from discharges, and for any of the minor components such as the benzenes or substituted benzenes which occur within these systems. The alteration minerals tend to record aspects of the fluid that pass them by. They tell us something about the pH, for example, but they may also trap some of the fluid as inclusions which can then be analysed. Some of these preserve high molecular weight hydrocarbons within them. So there are ways, from very limited amounts of rock information, preserved fluids in old systems, or new fluids in existing systems, that one can decode and work backwards down to the source region of hydrothermal systems.

Cowan: You mentioned the capacities of many of these minerals to absorb and concentrate the types of small molecules that are of interest in developing biogenesis. Can you expand on that in terms of the range of absorptive capacities, the concentrations one might expect and the sort of multilayers you would expect on different minerals?

Henley: I can't comment specifically on the numbers, but there is a very large literature on the cation exchange capacities of clay minerals and zeolites, and a lesser literature on the concentration of organic components in those. What struck me as most interesting, however, was the ability of these systems to produce complex microenvironments with large surface areas, such as in the silica precipitates, and, by virtue of the evaporation process, to increase the concentration of initially scarce ingredients. This would also enable you in some circumstances to initiate the development of films or membranes in those trapped environments. That's something I would like to see explored.

Cowan: Presumably these surfaces have a role as catalytic surfaces as well as just absorptive surfaces.

Henley: That's right. Many publications have explored the idea of clays acting as templates for biogenesis. Kauffman (1995) has recently pointed out that it is easier to set up a complex reactor sequence on a surface, because it is two-dimensional, than in a volume, which is three-dimensional. If you're working on any kind of high-surface-area precipitate around a hot spring system, you are maximizing the probability for complex reactions to become self-sustained and self-organized.

Cady: The idea of capillary action as a means of concentrating nutrients at the accretion surfaces of these delicate forms of geyserite dovetails nicely with our ideas regarding processes that could explain the presence of biofilms near the tops of protruding spicules of geyserite. Have you observed biofilms on the surfaces of your samples, or have you looked at them in detail with the scanning electron microscope?

Henley: No. When we go to these sites, we tend to rush to the hot spring because it is rather beautiful and we want to take a sample out of it, and we don't stop to see some of these rather delicate structures. When we first observed them at Waiotapu (New Zealand), we actually called them 'coccolithosilicii', because of their resemblance to coccoliths.

Cady: I've also observed variation in the morphology of spicular geyserite, and in many cases, that variation can be correlated with distance from the edge of the effluent. Even within the biolithofacies zone of spicular geyserite, as defined by Walter (1976), spicules display a variety of shapes and spatial orientations. We have observed high concentrations of biofilms on spicules that form at the air–water interface, whereas spicules that form further out from the pool edge acquire what appear to be crystallographically controlled forms.

Henley: I would be interested to look at the range of metals that accumulate in those. We tend to forget the importance of metals in biological processes. In those environments you can go from a molybdenum content of parts per billion through to percentage levels. That may prevent biofilms from forming in the more distal parts.

Nisbet: To take up the metal theme again: to a geologist, one of the great puzzles is how metals got into proteins. Presumably, the most primitive organisms were not as geochemically skilled as organisms now are, so the metals must have essentially forced themselves into the biological processes. One of the interesting things to me is the specificity of the metals: for instance, hyperthermophiles use tungsten, where other organisms use molybdenum. Perhaps the selection of metals for use in proteins descends from the types of metal concentrations found in particular hydrothermal settings. Above subduction zones one tends to get copper and molybdenum deposits; in black smokers on mid-ocean ridges you get manganese. One thing that is mainly specific to the Archaean is Komatiitic eruptions. Komatiites were high temperature, magnesium-rich lavas with high nickel and chromium contents which would have formed ocean islands like Hawaii, but much lower and of a larger diameter ($\sim$1000 km). A big komatiite eruption would have probably created a huge hurricane above it which would have spread any organic cells around the world. Perhaps a metal like nickel was incorporated into biochemistry around a komatiite volcano. One might argue that each of the particular interesting metalloproteins came from specific hydrothermal settings where the metal forced its way into the biology.

Henley: Because it was at such high concentration?

Nisbet: Yes, perhaps it simply got in accidentally. Likewise, for the minerals, I mentioned earlier the analogy between the zeolites and the heat-shock proteins — they both look like doughnuts. Compare GroES with a zeolite cage. Perhaps some of these minerals have their fingerprints written in life today, but have been replaced by organic substitutes.

Russell: It might be the other way round: that is, that the organics forced themselves onto what were the mineral clusters. After all, Bonomi et al (1985) have managed to coassemble protoferrodoxins, which are iron–sulfur redox catalysts, from iron sulfide clusters and thioethanol in water. These had 80% of the activity of a true peptide-based ferrodoxin. I think it is the metals and sulfur first, and then the organics are made in that milieu.

Pentecost: You mentioned ammonia. Would you say something about the speciation of nitrogen in hot springs, because we hear very little about this?

Giggenbach: Ammonia is quite a common constituent of thermal waters, and also volcanic steam. For instance, we have springs in New Zealand where ammonium sulfate waters are discharged that contain several hundred ppm ammonium balanced by sulfate. Especially in steam-heated areas and on the flanks of volcanoes one can get very high concentrations of ammonia and H_2S which then gets oxidized to sulfate and forms ammonium sulfate. Crater lakes, for instance, have high concentrations of ammonium, in the order of dozens of ppm.

Pentecost: Do you know the origins of the ammonia?

Giggenbach: Inorganic synthesis or from the interaction of volcanic volatiles with igneous rock.

Russell: What about cyanide?

Giggenbach: There was an eruption at Dieng, Indonesia, where people walking through the eruption cloud died. Also, after the Lake Nyos eruption, people reported the smell of cyanide and so a major effort was made to look for it, without success. It was probably once produced in the atmosphere.

Stetter: Is there evidence for nitrogen gas in hydrothermal systems?

Henley: The deeper parts of the system contain a wide range of gases, including nitrogen and hydrogen.

Stetter: Some hyperthermophiles are able to use N_2 as their sole nitrogen source in the laboratory; therefore there should be some N_2 present in these systems.

Giggenbach: At present the mantle which feeds mid-ocean ridge and other hydrothermal systems is very low in N_2. But N_2 is very high in subduction-related volcanoes and along convergent plate boundaries.

Sakai: In many of the island arc volcanic gases, N_2 concentration is very high, but the N/Ar ratio is much higher than the atmospheric N/Ar ratio, which means that N comes from the decomposition of organic matter in crustal materials.

Dick Henley showed that phosphate in acid thermal waters is higher than in neutral ones. Only limited data on phosphate concentration in Japanese acid thermal waters are available, but as far as I know there does not seem to be any clear relationship between water acidity and phosphate concentration.

Henley: This is one of the areas that I specifically tried to follow up for this meeting. You are right: there's very little information about phosphorus in hot springs. In Yellowstone National Park, Stauffer & Thompson (1978) showed that there was a slight increase of phosphate concentration up to about 10 ppm in acid hot springs. The concentration is generally lower than that. These concentrations are probably controlled by the solubility of apatite. But at Waimangu, in New Zealand, where you have an acidic hot lake, the fluid contains 100–300 ppm phosphate. The precipitate that forms is a silica, iron, manganese hydroxide precipitate, very high in phosphorus. So under some circumstances that we don't really understand, we can get high phosphorus concentrations beyond the solubility of apatite. If we can increase that concentration by evaporation locally, where biofilms are also produced, then we have created a nice opportunity for biogenic or biochemical reactions to take place with the involvement of elevated phosphorus and metal concentrations.

Pentecost: Levels of phosphate in the order of 1 ppm are actually quite high. Plants are incredibly efficient at mopping up phosphate. A 1 ppm phosphate system would be productive: phosphate would not be limiting.

Nisbet: We haven't really mentioned the extreme alkali volcanics, such as those which occur today, for instance in East Africa. They were more widespread in the Archaean; nowadays they are very rare. Very often they have large deposits of phosphate rock around them. For instance, Zimbabwe gets its phosphate fertilizer from digging up one such deposit. The problem with phosphorus biologically is that it doesn't travel — it precipitates, and so unless the unskilled early biology grabbed hold of it right there in the volcano, there probably wasn't very much of it around in the environment in an accessible form.

Going back to nitrogen, Jim Lovelock made the point that lightning would have probably fixed most of the nitrogen, which would have gone straight into the sea, and there wouldn't have been much N_2 around. Perhaps that is a limiting step.

Henley: An interesting issue here concerns how much the atmospheric concentration is topped up by eruptive volcanoes.

Nisbet: You can argue that volcanoes above subduction zones, like those that occur in Japan today, are basically just recycling surface nitrogen via the plate tectonic system. There isn't that much coming from deep down, so we're just recycling our store. What was it like in the Archaean? It is not obvious there was much N_2 around.

Giggenbach: Where did the N_2 come from then?

Nisbet: Prebiological N_2 might have degassed over time, or it might have come out early.

Des Marais: A key characteristic of N_2 is that it is not easily subducted: in that respect it behaves similarly to a noble gas. Consequently, the mantle is not being recharged with N_2 as it is with carbon and some of the other more soluble elements. In one model that I've seen N_2 behaves similarly to ^{36}Ar during outgassing (Zhang & Zindler 1993). Prior to 4.3 Ga, the upper mantle was pretty much depleted of N_2. We can't detect outgassing today above the noise level, so N_2 is largely in the atmosphere.

Davies: What's the situation on Mars with N_2?

Jakosky: It's a couple of percent, and there isn't any evidence for significant nitrate ground reservoirs. On the other hand, isotopic evidence for Mars N_2 suggests that about 90–95% N_2 that was there has been lost to space.

Carr: The noble gases are also very low on Mars. These inert gases that stay in the atmosphere were very vulnerable to impact erosion early in the planet's history: a lot of the N_2 and noble gases may have been lost at that time. That remaining has since been fractionated by very modest losses to space.

I'm intrigued — from the perspective of its implications for Mars — with the suggestions that the types of hydrothermal deposits depend on the tectonic environment. Many of these environments don't exist on Mars because there hasn't been any plate tectonics. What are the characteristics of purely basaltic hydrothermal deposits that are totally dissociated from subduction? These are the ones that are most likely on Mars.

Henley: The diversity of hydrothermal springs I described is basically common to all hydrothermal systems regardless of the nature of the volcanics, the alteration minerals, the type of fluid chemistry and so forth. There may be some more exotic ones, the more alkaline ones, but in general they're pretty much the same. It is because of this that we're able to use chemical ratios to identify source region temperatures, fluid flow paths and so forth. One principle variable is the total gas content: those springs around a subduction margin tend to be much higher in gas — CO_2 particularly. Another variable is salinity, because this is controlled largely by the ambient concentration of chloride. If you are looking at a black smoker, you are dealing with a system which cycles ocean water, so you have got 20 000 ppm chloride before you start. If you are looking at an inland terrestrial system the only source of chloride is magmatic gas, and you tend then to find only up to 1000 ppm chloride in the hot springs because of dilution by groundwater in the deep system. Most of the things I have described are common to all hydrothermal systems, and they would probably be common to the Martian systems.

Walter: We shouldn't forget that there are hydrothermal systems in sedimentary basins as well, in amagmatic settings where there are deeply circulating fluids. We don't necessarily have to restrict our thinking to association with a magmatic environment.

Horn: There's one environment which you might want to consider in Ontario. At Sudbury there's a well established impact structure. The result of the impact was an igneous complex. Lying above the complex is a sedimentary basin, within which there are hydrothermal deposits on faults. It is reasonable to assume those hydrothermal deposits are driven by the heat from the intrusion.

Walter: What sort of hydrothermal deposits are they?

Horn: Cu/Zn.

Walter: Are only the subterranean parts preserved?

Horn: No. They are exposed at surface. It's certainly the type of thing you want to think of.

Davies: Richard Henley, if I understood you correctly, you said that a 2D chemical system is more favourable for biogenesis than a 3D one; that things on surfaces would tend to facilitate the production of complex molecules. Why is this?

Henley: It's a suggestion of Stuart Kauffman's. His argument is that you are restricting the search strategy of a molecule to a much more confined region.

Davies: Is that good? Doesn't it restrict the possibilities?

Henley: It increases the possibility of molecules bumping into each other.

Des Marais: Also, larger molecules are stabilized on surfaces.

Davies: But this wouldn't necessarily have to be a surface exposed to the atmosphere. How large a surface are we talking about here? Could we be talking about porous rocks or are we talking about great slabs?

Henley: That's an interesting question. One of the most important things about the subaerial discharges is the variety of environments that they can form compared with submarine ones, and the variety of very complex high-surface-area precipitates, alteration minerals and so forth that they can form. But the important thing to remember about the volcanic systems in particular is that they're continuously evolving. Consequently, the large-surface-area microenvironment that you formed on Monday, by Tuesday may be covered by 100 m of a new volcanic deposit, so you're continuously producing these high surface area sites and *then* they may become subterranean. I'm intrigued about the question as to whether life evolves in the subterranean environment and then it escapes, or whether it evolves in the surface region, where the diversity of discharges is greater, and it then migrates underground.

Davies: It could be either way. But I'm interested to know whether we can identify surfaces sufficiently large, down below, for this process to take place, or whether it must occur on the surface.

Henley: Certainly, alteration clay minerals have very large surface areas.

Des Marais: Clays have surface areas of square metres per gram, as opposed to rocks, which might be only square centimetres per gram. So just having a clay in the subsurface and a predisposition towards reactions on surfaces would make hydrothermal systems chemically reactive.

Sakai: The acidity of the present volcanic thermal waters is created by addition of hydrogen chloride, by surficial oxidation of hydrogen sulfide, by disproportionation of SO_2 at a deeper level, or by a combination of these three processes. But in the Archaean volcanic systems, acidity would have been controlled mostly by hydrogen chloride, not by sulfur species. The amount of SO_2 relative to H_2S in volcanic gases would also have been lower in the Archaean than it is at present because of the lower oxygen pressure.

Walter: Yes, but we would still have had volcaniclastic sediments, for instance, with a high initial porosity. At least in many of these systems there would have been that source of porosity and permeability.

Henley: Just to follow up on that, if you have a volcano that is discharging SO_2, disproportionating to H_2S and sulfuric acid, you are producing strongly leaching acid high in the volcano. Therefore you are able to transport the condensate with its derived solutes to lower levels, where they may interact with other systems. That's one of the key things I was trying to get at in looking at microsystems within the bigger systems. The only environment of the set that I showed you that is definitely controlled by the atmosphere is the development of sulfur in steam-heated environments. Whether that process would occur or not becomes a matter of faith as to what you set the oxygen level at in the early Archaean.

References

Bonomi F, Werth MT, Kurtz DM 1985 Assembly of $Fe_nS_n(SR)^{2-}$ (n = 2,4) in aqueous media from iron salts, thiols and sulfur, sulfide, thiosulfide plus rhodonase. Inorg Chem 24:4331–4335

Kauffman SA 1995 At home in the Universe. Viking Press, London

Stauffer RE, Thompson JM 1978 Phosphorus in hydrothermal waters of Yellowstone National Park, Wyoming. J Res US Geol Survey 6:755–763

Walter MR (ed) 1976 Stromatolites. Elsevier, Amsterdam

Zhang Y, Zindler A 1993 Distribution and evolution of carbon and nitrogen in Earth. Earth Planet Sci Lett 117:331–345

Stable light isotope biogeochemistry of hydrothermal systems

David J. Des Marais

Space Science Division, NASA Ames Research Center, Moffett Field, CA 94035-1000, USA

Abstract. The stable isotopic composition of the elements O, H, S and C in minerals and other chemical species can indicate the existence, extent, conditions and the processes (including biological activity) of hydrothermal systems. Hydrothermal alteration of the $^{18}O/^{16}O$ and D/H values of minerals can be used to detect fossil systems and delineate their areal extent. Water–rock interactions create isotopic signatures which indicate fluid composition, temperature, water–rock ratios, etc. The $^{18}O/^{16}O$ values of silica and carbonate deposits tend to increase with declining temperature and thus help to map thermal gradients. Measurements of D/H values can help to decipher the origin(s) of hydrothermal fluids. The $^{34}S/^{32}S$ and $^{13}C/^{12}C$ values of fluids and minerals reflect the origin of the S and C as well as oxygen fugacities and key redox processes. For example, a wide range of $^{34}S/^{32}S$ values which are consistent with equilibration below 100 °C between sulfide and sulfate can be attributed to sulfur metabolizing bacteria. Depending on its magnitude, the difference in the $^{13}C/^{12}C$ value of CO_2 and carbonates versus organic carbon might be attributed either to equilibrium at hydrothermal temperatures or, if the difference exceeds 1% (10‰), to organic biosynthesis. Along the thermal gradients of thermal spring outflows, the $^{13}C/^{12}C$ value of carbonates and ^{13}C-depleted microbial organic carbon increases, principally due to the outgassing of relatively ^{13}C-depleted CO_2.

1996 Evolution of hydrothermal ecosystems on Earth (and Mars?). Wiley, Chichester (Ciba Foundation Symposium 202) p 83–98

This review summarizes some approaches by which stable isotopes can assist both in delineating modern and ancient hydrothermal systems and in understanding the components and conditions associated with their development. Special attention will be paid to biological features.

Isotopic abundances are expressed as their difference, in parts per thousand, from a standard material, as follows:

$$\delta^{m}E = ([R_{sam}/R_{std}] - 1)1000$$

where m is the mass of the heavier of two isotopes of element E whose ratio, R, is being measured; $R_{sam} = {}^{m}E/{}^{n}E$ of a sample, where n is the mass of the lighter isotope; $R_{std} = {}^{m}E/{}^{n}E$ of the standard. Standards are as follows: oxygen and hydrogen, 'Vienna standard mean ocean water'; carbon, Vienna Pee Dee Belemnite carbonate; sulfur, Canon Diablo meteoritic troilite; and nitrogen, atmospheric dinitrogen.

Oxygen and hydrogen

As water interacts with rock in hydrothermal systems, some minerals approach chemical equilibrium with water and others react to form new minerals. Water in equilibrium with silicates or carbonates has an isotopic composition which is typically enriched in D and depleted in ^{18}O, relative to these minerals. This difference is called the isotopic fractionation factor (α) and is equal to the quotient of the heavy-to-light isotope ratios of the two equilibrated compounds. The magnitude of this factor, α, varies inversely with temperature; therefore it can indicate the temperature at which equilibration occurred. Accordingly, the isotopic compositions of both water and minerals will be altered by hydrothermal activity to reflect ambient conditions (temperature, water/rock mass ratio). The $\delta^{18}O$ and δD values of newly-created minerals will reflect the environment of their formation. Accordingly, isotopic measurements can help to detect hydrothermal activity, delineate the conditions and mechanisms of mineral formation, and characterize sources of magma, sediment or water.

Figure 1 illustrates how the isotopic composition of a volcanic rock in a continental hydrothermal system is affected by exchange with water. The 'meteoric water line' at upper left describes the range of $\delta^{18}O$ and δD values in rainfall, the dominant source of water in hydrothermal systems (e.g. Craig 1963). For the purposes of this discussion, the letters 'W' and 'R' indicate the initial 'unaltered' isotopic compositions of the water and rock, respectively (Fig. 1). The curves describe how these isotopic compositions are affected at different temperatures and water/rock values (w/r), where w/r equals the abundance ratio of exchanged oxygen and hydrogen atoms in the water versus those in the rock. Thus, at $w/r = 0.01$, rock will dominate the interaction; whereas, at $w/r = 10$, water dominates.

Several features merit comment. First, the differences in isotopic composition between water and rock decrease as the temperature increases. Second, rocks typically have much less hydrogen than oxygen available to buffer against changes in isotopic composition; therefore, as w/r increases, rock δD values are significantly altered earlier than are $\delta^{18}O$ values. Third, minerals are typically depleted in D and enriched in ^{18}O, relative to the water with which they have equilibrated.

These isotopic features, together with other aspects such as redox state, mineral composition and rock textures, create a signature of hydrothermal alteration which can be used to map the intensity and extent of ancient hydrothermal systems. This is illustrated by the 55 Ma old Skaergaard intrusion and its country rocks, the hydrothermal alteration of which was delineated by mapping $\delta^{18}O$ and δD values (e.g. Fig. 2, Taylor & Forester 1979). These investigators estimated, among other things, the lateral extent of this continental system, the isotopic composition of the hydrothermal fluid, spatial variations in w/r and permeability, temperatures of alteration, and the

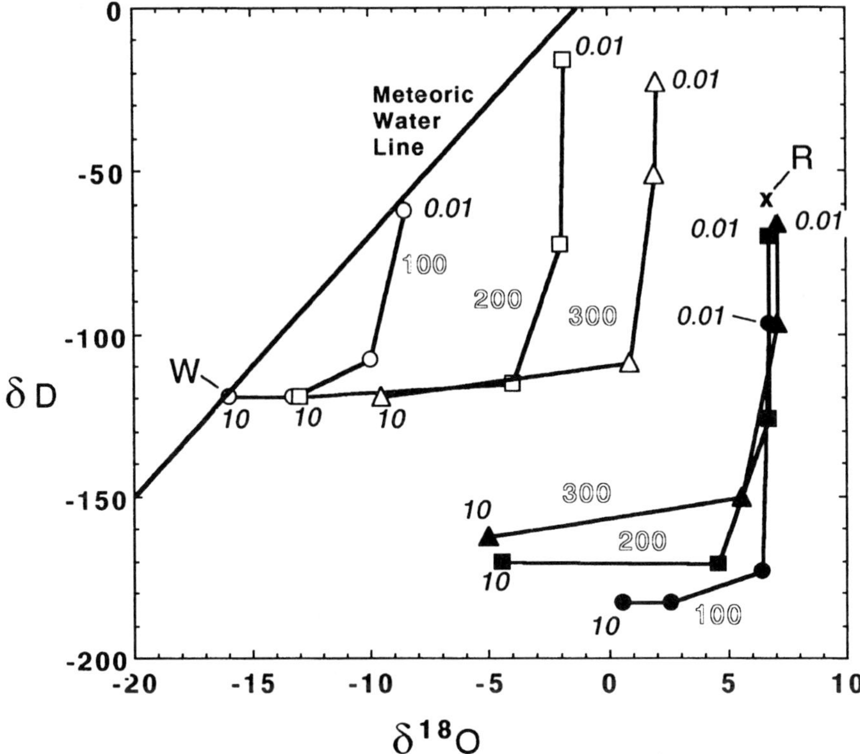

FIG. 1. Hydrogen and oxygen isotopic compositions of water and rock which have equilibrated at hydrothermal temperatures in a continental hydrothermal system (adapted from Field & Fifarek 1985). Temperatures (100, 200 and 300 °C) are indicated in outline lettering adjacent to the curves. Initial isotopic compositions of unreacted water and rock are indicated by the letters 'W' and 'R', respectively. Open data points show values for water; filled points show values for rock. Individual data points correspond to water/rock ratios which vary by one order of magnitude between the points on a given curve (only w/r values 0.01 and 10 are shown). Note that, at low w/r values, equilibrated rock isotope values are similar to R. At high w/r values, values for equilibrated water approach W. The meteoric water line at upper left is shown for reference.

sequence of flow regimes which occurred over the lifetime of the hydrothermal circulation. The outer boundaries of hydrothermal systems have been found, using isotopic measurements, to extend beyond the limits detected by mineral alteration (Field & Fifarek 1985). An excellent correlation between $\delta^{18}O$ and chlorite alteration, together with characteristic absorption bands for chlorite in near-infrared wavelength range, allows hydrothermal $\delta^{18}O$ alteration to be mapped by remote sensing methods (Gillespie & Criss 1984). Isotopic methods also have been applied to marine hydrothermal systems exumed by uplift and erosion (Schiffman & Smith 1988).

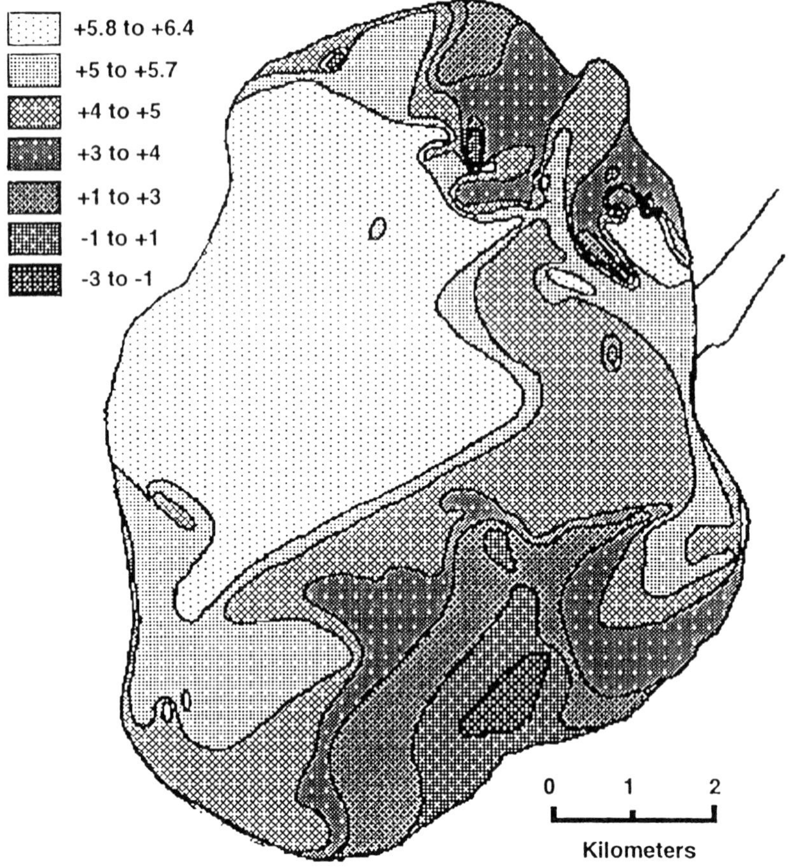

FIG. 2. Map of $\delta^{18}O$ plagioclase from outcrop localities in the Skaergaard intrusion (Taylor & Forester 1979). The most intense hydrothermal alteration corresponds to the lower $\delta^{18}O$ values, depicted at lower right.

Because silica dissolves at high temperatures and is redeposited downstream in cooler zones (White et al 1956), it helps to preserve a record of volatiles and biota in ancient systems (e.g. Vityk et al 1993, Rice et al 1995), particularly at temperatures below 250 °C (Sturchio et al 1990). Because the isotopic fractionation factor for $^{18}O/^{16}O$ between silica and water decreases with temperature (Kita et al 1985), silica $\delta^{18}O$ values record the thermal gradient between subsurface veins and subaerial sinters in fossil systems (Clayton & Steiner 1975, Ewers 1991). It remains to be demonstrated whether the thermal gradient within surface spring outflow deposits can be similarly detected by $\delta^{18}O$ measurements.

Oxygen isotope systematics in ancient hydrothermal deposits have been used to infer that the $\delta^{18}O$ value of Archean seawater was identical to today's value (e.g. de

Ronde et al 1994, Hoffman et al 1986), an inference which is supported by the proposal that seawater has been buffered isotopically by submarine hydrothermal circulation (Muehlenbachs 1986). Thus, in addition to preserving a record of early hydrothermal systems, silica deposits have recorded at least one key aspect of the global marine environment.

Sulfur

The key role played by sulfur in hydrothermal systems arises from both its substantial abundance and its participation in redox reactions. Economically important metals such as gold are transported by bisulfide complexes which, upon oxidation, deposit the metals in ore zones. Bacterially-sustained ecosystems derive energy from sulfur redox reactions (Jannasch 1995). Marine hydrothermal sulfur minerals can record the composition and oxidation state of ancient oceans (Eastoe et al 1990).

Oxidized sulfur compounds are enriched in ^{34}S, relative to the reduced compounds with which they have equilibrated (Sakai 1957). Magmatic mantle-derived sulfur has a $\delta^{34}S$ value close to 0 (Fig. 3), similar to meteoritic sulfur. Hypogene sulfates which formed at higher temperatures in equilibrium with magmatic sulfide will be relatively ^{34}S enriched (Fig. 3; Ohmoto 1972). Aqueous or gaseous sulfide which ascends to the supergene environment can form sulfides or, upon oxidation, sulfates which are relatively ^{34}S-depleted (Field & Fifarek 1985). Equilibration between sulfides and sulfates becomes sluggish below 350 °C (Ohmoto & Rye 1979), therefore the isotopic composition of these minerals can possibly reflect processes occurring deep in the system.

The widest range of $\delta^{34}S$ values observed between sulfates and sulfides in hydrothermal systems ultimately derives from bacterial processes (Ohmoto & Rye 1979). Bacterial sulfate reduction has created both a wide range of $\delta^{34}S$ values in sedimentary sulfides and ^{34}S-enriched seawater sulfate (Kaplan 1975). The thermal circulation of fluids can entrain sulfur from deeply-buried sediments and from seawater, thereby importing this bacterial isotopic signature into hydrothermal mineral assemblages (Fig. 3). One consequence of this is that sulfates and sulfides from continental hydrothermal systems are generally depleted in ^{34}S, relative to their counterparts in marine volcanogenic massive sulfide deposits, because the latter derive their sulfur largely from ^{34}S-enriched seawater sulfate (Field & Fifarek 1985).

The tendency for volcanogenic massive sulfides to assimilate isotopically heavy marine sulfate has been observed in deposits as old as those in the mid-Proterozoic McArthur Basin (Muir et al 1985). In contrast, sulfides in early Proterozoic and Archaean deposits have $\delta^{34}S$ values which lie within a few permil of 0, implying that the ancient deep ocean was anoxic and also had very low sulfate concentrations (Eastoe et al 1990). The relatively low ^{34}S enrichments (a few permil) occasionally observed in Archaean deposits (Fralick et al 1989, Hattori & Cameron 1986) are expected at equilibrium (Ohmoto 1972) and can be attributed to the oxidation of magmatic sulfide by water (Hattori & Cameron 1986).

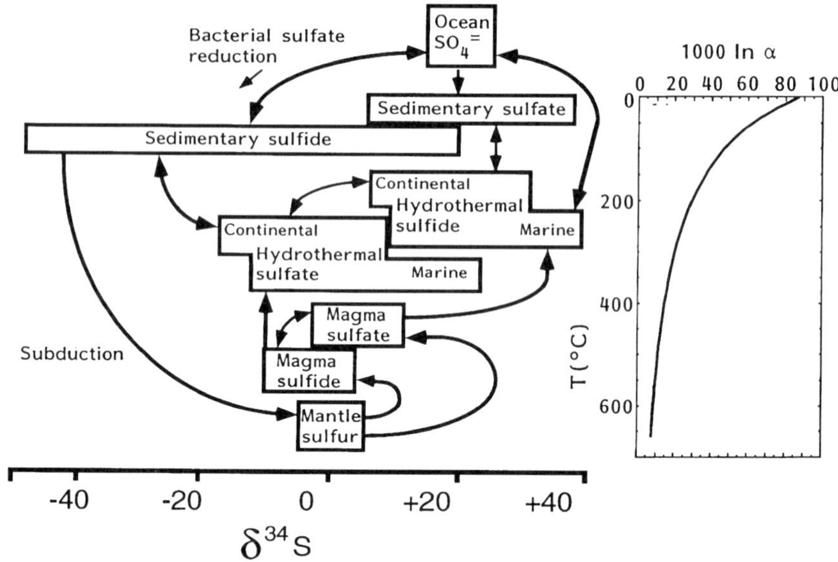

FIG. 3. Sulfur isotopic composition of crustal sulfur reservoirs, with special emphasis on hydrothermal systems. Boxes and arrows depict reservoirs and processes linking these reservoirs, respectively. The isotopic composition of hydrothermal sulfur species can reflect both mantle sulfur and contamination from crustal sources of sulfides and sulfates. Diagram at right shows how sulfur isotopic discrimination between hydrogen sulfide versus sulfate vary with temperature. Note how the magnitude of this discrimination correlates with the range of $\delta^{34}S$ values observed at various depths (temperatures) in the crust.

Vent ecosystems derive energy principally from redox reactions of sulfur; bacterial aerobic sulfide oxidation being the most important (see Jannasch 1995 for review). These bacteria grow either in suspension within the cooler zones of hydrothermal systems or as mats bathed in the spring outflow. Thus they provide organic carbon for filter-feeding and grazing vent animals (Kennicutt & Burke 1995).

The importance of hydrothermal sulfides in sustaining vent ecosystems was first demonstrated by a sulfur isotopic measurement. Because isotopic discrimination during assimilatory uptake of sulfur is small ($<3‰$), $\delta^{34}S$ values of marine phytoplankton and their food chain dependents are typically $+15$ to $+20‰$, close to the value for marine sulfate (+20). In contrast, vent clams had $\delta^{34}S$ values near 0, essentially identical to vent sulfide $\delta^{34}S$ values (Fry et al 1983). Greater variability in the $\delta^{34}S$ values of vent biota has been documented subsequently; however, this apparently reflects isotopic variations in the $\delta^{34}S$ value of the vent's sulfide source (Kennicutt & Burke 1995). These variations can be due either to changes in the proportion of sulfur derived from magmatic versus seawater sources, or to changes in the redox reactions leading to production and/or consumption of sulfates versus sulfides.

An isotopic signature of life which can be sought in fossil deposits is the extreme variability in $\delta^{34}S$, which arises from bacterial sulfate reduction (Kaplan 1975). For sulfides and sulfates, an overall variability of $\delta^{34}S$ values exceeding 100‰ has been observed. In contrast, overall variations attributable to abiogenic redox reactions are expected to be less than 20‰ (Ohmoto 1972). This is because non-biological equilibration between sulfate and sulfide is exceedingly slow below 350 °C and, above 350 °C, the $\delta^{34}S$ values of oxidized and reduced sulfur species at equilibrium differ by less than 20‰ (Sakai 1957). Thus massive sulfides, sulfates or other hydrothermal sulfur species having $\delta^{34}S$ values which are highly variable and/or differ substantially from 0‰ have probably incorporated the isotopic signature of ancient sulfate-reducing bacteria (Cameron 1982).

Carbon

Carbon species are principal components in geothermal emanations and, like sulfur, they can participate in redox reactions. A mantle component of CO_2 is virtually ubiquitous in marine and continental hydrothermal emanations, and its $\delta^{13}C$ value lies near -5 (Des Marais & Moore 1984, Ohmoto & Rye 1979). Methane occurs in mid-ocean ridge emanations at concentrations typically about 0.1 to 1% of the CO_2 concentration (Craig et al 1980). Compelling evidence for an abiotic origin of this CH_4 includes the following: it occurs in fluids too hot for bacteria; it emanates from mid-oceanic hydrothermal systems which are free of sedimentary organic matter; its $\delta^{13}C$ value lies in the range -15 to -17, which is much greater than values observed for CH_4 produced by biota or by the thermal decomposition of biological organic matter; CH_4 also occurs in the vesicles of basalts erupted from hydrothermal systems; and CH_4 appears to be in chemical equilibrium with CO_2 at temperatures (550–700 °C) which are consistent with equilibration within the hydrothermal fracture zone adjacent to the magma body (Welhan & Craig 1983).

The organic carbon and carbonate reservoirs in sediments and also the bicarbonate in seawater can become incorporated into hydrothermal fluids (Fig. 4). The $\delta^{13}C$ values of these reservoirs are controlled ultimately by biological isotope discrimination, which causes organic carbon to be depleted by typically 20–40‰ relative to carbonate (Kaplan 1975). These isotopic patterns can be incorporated into hydrothermal fluids. For example, methane and higher hydrocarbons are considerably more abundant in continental than in sediment-depleted marine hydrothermal systems, and their abundances and depleted $\delta^{13}C$ values indicate that they are products of the thermal decomposition of sedimentary organic matter (Des Marais et al 1981). Hydrothermal activity can remobilize sedimentary organic matter to form distinctive petroleums (e.g. Clifton et al 1990). Organic carbon also can become oxidized and remobilized in some magmatic and hydrothermal environments, creating CO_2 emanations and carbonates that can be as isotopically light as -10 to -15‰ (Field & Fifarek 1985). The unusual mineral whewellite ($CaC_2O_4.H_2O$), which has been identified in low-temperature hydrothermal veins, was derived from sedimentary organic matter, as evidenced by

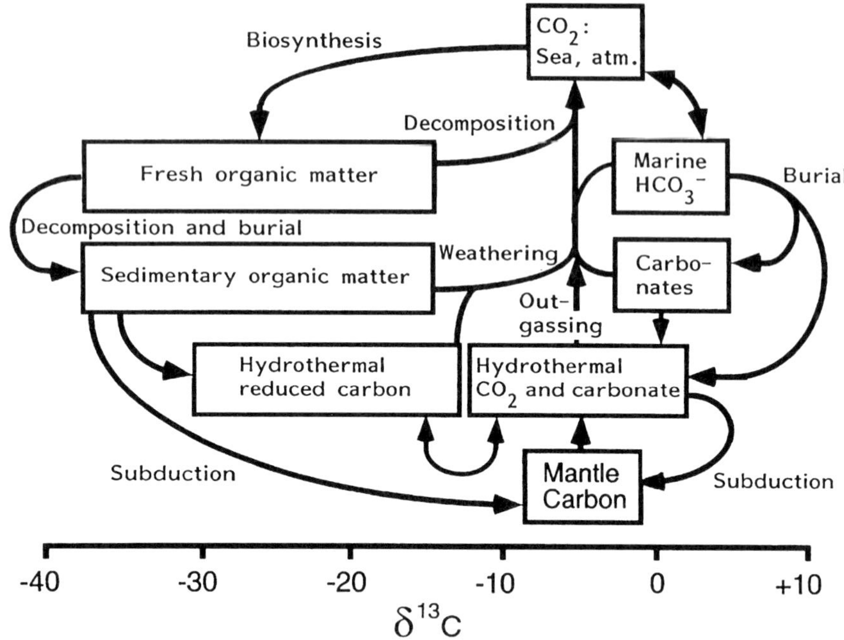

FIG. 4. Carbon isotopic composition of crustal carbon reservoirs, with special emphasis on hydrothermal systems. Boxes and arrows depict reservoirs and processes linking these reservoirs, respectively. The isotopic composition of hydrothermal carbon species can be affected both by mantle carbon and by contamination from crustal sources of organic and carbonate carbon.

its low $\delta^{13}C$ value (-28 to -32; Zak & Skala 1993). The thermal decomposition of sedimentary carbonates can create hydrothermal CO_2 and carbonates heavier than 0‰ (Kerrich 1990). Thus, in principle, hydrothermal carbon species remobilized from sedimentary reservoirs might vary isotopically by as much as 20‰ or more, e.g. from $+4$ to -20, whereas mantle carbon might vary by only a few permil, e.g. from -3 to -7 (Burrows et al 1986, Kerrich 1990, Des Marais & Moore 1984). Thus the presence of isotopically variable carbonates in ancient hydrothermal deposits might be a legacy of sedimentary organic matter and carbonates isotopically fractionated by an ancient biosphere. The carbon isotope systematics of Archaean Au–Ag vein deposits in the Superior Province, Canada support this hypothesis (Kerrich 1990).

In situ bacterial production is an additional source of organic matter in hydrothermal systems (Jannasch 1995). Much of the evidence for the carbon isotopic discrimination by marine vent bacteria has been obtained indirectly from analyses of vent fauna, such as bivalves, crustaceans and vestimentiferans. Both methanogenic and sulfide-oxidizing chemolithotrophic bacteria produce organic matter having δ values which are typically low, relative to those observed in communities sustained by marine phytoplankton (Kennicutt & Burke 1995). Sulfur-oxidizing bacteria produce organic matter that is typically 10‰ depleted, relative to planktonic organic carbon (Ruby et al 1987). As these bacteria utilize the CO_2-fixing enzyme rubisco (ribulose-1,5-bisphosphate carboxylase/oxygenase) (Nelson 1989), the substantial discrimination observed indicates that chemosynthesis in this environment proceeds in the presence of excess CO_2. Both the production (Jones et al 1983) and the aerobic consumption (Jannasch & Wirsen 1981) of methane occur in hydrothermal vents and can sustain bacterial organic synthesis. The stable isotope systematics of methane-producing and -consuming bacteria have not been studied specifically in hydrothermal vents, however methylotrophy does sustain highly ^{13}C-depleted communities in cold sedimentary methane seeps (e.g. Paull et al 1985). An additional, ^{13}C-enriched (−8 to −14‰) source of organic carbon has been identified in large vestimentiferan worms and among non-symbiont-containing (NSC) fauna which presumably graze upon free-living bacteria (Kennicutt & Burke 1995). The source of this ^{13}C-enriched carbon for the NSC fauna has not yet been identified.

Continental thermal springs frequently host highly productive, diverse photosynthetic bacterial communities. This diversity is sustained by the range of temperatures, pH, and concentrations of aqueous dissolved species. The isotopic consequences of this diversity have been explored (Estep 1984, Des Marais et al 1992). Mats with mixed communities of *Chloroflexus* and cyanobacteria occur above 45 °C and are typically depleted in ^{13}C, relative to inorganic carbon, by 17–24‰ (Estep 1984). This ^{13}C-depletion is less (0–15‰) in cyanobacterial mats growing at lower temperatures and/or lower CO_2 concentrations (Estep 1984, Des Marais et al 1992). A clear correlation exists between dissolved CO_2 concentrations versus discrimination against ^{13}C (Estep 1984, Des Marais et al 1992). Within a given spring channel, a correlation also exists between discrimination and temperature (Des Marais et al 1992). This temperature correlation arises due to CO_2 outgassing along the stream course, which both lowers the CO_2 concentration and enriches the remaining dissolved inorganic carbon in ^{13}C (Chafetz & Lawrence 1994). Accordingly, carbonates deposited along the spring run also become progressively ^{13}C-enriched downstream (Chafetz & Lawrence 1994).

Fossiliferous ancient sinters have been identified (White et al 1989, Rice et al 1995) which contain organic matter that could be analysed isotopically to search for parallels to the systematics observed in modern thermal spring mats. Such analyses should help to confirm not only the presence of hot spring deposits, but also the availability of CO_2 for biosynthesis in ancient spring environments.

Value of stable isotopic measurements

The patterns of isotopic abundance in hydrothermal fluids and deposits offer perspectives about the surrounding environment and about the key processes within hydrothermal systems, including the biota. The incorporation of seawater and crustal components into hydrothermal deposits can be used, through isotopic measurements, to monitor changes in the composition and oxidation state of sediments, the ocean and the atmosphere. Evolutionary changes in hydrothermal ecosystems have been only minimally explored, and isotopic measurements can offer perspectives about both the evolution of metabolism of microbes and its responses to environmental change.

References

Burrows DR, Wood PC, Spooner ETC 1986 Carbon isotope evidence for a magmatic origin for Archaean gold-quartz vein ore deposits. Nature 321:851–854

Cameron EM 1982 Sulphate and sulphate reduction in early Precambrian oceans. Nature 296:145–148

Chafetz HS, Lawrence JR 1994 Stable isotopic variability within modern travertines. Geogr Phys Quaternair 48:257–273

Clayton RN, Steiner A 1975 Oxygen isotope studies of the geothermal system at Wairakei, New Zealand. Geochim Cosmochim Acta 39:1179–1186

Clifton CG, Walters CC, Simoneit BRT 1990 Hydrothermal petroleums from Yellowstone National Park, Wyoming, USA. Appl Geochem 5:169–191

Craig H 1963 The isotopic geochemistry of water and carbon in geothermal areas. In: Tongiorgi E (ed) Nuclear geology on geothermal areas. Consiglio Nationale Delle Recherche, Pisa, p 17–53

Craig H, Welhan JA, Kim K, Poreda R, Lupton JE 1980 Geochemical studies of the 21° N EPR hydrothermal fluids. Eos (Trans Am Geophys Union) 61:1–88

de Ronde CEJ, de Wit MJ, Spooner ETC 1994 Early Archean (> 3.2 Ga) Fe-oxide-rich, hydrothermal discharge vents in the Barberton greenstone belt, South Africa. Geol Soc Am Bull 106:86–104

Des Marais DJ, Moore JG 1984 Carbon and its isotopes in mid-oceanic basaltic glasses. Earth Planet Sci Lett 69:43–57

Des Marais DJ, Donchin JH, Nehring NL, Truesdell AH 1981 Molecular carbon isotopic evidence for the origin of geothermal hydrocarbons. Nature 292:826–828

Des Marais DJ, Bauld J, Palmisano AC, Summons RE, Ward DM 1992 The biogeochemistry of carbon in modern microbial mats. In: Schopf JW, Klein C (eds) The Proterozoic biosphere: a multidisciplinary study. Cambridge University Press, New York, p 299–308

Eastoe CJ, Gustin MS, Hurlbut DF, Orr RL 1990 Sulfur isotopes in early Proterozoic volcanogenic massive sulfide deposits: new data from Arizona and implications for ocean chemistry. Precamb Res 46:353–364

Estep MLF 1984 Carbon and hydrogen isotopic compositions of algae and bacteria from hydrothermal environments, Yellowstone National Park. Geochim Cosmochim Acta 48:591–599

Ewers GR 1991 Oxygen isotopes and the recognition of siliceous sinters in epithermal ore deposits. Econ Geol 86:173–178

Field CW, Fifarek RH 1985 Light stable-isotope systematics in the epithermal environment. Rev Econ Geol 2:99–128

Fralick PW, Barrett TJ, Jarvis KE, Jarvis I, Schnieders BR, Vande Kemp R 1989 Sulfide-facies iron formation at the Archean Morley occurrence, northwestern Ontario: contrasts with oceanic hydrothermal deposits. Can Mineral 27:601–616

Fry B, Gest H, Hayes JM 1983 Sulphur isotopic compositions of deep-sea hydrothermal vent animals. Nature 306:51–52

Gillespie AR, Criss RF 1984 Correlation of infrared reflectance and ^{18}O profiles across a fossil hydrothermal system in the Idaho batholith. Eos (Trans Am Geophys Union) 65:290

Hattori K, Cameron EM 1986 Archaean magmatic sulphate. Nature 319:45–47

Hoffman SE, Wilson M, Stakes DS 1986 Inferred oxygen isotope profile of Archaean oceanic crust, Onverwacht Group, South Africa. Nature 321:55–58

Jannasch HW 1995 Microbial interactions with hydrothermal fluids. In: Humphris SE (ed) Seafloor hydrothermal systems: physical, chemical, biological, and geological interactions. American Geophysical Union, Washington DC, p 273–296

Jannasch HW, Wirsen CO 1981 Morphological survey of microbial mats near deep-sea thermal vents. Appl Environ Microbiol 41:528–538

Jones WJ, Leigh JA, Meyer F, Woese CR, Wolfe RS 1983 *Methanococcus jannaschii* sp. nov., an extremely thermophilic methanogen from a submarine hydrothermal vent. Arch Microbiol 136:254–261

Kaplan IR 1975 Stable isotopes as a guide to biogeochemical processes. Proc R Soc London B Biol Sci 198:183

Kennicutt MC II, Burke RA Jr 1995 Stable isotopes: clues to biological cycling of elements at hydrothermal vents. In: Karl DM (ed) The microbiology of deep-sea hydrothermal vents. CRC Press, Boca Raton, p 275–287

Kerrich R 1990 Carbon-isotope systematics of Archean Au-Ag vein deposits in the Superior Province. Can J Earth Sci 27:40–56

Kita I, Taguchi S, Matsubaya O 1985 Oxygen isotope fractionation between amorphous silica and water at 34–93 °C. Nature 314:83–84

Muehlenbachs K 1986 Alteration of the oceanic crust and the ^{18}O history of seawater. In: Taylor HP Jr, O'Neil JR, Valley JW (eds) Stable isotopes in high temperature geological processes. Mineralogical Society of America, Washington DC, p 425–444

Muir MD, Donnelly TH, Wilkins RWT, Armstrong KJ 1985 Stable isotope, petrological, and fluid inclusion studies of minor mineral deposits from the McArthur Basin: implications for the genesis of some sediment-hosted base metal mineralization from the Northern Territory. Aust J Earth Sci 32:239–260

Nelson DC 1989 Physiology and biochemistry of filamentous sulfur bacteria. In: Schlegel HG, Bowien B (eds) Autotrophic bacteria. Springer-Verlag, Berlin, p 219–238

Ohmoto H 1972 Systematics of sulfur and carbon isotopes in hydrothermal ore deposits. Econ Geol 67:551–578

Ohmoto H, Rye RO 1979 Isotopes of sulfur and carbon. In: Barnes HL (ed) Geochemistry of hydrothermal ore deposits. Wiley, New York, p 509–567

Paull CK, Jull AJT, Toolin LJ, Linick T 1985 Stable isotope evidence for chemosynthesis in an abyssal seep community. Nature 317:709–711

Rice CM, Ashcroft WA, Batten DJ et al 1995 A Devonian auriferous hot spring system, Rhynie, Scotland. J Geol Soc 152:229–250

Ruby EG, Jannasch HW, Deuser WG 1987 Fractionation of stable carbon isotopes during chemoautotrophic growth of sulfur-oxidizing bacteria. Appl Environ Microbiol 53:1940–1943

Sakai H 1957 Fractionation of sulfur isotopes in nature. Geochim Cosmochim Acta 12:150–169

Schiffman P, Smith BM 1988 Petrology and oxygen isotope geochemistry of a fossil seawater hydrothermal system within the Solea Graben, northern Troodos ophiolite, Cyprus. J Geophys Res 93:4612–4624

Sturchio NC, Keith TEC, Muehlenbachs K 1990 Oxygen and carbon isotope ratios of hydrothermal minerals from Yellowstone drill cores. J Volcanol Geotherm Res 40:23–37

Taylor HP Jr, Forester RW 1979 An oxygen and hydrogen isotope study of the Skaergaard intrusion and its country rocks: a description of a 55-million year-old fossil hydrothermal system. J Petrol 20:355–419

Vityk MO, Krouse HR, Demihov YN 1993 Preservation of $\delta^{18}O$ values of fluid inclusion water in quartz over geological time in an epithermal environment: Beregovo deposit, Transcarpathia, Ukraine. Earth Planet Sci Lett 119:561–568

Welhan JA, Craig H 1983 Methane, hydrogen and helium in hydrothermal fluids at 21°N on the East Pacific Rise. In: Rona PA, Bostrom K, Laubier L, Smith KL (eds) Hydrothermal processes at seafloor spreading centers. Plenum, New York, p 391–409

White DE, Brannock WW, Murata KJ 1956 Silica in hot-spring waters. Geochim Cosmochim Acta 10:27–59

White NC, Wood DG, Lee MC 1989 Epithermal sinters of Paleozoic age in north Queensland, Australia. Geology 17:718–722

Zak K, Skala R 1993 Carbon isotopic composition of whewellite (CaC_2O_4 H_2O) from different geological environments and its significance. Chem Geol 106:123–131

DISCUSSION

Jakosky: It is striking that there is a trend in $\delta^{13}C$ for organic carbon, but not for the carbonate. You mentioned the change in the value between them with time. Why would the carbonate be so constant?

Des Marais: There's a good quantitative explanation (Des Marais et al 1993). During the Proterozoic, two trends occurred, each of which affected the carbon isotopic composition of organic matter and carbonate. One is that the isotopic difference between organics and carbonates decreased. This would tend to make organics isotopically heavier and the carbonates lighter. The other effect is that the fraction of carbon buried as organic matter increased, relative to carbonate. This would have made the organic carbon heavier *and* the carbonate heavier. Thus these effects tend to reinforce each other with respect to their effect on the organic carbon isotopes (they both cause an increase in $^{13}C/^{12}C$), and they tend to cancel each other with respect to their effects on the carbonate carbon.

Jakosky: It's clear that it would work quantitatively, but does that mean that it is then coincidence that the carbonate stays constant?

Des Marais: It doesn't have to be as precise a coincidence as you might think, because there's a fair amount of scatter in the carbonate data, in the order of several permil, and since carbonate constitutes about 80% of the carbon in the system, a small change in its isotopic composition causes an important shift in the mass balance. A sample calculation illustrates why the observed trends can be explained in a reasonable fashion. If you assume that the isotope difference between organics and carbonates decreased as is observed, and if you assume that the crustal organic carbon inventory increased in order to balance the appearance of modern-sized O_2 and sulfate reservoirs during the Proterozoic, then you estimate that the organic reservoir's $\delta^{13}C$ value increased substantially, and that the carbonate $\delta^{13}C$ value changed by less than 2‰. Compared with the 'noise' in the carbonate $\delta^{13}C$ data, a 1–2‰ change would be hard to see.

Nisbet: We are also making hidden assumptions about the mantle source of carbon, and the isotopic ratio of the carbon which the mantle is putting into the system. Would you like to comment on this?

Des Marais: We have made a number of measurements along the length of the mid-ocean ridge system, and we see very constant values. The value I would favour for the mantle average is more like -4 or -5‰. If any isotopic discrimination has occurred between crustal and mantle exchange, then its effect would be accumulated in the crustal sediments, such that you would see a change in the average value for the total crustal carbon reservoir, relative to the mantle. In fact, the crustal average is almost identical to the mantle value. Consequently, I don't think there has been a major drift in the isotopic composition of crustal material relative to mantle through time. Therefore I would place a lot of credibility in the idea that the average isotopic combination of crustal carbon hasn't changed that much over time.

Sakai: I would like to comment on the sulfur isotope ratio in the Archaean age. David Des Marais has shown us that in the Phanerozoic and late Proterozoic sedimentary rocks, sulfate and sulfide sulfur show large isotopic fractionation owing to sulfate-reducing bacteria. This fractionation, however, should diminish as we go back to the Precambrian, prior to the appearance of the sulfate reducers. The oldest sedimentary sulfate so far analysed is barite, barium sulfate, in a 3.5 Ga old sedimentary formation in Africa, which indicates no substantial frationation. Data for Precambrian sulfate and sulfide are scarce and some claim that fractionation only becomes notable after 2.2 Ga, while others suggest that sulfur isotopic ratios were already affected at 3.5 Ga. What does the phylogenetic tree of life tell us about the time gap between the appearance of photosynthetic and sulfate-reducing microorganisms?

Stetter: Some archaeal sulfate reducers are rather deep in the phylogenetic tree. Nowadays these are only found associated with hydrothermal vent systems.

Des Marais: The major divergence that gives rise to cyanobacteria on one side and sulfate reducers on the other indicates that their time of origin in the eubacteria was very similar. I don't have any problem reconciling that with the isotope data, because to express a large discrimination you need a large substrate pool. So what we might be seeing here is an ecosystem where you have primary producing cyanobacteria and sulfate reducers existing as early as 2.7 Ga, but there's a lot of bicarbonate and CO_2 around, so you can get a big carbon isotopic discrimination, but the concentration of seawater sulfate is very low. Thus even though sulfate-reducing bacteria existed by the late Archaean, you don't observe a substantial isotope effect. Thus the time of appearance of large isotopic contrasts in the oxidized and reduced reservoirs of carbon and sulfur might reflect the evolution of the sizes of these various reservoirs.

Shock: With respect to the change in sulfur isotope fractionation with time, is it a unique solution to call upon sulfate reduction, or can the appearance of sulfide oxidation later on also explain the curve?

Des Marais: Sulfide oxidation gives rise to sulfate that is actually isotopically lighter than the sulfide, so it's going the wrong way. However, there is now evidence in sediments that perhaps part of the reason you see such a nice isotope spread between sulfate and sulfide is that there's a consortium of bacteria that facilitates the exchange between sulfate and sulfide. Of course, that's not sulfate reduction but sulfate oxidation also, and thiosulfate disproportionation becomes very important (Canfield & Thamdrup 1994). You can get bidirectional reaction of sulfur between sulfate and sulfide that allows equilibrium isotope values to be approached, even at temperatures so low that the sulfate and sulfide could never equilibrate abiologically.

Shock: If the sulfate concentration is going up, and sulfide oxidation sends the isotopes in the wrong direction, where is the sulfate coming from later on in geological time? Is it from weathering of continents?

Des Marais: Yes, it is coming from the attack of pre-existing sediments and volcanic rocks by oxygen or some other oxidizing agent.

Shock: So it is dependent on the appearance of a more-oxidizing atmosphere.

Des Marais: Yes, and the ability of the Earth's crust to sequester reduced carbon such that there is excess oxygen left over. All this oxygen and organic matter is produced by photosynthesis, but if you don't hide the organic matter in sediments, it will react with and destroy all of the oxygen. The primary productivity on the earth fixes enormous quantities of carbon, but so much of it is re-oxidized that only 0.03% of the carbon that is formed escapes to sediments; the rest is destroyed. The same then applies for oxygen. So if you want to create a net oxidation over time, this reduced carbon has to be removed. If this is done, you have this oxygen which can then go around oxidizing sulfur and iron in sediments and building up their crustal reservoirs.

Sakai: An important fact is that even in the most basaltic magma, some sulfate always exists together with sulfide, the relative abundance of which depends on the redox potential of the magma. For mid-ocean ridge and Kilauea submarine basalt, for instance, I have measured sulfate levels of 5–20% of total sulfur.

Des Marais: This could be one of the sources of the sulfate in the Archaean barites.

Trewin: I wanted to ask about the oxygen isotopes in the modern sinters. You have been measuring values from modern sinter surfaces. Do you think they will survive and keep the detailed signatures into the subsurface? Modern sinters are roughly 50–80% porosity, and those spaces all have to be filled in by later deposits which are probably going to have different isotopic signatures. Can we extrapolate the modern sinter oxygen isotope data into subsurface preservation?

Des Marais: The story with sinters is every bit as complicated as the story with modern carbonates, where there are open textures initially which are later infilled. I've done oxygen isotope measurements on young sinters and I don't see a clear isotope trend across a range of textures that represent a range of depositional temperatures in a surface deposit. But the expected isotope trend with temperature might have been obscured by the addition of these other silica phases that fill in the porosity. This is not to say it's a hopeless situation: you can drill certain phases within the sinters that are as primary as we can hope to find and see if there's a pattern within those particular phases, but we don't have the data yet.

Trewin: We don't appear to have a decent survey of recent to subrecent sinters to see whether isotopic values change or not in the early stages of preservation of sinters.

Des Marais: We lack a detailed study in which very discrete phases have been sampled in order to reconstruct the stages of burial and preservation of sinters.

Sakai: I would like to ask the biologists here about the carbon isotope fractionation during photosynthesis. There are two types of the photosynthetic pathway. One is the Calvin–Benson cycle, or C3 cycle, which strongly enriches ^{12}C into plants. The other is the C4 cycle which involves much less fractionation. C3 plants exhibit carbon isotopic ratios of −20 to −30‰, whereas those of C4 plants range from −10 to −15‰. The fact that reduced carbon in Archaean sedimentary formations as old as 3.5 Ga shows carbon isotopic ratios of −30‰ or so has been taken to indicate the photosynthetic or chemosynthetic organisms using the C3 pathway had already evolved. Might not microorganisms using the C4 pathway have evolved prior to C3 organisms? It has been proposed that some photosynthetic green bacteria and anaerobic bacteria have C4 pathways (Schdlowski et al 1983).

Des Marais: There's been a lot of confusion over the years about this. The consensus is that the C4 mechanism is exclusive to vascular plants, because it involves a multi-cellular architecture, where fixation of CO_2 occurs in one cell type and the C4 compound is translocated to another cell.

Sakai: Wong & Sackett (1978) also found that a marine diatom, *Nitzschia* sp., is capable of utilizing both pathways.

Des Marais: I think they observed ^{14}C showing up in glutamic acid and some of the amino acids associated with the Krebs cycle. It is not clear to me that this represents the C4 mechanism as it is expressed in plants. All organisms will fix some ^{14}C to the Krebs cycle during growth. The reverse TCA cycle which some bacteria exploit could mimic this initial labelling of ^{14}C into amino acids around the Krebs cycle. The reverse Krebs cycle is employed by green bacteria such as *Chloroflexus* or *Chlorobium.*

Knoll: I was going to add another level of complexity to the isotopic interpretation. It is becoming apparent that rubisco, the enzyme involved in CO_2 fixation in at least those upper branches of the bacteria, comes in at least two different forms, one of which is an octomer and the other a heterodimer. Sometimes within a single bacterium both genes are present and only one will be active at any one time. The point is that these two forms of rubisco have very different fractionations associated with them, so that some ecological cases that have traditionally been described as CO_2 limitation are now being re-interpreted as organisms that are using alternate forms of rubisco.

What kind of fractionation is known to be associated with non-cyanobacterial autotrophs?

Des Marais: Purple bacteria discriminate against ^{13}C during CO_2 fixation by about 20–25‰ (2–2.5%), which is somewhat less than is seen for cyanobacteria. The green bacteria that have the reverse TCA cycle have a much smaller discrimination, in the order of 10‰. The methanogens have a discrimination during organic synthesis in the 20‰ range. The bottom line is that when you look at the fractionation going from CO_2 into organic material in almost all organisms, the cyanobacteria and the green plants in particular have the largest discrimination associated with CO_2 assimilation. The methanogens have the additional complication that they're not only making organic matter, they're making this isotopically light methane, which leads to an isotopic competition for ^{12}C within the organism. We've done experiments where the organic matter ends up heavier than the CO_2 that the organisms are using: because of the way that carbon flows are partitioned, the competition for ^{12}C is won by the methane production process and the cellular material ends up with a lot of the ^{13}C. The general rule is that if you are looking at organic matter, the biggest fractionators are the photosynthetic organisms, specifically the green plants. Methanotrophs are another kind of bacteria — these are methane oxidizing bacteria, and the methane they use has very isotopically light carbon that could find its way into the biomass of methane oxidizing bacteria, so you could argue that their biomass could be quite light.

Shock: What is the carbon isotopic fractionation of the hyperthermophiles like at the temperature that Karl Stetter is growing them in the laboratory?

Des Marais: Certain methanogens that utilize CO_2 and tend to grow at 50 °C discriminate against ^{13}C less than the methanogens that grow at lower temperatures. Isotope discrimination hasn't been investigated in hyperthermophiles

Shock: I think it would be an excellent thing for someone who is concerned about unravelling the isotopic record in some of these rocks to investigate.

References

Canfield DE, Thamdrup B 1994 The production of ^{34}S-depleted sulfide during bacterial disproportionation of elemental sulfur. Science 266:1973–1975

Des Marais DJ, Strauss H, Summons RE, Hayes JM 1993 Proterozoic carbon cycle—reply. Nature 362:118

Schdlowski M, Hayes JM, Kaplan TR 1983 Isotopic influence of ancient biochemistry: carbon, sulfur, hydrogen, and nitrogen. In: Schopf JW (ed) Earth's earliest biosphere: its origin and evolution. Princeton University Press, Princeton, NJ, p 149–186

Wong WW, Sackett WM 1978 Fractionation of stable carbon isotopes by marine phytoplankton. Geochim Cosmochim Acta 42:1809–1815

High temperature ecosystems and their chemical interactions with their environment

Allan Pentecost

Division of Life Sciences, King's College London, Campden Hill Road, London W8 7AH, UK

Abstract. Phototrophic thermal ecosystems consist of microbial mats whose composition is largely determined by water temperature, dissolved oxygen, sulfide and pH. Mats exposed to sunlight consist of an upper zone of phototrophic bacteria and cyanobacteria and an undermat of heterotrophic bacteria. There is little or no net accumulation of reduced carbon and a quasi-equilibrium is established between the synthesis and oxidation of reduced carbon. The flux of carbon and other metabolites induces chemical change in the interstitial water which may assist the deposition of hydrothermal minerals. Uptake of carbon dioxide by phototrophs is favourable to calcium carbonate (travertine) deposition. Thermal systems also contain a range of chemolithotrophs and sulfate reducers potentially capable of depositing carbonate. Acid production by sulfate reducers may have the ability to precipitate silica from alkaline thermal waters but has not yet been demonstrated *in vivo*. Deposition of thermal ochre is also possible via bacterial oxidation of reduced iron and manganese. It appears that bacteria play a minor role in the deposition of hydrothermal minerals through chemical interaction. However, they may play a more important physical role by providing a large surface area suitable for mineral nucleation. If hydrothermal deposits occur on Mars, the distribution of travertine is likely to be restricted if there is a lack of pre-existing sedimentary carbonate. Less biologically interactive deposits of silica and ochre may predominate.

1996 Evolution of hydrothermal ecosystems on Earth (and Mars?). Wiley, Chichester (Ciba Foundation Symposium 202) p 99–111

High temperature ecosystems may be defined as those developing at temperatures in excess of 60 °C, but it is important to realise that some community members such as *Fischerella (Mastigocladus) laminosus* have a wide temperature tolerance and grow well below 60 °C. In addition, hot-spring deposits can form at 10 °C or lower. Temperatures >60 °C are attained for prolonged periods on the Earth's surface only in hot springs and related phenomena associated with vulcanism and tectonic activity. Such extremes are rarely encountered through solar heating, where they are sustained for just a few hours in the day.

It is important to distinguish between chemical and physical interactions between the microbes and their environment. The former may produce chemical changes through molecular flux and the latter through physicochemical properties of the cell surface. The last will not be considered here, but is often important in promoting the nucleation of minerals (Thompson & Ferris 1990).

There has been much interest in high temperature ecosystems and they have been well studied in parts of North America, particularly at Yellowstone, Wyoming. Elsewhere they have received scant attention. Although hot springs $>60\,^{\circ}C$ account for less than one in 10^6 of the earth's springs, they are widely distributed and found on all of the continents, including Antarctica. This discussion is limited to aquatic-continental systems, as opposed to marine systems.

The ecosystems

The communities are dominated by prokaryotes — the bacteria and archaea — a diverse group of microorganisms with a simple internal structure. Eukaryotes rarely grow above 60 °C, although a few are known, e.g. *Cyanidium* (Brock 1978). The communities contain a comparatively small number of species and have been used widely to model more complex systems (e.g. Wiegert 1977). Where light is available, the primary producers are photoautotrophs utilizing H_2S or H_2O for reductant. In the absence of light, the primary producers obtain energy largely from sulfide oxidation (Castenholz 1984). At Octopus Spring, Yellowstone — one of the most intensively investigated sites — the production of organic carbon by photoautotrophs is balanced by decomposition, so there appears to be little or no net accumulation of organic matter (Anderson et al 1987). This seems to be true of most thermal ecosystems, and is important when considering chemical interactions. Losses via invertebrate grazing are minimal above 60 °C, but may become important below 50 °C, when Ostracoda and Diptera appear (Brock 1978). The thermal ecosystem may then be considered to be in a quasi-equilibrium with its environment, with the autotrophs removing exogenous carbon, as CO_2 plus small quantities of nutrients, while the heterotrophs, which generally form an undermat, use the reduced carbon and released nutrients for their own growth requirements. Ultimately, fully or partially oxidized carbon compounds and nutrients are released back to the environment. Only a small proportion of the reduced carbon (10% or less in the few measured examples) remains unoxidized and either becomes incorporated into hot-spring deposits or is washed downstream as soluble extracellular metabolite.

In most hot springs, the microbial communities occur as gelatinous mats a few millimetres in thickness. The mat surface consists of photoautotrophs (almost always cyanobacteria) with an undermat of mixed photoautotrophs and photoheterotrophs — the cyanobacteria and photosynthetic bacteria, particularly *Chloroflexus* (Ward et al 1987, Kullberg & Scheibe 1989). Beneath this, where there is insufficient light for photosynthesis, heterotrophic prokaryotes predominate (Fig. 1). This structure is almost universal in hot-spring communities but the species composition varies considerably depending on water temperature, water chemistry and even geographic location (Castenholz 1984).

With reference to ecosystem composition, thermal waters may be conveniently divided into three classes: acidic (pH 1–5), neutral (pH 6–7.5) and alkaline (pH > 7.5). Though this classification is a gross oversimplification of hot spring chemistry, it is conceptually useful with regard to their ecosystems and associated mineral deposits. Acidic springs tend to have a reduced microbial flora dominated by *Cyanidium caldarium*, contain toxic heavy metals and are corrosive. Neutral waters tend to be high in the alkaline earth elements and often deposit large amounts of travertine or ochre. They have a diverse microbial flora. The alkaline springs often deposit silica as sinter and also possess a rich flora. However, where mineral deposition is rapid, biomass declines as the microbes become smothered and irradiance is reduced. Many thermal waters contain sulfide which has a profound influence on microbial communities, suppressing cyanobacteria at temperatures above 55 °C, and selecting sulfide-utilizing eubacteria (Castenholz 1984, Giovannoni et al 1987).

Another important determinant of ecosystem composition is oxygen. Even the transient presence of oxygen resulting from oxygenic photosynthesis can influence microbial communities. Many prokaryotes cannot tolerate oxygen and will be excluded by its presence, and some microbial mats develop in the complete absence of oxygen.

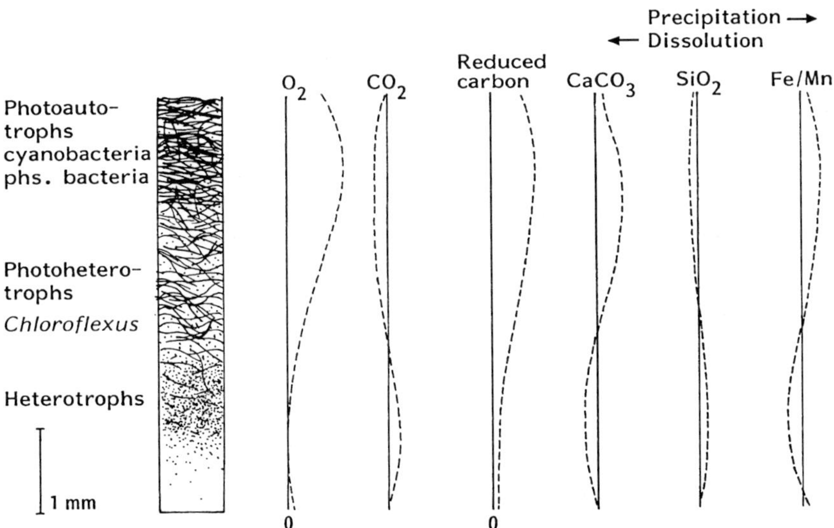

FIG. 1. Schematic illustrating a typical hot-spring microbial mat profile and some of the chemical gradients likely to be present. Relative levels of mineral precipitation and dissolution are shown.

The number of hot-spring photoautotrophs appears to be limited to a handful of species existing as a number of physiological strains. Geitler (1932) mentioned 98 thermal cyanobacteria in his treatise, but most are inadequately characterized. Several thermophilic non-filamentous photosynthetic bacteria have also been described, e.g. *Chlorobium tepidum* (Wahlund et al 1991). The heterotrophic bacteria and archaea metabolize reduced carbon and nitrogen by many routes. Some require oxygen, while others oxidize organic matter anaerobically. In the Octopus Spring mats, sulfate-reducing bacteria ferment a range of organic substrates, while the fermentation products themselves, which are only partially oxidized, may be consumed by methanogens (Ward et al 1984). Taken together, the diversity of thermal microbes would not appear to be large. About a dozen phototrophic and 50 heterotrophic microbes have been reasonably well characterized. However, recent work employing 16S rRNA sequencing (Ward et al 1990, Weller et al 1992, Barns et al 1996, Stetter 1996, this volume) suggests that the diversity is considerably greater.

Interactions between ecosystem and environment

One might suppose that environmental effects of chemical flux would cancel, since most of that which is biologically removed from the environment is soon returned to it, or internally cycled.

Nevertheless, chemical fluxes from these ecosystems have often been claimed to give rise to hot-spring deposits. Accumulations of travertine and, to a lesser extent, sinter and other hydrothermal minerals have often been thought to form through biological activity. Krumbein (1986) lists 59 bacterial 'biominerals', and more than 50 authigenic minerals occur in hot-spring deposits (author's unpublished results). The most important of these are calcite/aragonite (travertine), chalcedony and other hydrated silicas (sinters), and various oxides of iron and manganese (ochres). Microorganisms influence mineral deposition by altering their immediate chemical and physical environment such that minerals may precipitate upon their surface or immediately adjacent to them. Some of the metabolic processes leading to mineral formation are summarized in Table 1 and are considered in more detail below. Such deposits might be termed 'hydrobiothermal', but the significance of biological activity varies enormously and is rarely responsible for the majority of the deposit.

Microbial travertine deposition

Spring deposits of calcium carbonate are formed largely by the degassing of calcium bicarbonate-rich groundwaters. Those formed around hot springs are often spectacular, since the carbonate is deposited at a rate of several centimetres per year. The rapid rate of deposition results from the elevated water temperature (CO_2 less soluble with increasing temperature) and high springhead concentrations of Ca and CO_2, the latter often originating from thermal processes associated with vulcanism (Pentecost & Viles 1994, Pentecost 1995a).

TABLE 1 Potential chemical interactions between microorganisms and hot-spring deposits

Mineral	*Microbial reactions significant in mineral deposition/dissolution*	*Participating microbes*
Aragonite/calcite $CaCO_3$ (precipitation/dissolution)	$Ca(HCO_3)_2 \rightarrow CaCO_3 + H_2O + CO_2$	*Synechococcus*, *Oscillatoria terebriformis*, *Fischerella laminosus*, *Chloroflexus*, *Hydrogenobacter thermophilus*, *Methanobacterium*, *Thermothrix*, *Thermoproteus*
	$CaSO_4 + 2(CH_2O) \rightarrow CaCO_3 + H_2O + CO_2 + H_2S$	*Desulfovibrio*, *Clostridium thermohydrosulfuricum*
	$Ca(NO_3)_2 + 3H_2 + C$ (organic) $\rightarrow CaCO_3 + N_2 + 3H_2O$	*Bacillus thermophilum*
	$2NH_3 + 2H_2O + Ca(HCO_3)_2 \rightarrow CaCO_3 + (NH_4)_2CO_3 + 2H_2O$	Ammonia producers, ?*Pseudomonas*
	$(CH_2O)_n$(organic matter)$+ O_2 \rightarrow CO_2 + H_2O$	*Synechococcus*, *Oscillatoria*, *Thermus*
Silica (hydrated) $SiO_2\ nH_2O$ (precipitation/dissolution)	$CO_2 + H_2O \rightarrow HCO_3^- + H^+$	*Desulfovibrio*
	$CO_3^= + H_2O \rightarrow HCO_3^- + OH^-$	Photosynthetic alkalization
Goethite FeO OH	$Fe^{2+} \rightarrow Fe^{3+}$	Iron-oxidizers, ?*Thiobacillus*
	$Fe^{3+} + H_2O \rightarrow FeO\ OH$	
Gypsum $CaSO_4$	$Ca^{2+} + S + 2O_2 \rightarrow CaSO_4$	*Chromatium*, *Desulfovibrio*
Limonite $Fe_2O_3\ nH_2O$	$4Fe^{2+} \rightarrow 4Fe^{3+}$	Iron oxidizers ?*Thiobacillus*
	$4Fe^{3+} 3O_2 + H_2O \rightarrow 2Fe_2O_3 nH_2O$	
Manganite MnO OH	$Mn^{2+} \rightarrow Mn^{3+}$	
	$Mn^{3+} + H_2O \rightarrow MnO\ OH$	Manganese oxidizers (poorly characterized)
Pyrite FeS_2	$Fe^{2+} + 2S^- \rightarrow FeS_2$	*Desulfovibrio*

The removal of carbon dioxide by phototrophic and chemolithotrophic microbes is favourable to carbonate (travertine) deposition as indicated in equation 1 below:

$$Ca(HCO_3)_2 = CaCO_3 + H_2O + CO_2 \quad (1)$$

Estimates have been made of the likely extent of photoautotrophic calcium carbonate deposition in several cold and mesothermal travertine-depositing springs (e.g. Spiro & Pentecost 1991, Pentecost 1994) and indicate that a comparatively small amount of mineral could be formed in this way (up to 20% in favourable circumstances). The amount contributed in thermal ecosystems ($>60\,^{\circ}C$) has not been determined, but can be estimated approximately by comparing the net productivities of thermal and athermal ecosystems (Table 2). The data, though limited, suggest that the travertine-depositing potential of thermal systems is probably no greater than that occurring at ambient temperature.

Where photosynthetically available radiation is absent, energy for the reduction of carbon dioxide can be obtained from the oxidation of hydrogen and sulfur. At least a dozen well characterized thermophilic chemolithotrophs are known, for example *Hydrogenobacter thermophilus* which oxidizes H_2 under aerobic conditions, and archaea belonging to the genera *Methanobacterium* and *Thermoproteus* which also oxidize H_2 though some are facultative chemolithotrophs. Members of the genus *Thermothrix* oxidize sulfur to sulfuric acid and are again facultative chemolithotrophs and also facultative anaerobes. Species of *Thermothrix* are among the most frequently reported chemolithotrophs of hot-spring ecosystems, but they have not yet been shown to influence travertine formation. Several genera of thermophilic bacteria and archaea use oxidized sulfur compounds as terminal electron acceptors. In this group there are no autotrophs and all are anaerobic. A range of simple organic compounds are utilized for energy. In all cases, sulfide is produced from the reduction of sulfur compounds. These bacteria are particularly numerous in reducing environments containing high levels of sulfate (>2 mM/l). The sulfide may escape as H_2S, or react with reduced iron to form the minerals FeS and FeS_2, or might itself serve as H donor to a range of autotrophs. There are some well established examples of calcium carbonate precipitation associated with sulfate reduction in non-thermal sites (e.g. Jørgensen & Cohen 1977). The overall reaction can be summarized as:

$$CaSO_4 + 2(CH_2O) = CaCO_3 + CO_2 + H_2O + H_2S \quad (2)$$

(Krumbein 1979). However, this reaction is dependent upon the rate of sulfate reduction. Sulfate-reducing bacteria have been isolated from hot-spring mats and thermal travertines (e.g. Pentecost 1995b) but no mineralization has yet been attributed to them. Sulfite- and thiosulfate-reducing bacteria such as *Clostridium thermohydrosulfuricum* also occur in hot springs (Wiegel et al 1979).

Denitrifying bacteria utilize the oxygen in nitrate or nitrite for respiration. This could result in travertine deposition under favourable circumstances (Table 1). The microbes are facultative anaerobes and require fairly high levels of organic matter.

TABLE 2 Some net production estimates for flowing thermal and athermal waters

Location	*Net production* $(gC\,m^{-2}\,d^{-1})$	*Temperature* (°C)	*Reference*
Thermal waters			
Octopus Spring, Wyoming	6	50–60	Revsbech & Ward 1984
Ohana-Pecosh Hot Springs, Washington	0.5	39	Stockner 1967 ($CaCO_3$-depositing)
Drakesbad Springs, California	7–12	—	Lenn 1966 quoted in Wiegert & Fraleigh 1972
Mimbres hot spring	0.1	—	Duke 1967 quoted in Wiegert & Fraleigh 1972
Serendipity Springs, Wyoming	4	36–42	Weigert & Fraleigh 1972 (Si-depositing)
Athermal waters			
Waterfall Beck, UK	0.3	8	Pentecost 1991 ($CaCO_3$ depositing)
Itchen River, UK	5	10	Odum 1956 ($CaCO_3$-depositing) (gross production)
Blue River, Oklahoma	20	23–24	Duffer & Dorris 1966 (gross production)
Logan River, Utah	1	8	McConnell & Sigler 1959

Most nitrate-reducers have been isolated from soils but there are thermophilic forms, e.g. *Bacillus 'Dinitrobacterium' thermophilum* and *B. licheniformis* (Konohana et al 1993). Most of the nitrate reducers belong to the large genus *Pseudomonas*. Members of this group are frequently isolated from travertines (Pentecost & Terry 1989) but have not yet been shown to deposit carbonates in any freshwater environment.

Some bacteria excrete ammonia through the oxidative deamination of amino acids. The reaction provides energy, but a good supply of protein or amino acid is required. The ammonia then reacts with water to form the base 'ammonium hydroxide' capable of precipitating calcium carbonate (Table 1).

Ammonia production is easy to demonstrate with bacterial isolates grown on a suitable medium but, again, there is no evidence that it is important in any thermal system. However, K. Stetter (personal communication) has recently isolated an ammonia-forming thermophile from Indonesia.

Sinter (hydrated silica) and microorganisms

Microbes are often coated or even replaced by silica, but there is little evidence to indicate that microbial metabolism is a major factor in sinter deposition. Most sinter appears to be formed through the cooling and evaporation of Si-enriched thermal waters (Walter 1976, Cady & Farmer 1996, this volume). However, acidification of alkaline waters could lead to precipitation of hydrated silica and might be triggered by microbial acid production. Experiments with the sulfate reducer *Desulfovibrio* have demonstrated silica deposition *in vitro* (Birnbaum et al 1989), but to obtain large colonies *in vivo*, an organic substrate would be required and it is unclear where this could originate in hot, Si-depositing waters. Conversely, the alkalization accompanying photosynthesis may lead to silica dissolution.

Deposits of iron and manganese (ochre)

The oxidation of Fe^{2+} and Mn^{2+} liberates energy which may be used by some bacteria. The phenomenon has been well studied in the sulfur bacterium *Thiobacillus ferrooxidans* (Kelly & Harrison 1989). Several Fe- and Mn-encrusted bacteria are described in the literature but have yet to be isolated and studied *in vitro*. None of these bacteria are thermophilic, but Mn precipitation associated with bacteria has been reported from hot springs (Ferris et al 1987). Below pH 8.0, Mn^{2+} is oxidized slowly in the absence of bacteria and may be bacterially catalysed, although the existence of autotrophic Mn-oxidizing bacteria remains problematic (Erhlich 1978, 1995). Oxidized Mn minerals are common in hot-spring deposits though there have been no biological investigations to date. Reduced iron occurs in surface thermal waters only where sulfide is absent, and is rapidly oxidized on contact with the atmosphere. The free energy available is small (c. 10 kcal/mole at pH 2) so a considerable quantity of reduced iron would be needed for chemosynthesis, though autotrophic iron-oxidizers are well known (Ehrlich 1978).

Figure 1 illustrates a typical microbial mat with an upper photoautotrophic zone and a lower heterotrophic zone. Chemical gradients and fluxes will vary with depth and the figure illustrates regions where some of the above minerals are likely to precipitate or dissolve. Overall, the surface zone is a region of mineral deposition, with potential for dissolution below. As mineral accumulation proceeds, the net microbial mineralization caused by chemical interactions is minimal in most situations. This does not necessarily apply to deposition on microbial surfaces however, where the energy barrier to nucleation may be lowered on epitaxial layers.

Hydrobiothermal deposits on Mars?

Virtually all travertines so far investigated contain traces of organic matter, including microfossils. On earth, the vast majority of thermogene (hydrothermal) travertines are formed by the reaction between hot CO_2-rich groundwaters and limestones. Travertines formed from carbonatites, serpentinites and other non-sedimentary calcium-rich rocks are rare. In contrast, deposits of silica, iron and manganese do not depend upon a sedimentary carbonate source and may thus be more common on planets possessing hydrothermal circuits but a dearth of sedimentary carbonate. Whether there is sedimentary carbonate on Mars remains to be seen. There is no reason in principle why microbes could not exist in the hydrothermal deposits of other planets under an appropriate temperature and pressure, but deposits of ochre and silica, whose structure appears to be less influenced by biological processes may predominate.

References

Anderson KL, Tayne TA, Ward DM 1987 Formation and fate of fermentation products in hot spring cyanobacterial mats. Appl Environ Microbiol 53:2343–2352

Barns SM, Delwiche CF, Palmer JD, Dawson SC, Hershberger KL, Pace NR 1996 Phylogenetic perspectives on microbial life in hydrothermal ecosystems, past and present. In: The evolution of hydrothermal ecosystems on Earth (and Mars?). Wiley, Chichester (Ciba Found Symp 202) p 24–39

Birnbaum SJ, Wireman JW, Borowski R 1989 Silica precipitation induced by the anaerobic sulfate-reducing bacterium *Desulfovibrio desulfuricans*: effects upon cell morphology and implications for preservation. In: Crick RE (ed) Origin, evolution and modern aspects of biomineralization in plants and animals. Plenum, New York, p 507–516

Brock TD 1978 Thermophilic microorganisms and life at high temperatures. Springer-Verlag, New York

Cady SL, Farmer JD 1996 Fossilization processes in siliceous thermal springs: trends in preservation along the thermal gradient. In: Evolution of hydrothermal ecosystems on Earth (and Mars?). Wiley, Chichester (Ciba Found Symp 202) p 150–173

Castenholz RW 1984 Composition of hot spring microbial mats: a summary. In: Cohen Y, Castenholz RW, Halvorsen HO (eds) Microbial mats: stromatolites. Alan R. Liss, New York, p 107–119

Duffer WR, Dorris TC 1966 Primary production in a southern Great Plains stream. Limnol Oceanogr 11:143–151

Ehrlich HL 1978 Inorganic energy sources for chemolithoautotrophic and mixotrophic bacteria. Geomicrobiol J 1:65–83

Ehrlich HL 1995 Geomicrobiology. Marcel Dekker, New York

Ferris FG, Fyfe WS, Beveridge TJ 1987 Manganese oxide deposition in a hot spring microbial mat. Geomicrobiol J 5:33–42

Geitler L 1932 Cyanophyceae. In: Rabenhorst L (ed) Kryptogamen-Flora von Deutschland, Osterreich und der Schweiz 14. Akademische Verlagsgesellschaft, Leipzig

Giovannoni SJ, Revsbech NP, Ward DM, Castenholz RW 1987 Obligately phototrophic *Chloroflexus*: primary production in anaerobic hot spring microbial mats. Arch Microbiol 147:80–87

Jørgensen BB, Cohen Y 1977 Solar Lake (Sinai). 5. The sulfur cycle of the benthic cyanobacterial mats. Limnol Oceanogr 22:657–666

Kelly DP, Harrison AP 1989 Genus *Thiobacillus*. In: Staley JT, Bryant MP, Pfennig N, Holt JG (eds) Bergey's manual of systematic bacteriology, vol 3. Williams & Wilkins, Baltimore, MD, p 1842–1858

Konohana T, Murakami S, Nanmori T, Aoki K, Shinke R 1993 Nitrate reduction in *Bacillus licheniformis* in the presence of ammonium salt by shaking cultures. Biosci Biotechnol Biochem 57:1082–1086

Krumbein WE 1979 Calcification by bacteria and algae. In: Trudinger PA, Swaine DJ (eds) Biogeochemical cycling of mineral-forming elements. Elsevier, Amsterdam, p 47–68

Krumbein WE 1986 Biotransfer of minerals by microbes and microbial mats. In: Leadbeater BSC, Riding R (eds) Biomineralization in lower plants and animals. Clarendon, Oxford, p 55–72

Kullberg RG, Scheibe JS 1989 The effects of succession on niche breadth and overlap in a hot spring algal community. Am Midl Nat 121:32–31

McConnell WJ, Sigler WF 1959 Chlorophyll and productivity in a mountain river. Limnol Oceanogr 4:335–351

Odum HT 1956 Primary production of flowing water. Limnol Oceanogr 1:102–117

Pentecost A 1991 Algal and bryophyte flora of a Yorkshire (UK) hillstream: a comparative approach using biovolume estimations. Arch Hydrobiol 121:181–201

Pentecost A 1994 Formation of laminate travertines at Bagno Vignoni, Italy. Geomicrobiol J 12:239–252

Pentecost A 1995a Geochemistry of carbon dioxide in six travertine-depositing waters of Italy. J Hydrol 167:263–278

Pentecost A 1995b The microbial ecology of some Italian hot-spring travertines. Microbios 81:45–58

Pentecost A, Terry K 1988 Inability to demonstrate calcite precipitation by bacterial isolates from travertine. Geomicrobiol J 6:185–194

Pentecost A, Viles HA 1994 A review and re-assessment of travertine classification. Geogr Phys et Quaternaire 48:305–314

Revsbech NP, Ward DM 1984 Microelectrode studies of interstitial water chemistry and photosynthetic activity in a hot spring microbial mat. Appl Environ Microbiol 48:270–275

Spiro B, Pentecost A 1991 One day in the life of a stream: a diurnal inorganic carbon mass balance for a travertine-depositing stream (Waterfall Beck, Yorkshire). Geomicrobiol J 9:1–11

Stetter KO 1996 Hyperthermophiles in the history of life. In: Evolution of hydrothermal ecosystems on Earth (and Mars?). Wiley, Chichester (Ciba Found Symp 202) p 1–18

Stockner JG 1967 Observations of thermophilic algal communities in Mount Rainier and Yellowstone National Parks. Limnol Oceanogr 12:13–17

Thompson JB, Ferris FG 1990 Cyanobacterial precipitation of gypsum, calcite and magnesite from natural alkaline lake water. Geology 18:995–998

Wahlund TM, Woese CR, Castenholz RW, Madigan MT 1991 A thermophilic green sulfur bacterium from New Zealand hot springs, *Chlorobium tepidum* sp. nov. Arch Microbiol 156:81–90

Walter MR 1976 Geyserites of Yellowstone National Park: an example of abiogenic stromatolites. In: Walter MR (ed) Stromatolites. Elsevier, Amsterdam, p 87–112

Ward DM, Beck E, Revsbech NP, Sandbeck KA, Winfrey MR 1984 Decomposition of microbial mats. In: Cohen Y, Castenholz RW, Halvorson HO (eds) Microbial mats: stromatolites. Alan R. Liss, New York, p 191–214

Ward DM, Tayne TA, Anderson KL, Bateson MM 1987 Community structure and interactions among community members in hot-spring microbial mats. In: Fletcher M, Gray TRG, Jones JG (eds) Ecology of microbial communities 41:179–210

Ward DM, Weller R, Baterson MM 1990 16S rRNA sequences reveal uncultured inhabitants of a well-studied thermal community. FEMS Microbiol Rev 75:106–116

Weller R, Bateson MM, Heimbuch BK, Kopczynski ED, Ward DM 1992 Uncultivated cyanobacteria, *Chloroflexus*-like inhabitants and *Spirochaete*-like inhabitants of a hot spring microbial mat. Appl Environ Microbiol 58:3964–3969

Wiegel J, Ljungdahl LG, Rawson JR 1979 Isolation from soil and properties of the extreme thermophile *Clostridium thermohydrosulfuricum*. J Bacteriol 139:800–810

Wiegert RG 1977 A model of a thermal spring food chain. In: Hall CAS, Day JW (eds) Ecosystem modelling in theory and practice. Wiley, New York, p 290–315

Wiegert RG, Fraleigh PC 1972 Ecology of Yellowstone thermal effluent systems: net primary production and species diversity of a successful blue–green algal mat. Limnol Oceanogr 17:215–218

DISCUSSION

Jakosky: There has been a lot of discussion of carbonate–limestone deposits on Mars. One of the big questions is: given the idea that there has been CO_2 in the atmosphere in the past, where is the CO_2 now? We don't have the data to answer that. However, in the Martian meteorites, there are carbonate deposits.

Farmer: I wonder if Everett Shock would be willing to comment on his model-based work regarding the sequestering of carbonates by geothermal systems on Mars?

Shock: We published some model calculations last year on reacting potentially molten ice or water from near-surface on Mars, with rocks like basalts (Griffith & Shock 1995). We evaluated the extent to which carbonate minerals might precipitate in hydrothermal deposits that form on Mars and how much of the carbonate could be sequestered below the surface. These results place constraints on the temperatures and partial pressures of CO_2 that are necessary for formation of carbonate minerals.

Walter: I was wondering about the isotopic signals that might result from the very different pathways of carbonate deposition that occur. There might be some significant signals contained in these that would allow us to distinguish some of the broadly different groups of organisms.

Des Marais: Carbonates coming from an alkalization due to photosynthesis would tend to be enriched in ^{13}C relative to the source bicarbonate. If you get into the mat and start looking at diagenetic reactions it gets a little complex. If carbonates are produced as a consequence of sulfate reduction, alkalinity is produced in the sulfate reduction reaction, and bicarbonates coming from that organic matter would form carbonates that are low in $^{13}C/^{12}C$. On the other hand, there is methanogenesis: to the extent that the CO_2 that is being used in methanogenesis becomes enriched in ^{13}C because the light carbon is being reacted to form the methane, carbonates would form that are enriched in ^{13}C. Perhaps the most dependable isotopic feature one would get from biologically influenced carbonate deposition would be a scatter in isotope values, ranging from heavier to lighter than one would expect under equilibrium conditions. The net isotope effect of biological activity depends on the balance of all these processes that could occur within the milieu of the microbial mat.

Knoll: The good news is that all of the isotopic possibilities that David mentioned have actually been seen in rocks as old as the Archaean. There is a potentially very strong geochemical signal attached to this.

Pentecost: There are about 1500 stable isotope determinations on travertine, and they form two well-defined peaks. The mode of the 'heavy' peak for the hydrothermal travertines I've been talking about is about +4‰ and the 'lighter' relates to the meteogene travertines formed in karst environments where the carbon is coming from soil respiration, with a mode of about −10‰. There's another small group of travertines, which are exceedingly light (about −20‰), formed by the reaction between the atmospheric carbon dioxide and hydroxide waters produced through serpentenization (Pentecost & Viles 1994).

Horn: What was the carbon isotopic composition of the Martian meteorite material?

Jakosky: Carbonate in meteorites has a wide range of values. But when you throw out those that seem to be contaminated by terrestrial carbon, it is around 50‰. That is probably heavily influenced by the evolution of the carbon isotopes by escape to space. Carbon can be removed by various non-thermal processes and the lighter isotope is preferentially removed, so the atmosphere is enriched in the heavier isotope. That's probably the signature that we're seeing in the carbonates and it is going to muck up any interpretation in terms of biological activity.

Cady: Have you identified whether or not the mechanism of carbonate deposition differs for different species?

Pentecost: In my paper, I was talking about chemical fluxes and the way they change the immediate environment. The surface of the microorganisms is also important in determining both the form of calcium carbonate and to some extent in how well it nucleates on the surface. One can see marked differences between carbonate structures associated with different microorganisms, particularly the bacteria. The bacteria may or may not be setting off the precipitation by removing CO_2, but they could be reducing the energy barrier to nucleation at their surface and thereby favouring precipitation on their cell walls.

Cady: Do the differences in the mechanisms of carbonate deposition have any impact on the potential to preserve the organisms?

Pentecost: They must do, because in some cases the carbonate deposit completely ensheathes the bacteria with calcite crystals, whereas in other cases it is very patchy.

Cady: Because of the number and types of minerals formed on the organisms?

Pentecost: Yes. Organisms like *Geitleria*, which is a cyanobacterium, produce a complete tube with crystals perpendicularly radiating off all the surfaces. Others such as *Calothrix* may just have one crystal. So there does appear to be some species specificity (Pentecost & Riding 1986).

Walter: Jack Farmer, in your work on the paragenesis of diagenetic mineral sequences in travertine, you can recognize different carbonate phases that have come out of different stages of infilling of the porosity of the travertines. You might be able to predict that any particular carbonate phase formed in a subterranean environment where there would have been a distinctive biological system operating. It might be possible to see through the isotopic complexity that Dave Des Marais was talking about, if you link isotopic observations with petrographic observations and try to systematically approach the petrographic system in a way that might reveal the operation of the chemical pathways of carbonate precipitation.

Farmer: The approach we are taking is to microsample different phases within the microfabric framework of ancient sinters to distinguish between primary phases and secondary infills.

Also of importance is that above 40 °C the primary phase precipitated is aragonite, which undergoes a radical change during diagenesis, recrystallizing to form calcite. However, below 40 °C, the primary phase is calcite, and primary signatures appear to survive diagenesis. The basic process at the lower temperatures is infilling. Recrystallization of calcite is minimal. Thus, 40 °C is the threshold temperature for preserving many aspects of primary microfabric, and perhaps geochemistry in travertine thermal spring systems.

Pentecost: Travertines can precipitate up to a metre a year, though this is unusual. But if they are precipitating fast enough, they have daily laminations, and you can distinguish night-time precipitations from daytime. You can also get seasonal laminations, and we have picked up a 1‰ difference in these. This is probably attributable to photosynthesis (Pentecost & Spiro 1990). It may be possible to pick up environmental changes in this way.

Des Marais: Don Canfield (Max Planck Institute for Marine Microbiology) has done a lot of modelling of diagenesis and the effects of sulfate reduction. One can observe either carbonate precipitation or dissolution due to sulfate reduction, depending upon porewater composition.

Walter: You said that 10% of the originally fixed carbon survives decomposition and is left at the bottom of the mat, but how much of that ends up in the rock?

Pentecost: Pretty much that amount, to begin with.

Walter: Do you have an estimate for the organic content of ancient travertines?

Pentecost: The Holocene travertines can have up to 3% organic content, but the usual figure is about 1% (Pentecost 1993). I don't know about older travertines.

Walter: That's more than enough organic matter to allow a lot of interesting geochemistry.

Farmer: Do you know how that varies along the thermal gradient?

Pentecost: No, I don't think there have been any measurements.

Stetter: I want to comment on biological ammonia formation in hot springs, which as far as I was aware had not been demonstrated so far. We have a recent result from samples I took from a hot spring in Kawah Candradimuka Crater, at the Dieng Plateau, Java (Stetter 1991). From there we have isolated an extreme thermophile that forms ammonia from nitrate and uses hydrogen. The organism is a strictly chemolithoautotrophic bacterium which we named *Ammonifex degensii* (Huber et al 1996). It grows up to 85 °C. In addition, the novel submarine archaeal hyperthermophile *Pyrolobus fumarius* is able to gain energy by reducing nitrate all the way through to ammonia at temperatures of up to 113 °C.

References

Griffith LL, Shock EL 1995 A geochemical model for the formation of hydrothermal carbonate on Mars. Nature 377:406–408

Huber R, Rossnagel P, Woese CR, Rachel R, Langworthy TA, Stetter KO 1996 Formation of ammonium from nitrate during chemolithoautotrophic growth of the extremely thermophilic bacterium *Ammonifex degensii* gen. nov. sp. nov. Syst Appl Microbiol 19:40–49

Pentecost A 1993 British travertines: a review. Proc Geol Assoc 104:23–39

Pentecost A, Riding R 1986 Calcification in cyanobacteria. In: Leadbeater BSC, Riding R (eds) Biomineralization of lower plants and animals. Oxford University Press, Oxford, p 73–90

Pentecost A, Spiro B 1990 Stable carbon and oxygen isotope composition of calcites associated with modern freshwater cyanobacteria and algae. Geomicrobiol J 8:17–26

Pentecost A, Viles HA 1994 A review and reassessment of travertine classification. Geogr Phys et Quaternaire 48:305–314

Stetter KO 1991 Life at the upper temperature border. In: Trân Tranh Vân J, Mounolou K, Schneider J, McKay C (eds) Colloque Interdisciplinaie du Comité National de la Recherche Scientifique, Editions Frontiéres. Gif-sur-Yvette, France

Ancient hydrothermal ecosystems on Earth: a new palaeobiological frontier

M. R. Walter

School of Earth Sciences, Macquarie University, NSW 2109 and Rix & Walter Pty Ltd, 265 Murramarang Road, Bawley Point, NSW 2539, Australia

Abstract. Thermal springs are common in the oceans and on land. Early in the history of the Earth they would have been even more abundant, because of a higher heat flow. A thermophilic lifestyle has been proposed for the common ancestor of extant life, and hydrothermal ecosystems can be expected to have existed on Earth since life arose. Though there has been a great deal of recent research on this topic by biologists, palaeobiologists have done little to explore ancient high temperature environments. Exploration geologists and miners have long known the importance of hydrothermal systems, as they are sources for much of our gold, silver, copper, lead and zinc. Such systems are particularly abundant in Archaean and Proterozoic successions. Despite the rarity of systematic searches of these by palaeobiologists, already 12 fossiliferous Phanerozoic deposits are known. Five are 'black smoker' type submarine deposits that formed in the deep ocean and preserve a vent fauna like that in the modern oceans; the oldest is Devonian. Three are from shallow marine deposits of Carboniferous age. As well as 'worm tubes', several of these contain morphological or isotopic evidence of microbial life. The oldest well established fossiliferous submarine thermal spring deposit is Cambro-Ordovician; microorganisms of at least three or four types are preserved in this. One example each of Carboniferous and Jurassic sub-lacustrine fossiliferous thermal springs are known. There are two convincing examples of fossiliferous subaerial hydrothermal deposits. Both are Devonian. Several known Proterozoic and Archaean deposits are likely to preserve a substantial palaeobiological record, and all the indications are that there must be numerous deposits suitable for study. Already it is demonstrable that in ancient thermal spring deposits there is a record of microbial communities preserved as stromatolites, microfossils, isotope distribution patterns and hydrocarbon biomarkers.

1996 Evolution of hydrothermal ecosystems on Earth (and Mars?). Wiley, Chichester (Ciba Foundation Symposium 202) p 112–130

Hydrothermal ecosystems now are often thought of as pinpoint anomalies against a background of ambient life—yet they carry our deepest insights into earliest life on Earth, and maybe even to the origin of life. A thermophilic lifestyle has been proposed for the common ancestor of extant life (Woese 1987, Stetter 1994) and life may have started in submarine hydrothermal vents (Baross & Hoffman 1985, Nisbet 1986, Russell et al 1994). This recognition has come to microbiologists during the last two decades, but it has yet to have any significant impact on the discipline of palaeobiology.

Microbial palaeobiologists like myself are working with paradigms that date to the early 1950s and before. Our efforts have been directed almost exclusively to shallow marine sediments, and occasionally—usually unintentionally—to lacustrine deposits. Although this has been a very successful approach in palaeobiology, perhaps the time has come for a broader perspective. It is suggested here that consideration of hydrothermal systems will provide such a perspective by elucidating the past distribution and the composition of high temperature biological communities.

Nearly 25% of Earth's total heat loss is by thermally driven flow of seawater through the sea floor. The mass of the oceans is circulated through the oceanic crust every 10^6–10^7 years. Hydrothermal flow at temperatures $<200\,^{\circ}C$ is ubiquitous in the deeper regions of the oceans, while high intensity activity, with high flow rates and temperatures of 200–400 °C, is abundant on spreading ridges, island arcs and intraplate volcanic centres. Most research has focused on the high-intensity examples. Individual high-intensity hydrothermal fields may last more than 10^6 years. About 139 of these are currently known (Lowell et al 1995, Rona & Scott 1993). Such systems would have been even more abundant on the early Earth, when the heat flow was higher. The macrobiota of the modern systems is now well known (Tunnicliffe 1991, Lutz & Kennish 1993) and the microbiota has been described by many authors (see review by Segerer et al 1993).

Thermal springs are also common on land. Some result from deeply circulating waters in sedimentary basins, but most are associated with igneous sources of heat. In the continuum of variation of hydrothermal systems associated with volcanism and having a subaerial component, two distinctive end-members can be distinguished: acid-sulfate types occur high in volcanic domes where meteoric waters interact with magmatic volatiles, and 'adularia–sericite' types which form in extensive meteoric lateral flow regimes high above an igneous heat source at depth. In the latter, neutral pH, alkali–chloride waters dominate (Henley & Ellis 1983, Heald et al 1987, Berger & Henley 1989). The range of microbially habitable environments that occurs in all these systems is described by Henley (1996, this volume). Diverse assemblages of bacteria and archaeabacteria occur in the subaerial parts of these systems (see, for instance, Ward et al 1992), but the biology of the subsurface parts is essentially unknown.

Figures 1–3 show some of the key physical and chemical features of subaerial and subaqueous hydrothermal systems (Henley & Ellis 1983, Heald et al 1987, Krupp & Seward 1987, Berger & Henley 1989, Humphris 1995, Henley 1996, this volume). One important fact is readily apparent: the deposits of the subaerial and submarine (or sublacustrine) surfaces, on which all current palaeobiological attention is focused, are in some systems much less voluminous than the potentially habitable subterranean environments. Microbial life must be expected in this environment (Deming & Baross 1993), just as it has been found in other subterranean settings (e.g. Pedersen 1993, Parkes et al 1994, Stetter et al 1993). Indeed, the presence of suspensions of bacteria in submarine hydrothermal plumes is taken as evidence for a subterranean microbiota (Jannasch 1995). Bacteria congregate on and attach to surfaces, forming biofilms, because of relatively high nutrient concentrations at surfaces and to maintain

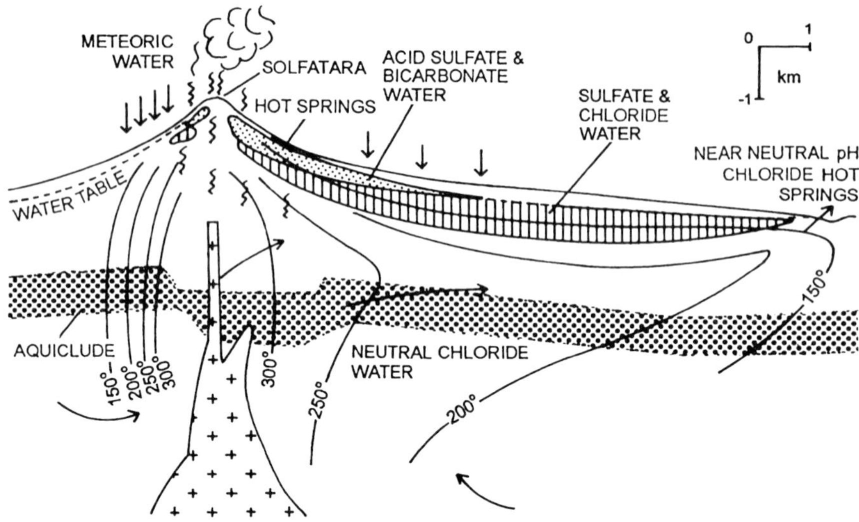

FIG. 1. Hydrothermal system of the acid-sulfate kind (after Henley & Ellis 1983). In this and Fig. 2, the 150 °C isotherm can be taken as delimiting the possible extent of the microbially habitable subterranean environment, which greatly exceeds in volume the size of the subaerial spring and solfatara environments.

themselves in a stable environment (Marshall 1976). Subterranean vadose and phreatic surfaces are very extensive in hydrothermal systems, and the associated mineral deposits are voluminous. It is inevitable (but currently unproven) that where these formed at temperatures of less than 110 °C, the deposition will have been influenced by, or will have entombed, microorganisms. This is a phenomenon familiar to engineers and microbiologists who deal with plumbing systems such as those of hydroelectric power stations or irrigation systems, where the 'scale' produced by iron- and manganese-oxidizing bacteria can be a major problem (Marshall 1976). Recent research has shown that deep subterranean biofilms (up to 1 cm thick) are the locus of the deposition of gypsum, hematite, siderite, silicates and other minerals, in microbially mediated reactions (Brown et al 1994, Doig et al 1995). Such subterranean deposits are an unexploited palaeobiological resource. There will be 'subterranean stromatolites' and subterranean microfossils. Given that the extreme conditions in the subterranean parts of active hydrothermal systems make any biological study very difficult, fossil systems may provide important insights into the biology of this environment. Not only that, but also the subterranean parts of hydrothermal systems are far more commonly preserved in the geological record than are the surficial parts (both because they are relatively much more voluminous and because they were protected from erosion), and thus represent an important palaeobiological opportunity. Subsurface deposits that are capable of preserving microfossils — and may well do so — are abundant, and include the cherty and ferruginous infillings of conduits and open spaces through which the water flowed. It

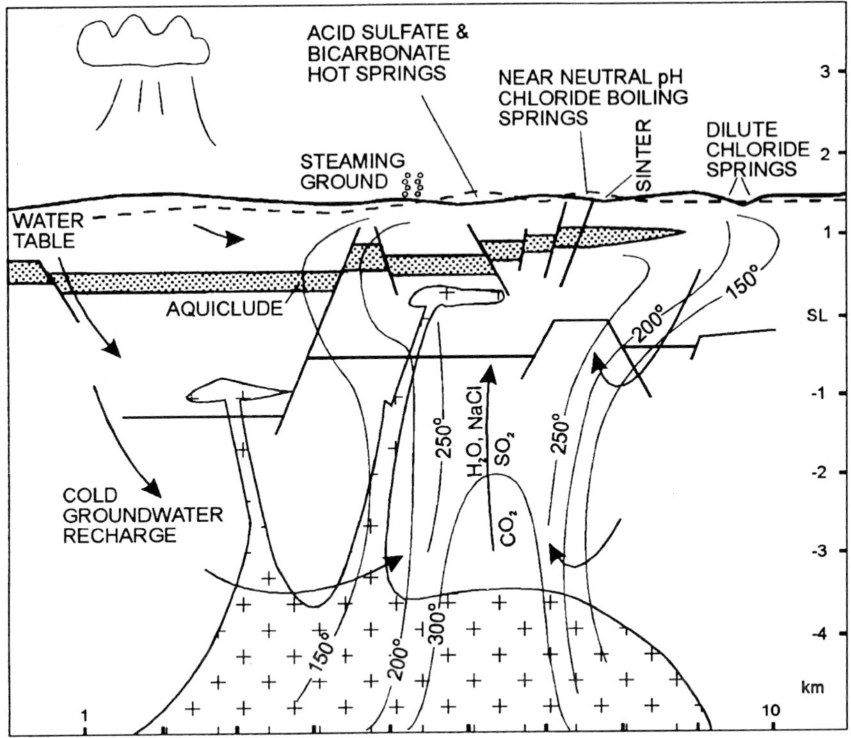

FIG. 2. Hydrothermal system of the adularia-sericite kind (after Henley & Ellis 1983, Heald et al 1987). SL, sea level.

is worth noting that in their study of the giant hydrothermal deposit of probable late Palaeozoic age at Mount Painter in South Australia, Lambert et al (1982) found vein calcite highly depleted in ^{13}C ($\delta^{13}C$, $-22.3‰$), which they interpreted as indicating a subterranean source of organic carbon.

There are several ways to estimate the original temperatures of the fluids involved in ancient hydrothermal systems. Tiny samples of the fluids were often trapped as inclusions in precipitating minerals; when these cooled, different phases in the fluids unmixed or precipitated, and there are well-established techniques for determining the minimum temperature at which a homogeneous fluid can be re-established. During mineral precipitation from hydrothermal fluids, fractionations of sulfur and oxygen isotopes are established between coeval minerals, allowing the fluid temperature to be calculated. Additionally, hot fluids chemically alter organic matter in sediments in a number of ways that allow the temperature of the fluids to be estimated. In every example it is necessary to determine whether the temperature estimate applies to the time of the hot spring event rather than some later time, and that is usually established by detailed petrographic studies. Because many of the spring deposits described below

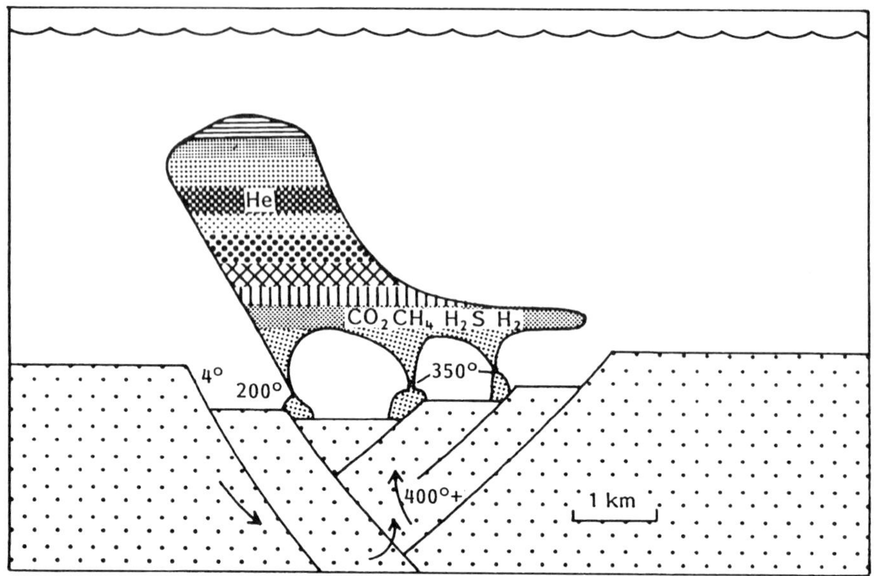

FIG. 3. Bathyal marine hydrothermal system with black and white smokers. Plumes of hot water eminating from springs carry bacteria (Jannasch 1995) and would spread the evidence of thermophiles over wide areas (for instance, as biomarkers in sediment).

are also ore deposits, their mineralogy and petrography are usually well known as they play an essential role in understanding the genesis of the ores. In fact, the mineralogy itself may often provide the best guide to palaeotemperatures, as the stability fields of the relevant minerals are well known.

Fossiliferous ancient hydrothermal systems

The discovery of the rich communities of worms and bivalves around modern black smokers in the mid-1970s quickly resulted in the successful search for their fossil counterparts. Several of these are in ophiolite complexes, which are fragments of the deep sea floor thrust up on to continental plate margins during later tectonic activity. Other examples are in shallow marine deposits. The fossils are in deposits of metal sulfides comparable with those of the modern hydrothermal vents, and the deposits have been mined for their metals. Fossils have also been found in ancient subaerial thermal spring deposits, mostly incidental to other studies. The opportunity now exists to make deliberate, focused searches.

Subaerial

There are numerous late Tertiary to Quaternary examples which are not discussed here.

Drummond Basin, Queensland

The Drummond Basin is a north–south trending back-arc basin of Late Devonian to Early Carboniferous age. Hydrothermal sinters and alteration mineral assemblages of the adulariasericite kind are well exposed in at least seven sites, and are interbedded with volcaniclastic sediments, lava and marsh sediments containing silicified plants (Cunneen & Sillitoe 1989, White et al 1989). They are very closely comparable with modern examples in Yellowstone National Park and elsewhere. Thirteen microfacies are recognizable in the field, ranging from high-temperature abiotic geyserite through various forms of stromatolitic sinter probably of cyanobacterial origin to ambient-temperature marsh deposits. Microfossils in the stromatolites are interpreted as cyanobacterial sheaths. By analogy with modern springs, microbial communities that lived at temperatures of up to about 70 °C can be inferred. Herbaceous lycopsids occur in the lower-temperature deposits (Walter et al 1996).

Rhynie chert, Scotland

This Early Devonian (~396 Ma) deposit in Scotland is famous as a rich source of fossils of early land plants, insects, cyanobacteria, algae, fungi and lichens. It was thought to be a thermal spring deposit when it was discovered early this century, and that has now been demonstrated by detailed mapping, drilling, petrography and geochemistry. It is of the adularia-sericite kind. The springs developed as the final stage of andesitic volcanism which had earlier built volcanic cones on a flood plain located adjacent to a fault zone in a terrestrial environment. The cherts were deposited in small ephemeral pools on an alluvial flood plain adjacent to the extinct volcanoes. The spring water was low salinity and dominantly meteoric, and had been heated by circulating in contact with the volcanic rocks. The climate varied from pluvial to semi-arid, and occasionally the pools dried out. The plants grew on sandy substrates at the margins of the pools. The higher-temperature parts of the ecosystem are poorly known, but there is high-temperature geyserite, and the fossil cyanobacteria described (Croft & George 1959, Edwards & Lyon 1983) probably represent a higher-temperature environment. The wealth of information that has come from this deposit (Trewin 1994, Rice et al 1995 and earlier references therein) is all the more remarkable in that there is almost no outcrop and most palaeontological work has been done on loose rocks from farmers' fields and walls.

Tambrey, Western Australia

The 2.7–2.8 Ga Maddina Basalt of the Hamersley Basin, Western Australia at a locality called Tambrey contains numerous large chert bodies that have been interpreted as hydrothermal deposits (Smith 1979). At Tambrey the basalt is flat lying and chert masses 520 m thick and 20–300 m wide are scattered over a plain along a zone 5 km wide by 20 km long. They are saucer-shaped, with inward dips of 7–15°. There are probable stromatolites in some of them (Packer & Walter 1986, Packer 1991). There is considerable debate as to whether these are thermal spring deposits. Packer (1991) concluded that the cherts are not hydrothermal but instead are lacustrine or marine. However, she was unable to satisfactorily explain the saucer-shaped outcrop patterns, the presence of breccia pipes, the presence of crystal pseudomorphs in the cherts and the presence of quartz geodes, all features potentially explicable as having a hydrothermal origin.

Sub-lacustrine

Scots Bay formation, Nova Scotia

In early Jurassic times, subtropical lakes formed in a rift valley setting in eastern North America. Richly fossiliferous muds were deposited in the lakes, along with sands and basalt flows. Alkaline silica-rich hydrothermal springs vented onto the floor of one of these lakes in Nova Scotia, depositing a siliceous tufa that mantles the irregular surface of a basalt flow and surrounds former fluid conduits in the basalt. A specifically hydrothermal biota has not been recognized, but a close and regular spatial association between the tufa and stromatolitic limestones is reported (De Wet & Hubert 1989), raising the possibility that benthic mats of thermophilic organisms were present.

East Kirkton, Scotland

Carboniferous limestone at this locality is famous for its assemblage of fossils, including a terrestrial biota. The limestone was deposited in an evaporitic playa lake. Most of the deposit is flat-laminated limestone and tuff, but there are two massive lenses of cherty, pyritic limestone that are probably spring vent deposits. Probable stromatolites occur in these and in the surrounding sediments (Rolfe et al 1994; and author's observations).

Barney Creek formation, McArthur Basin, Australia

One of the world's largest known stratiform Pb-Zn sulfide deposits occurs in the HYC Shale Member of the 1.69 Ga (Palaeoproterozoic) Barney Creek formation. These organic-rich shales were deposited in a lacustrine or lagoonal environment that was possibly evaporitic, but deeper than the photic zone. The metal sulfides were deposited from solutions that suffused the sediments and probably entered the overlying water body (Solomon et al 1994); temperatures as determined by sulfur isotope systematics were 100–260 °C, with the higher temperatures being in the subterranean feeder zone (Williams 1980, Logan et al 1990). An unusual assemblage of microfossils occurs in chert lenses in the shale. It is dominated by pyritized filamentous bacteria and organically preserved coccoid (and rarely filamentous) cyanobacteria; 21 species in 16 genera are described. The bacteria include branching and budding forms. It is considered to be a benthic heterotrophic community with some allochthonous planktonic forms. The shale contains up to 4.8% organic carbon by weight, with a $\delta^{13}C$ of −26.6‰ to −31.0‰. Sulfur isotope patterns in the pyrite in the shale indicate the action of sulfate-reducing bacteria. At least some of the base-metal sulfides formed penecontemporaneously with the silica that preserves the microfossils (Oehler 1977, Oehler & Logan 1977). Kerogen in the formation is in the form of 'lamalginite', interpreted as remains of benthic mats of filamentous organisms (Crick et al 1988). The environment is interpreted as deeper than the photic zone because the only probable photosynthesisers amongst the microfossils are differently preserved from the benthic microfossils and are probably allochthonous; deposition in more than 40 m of water is also required by the high temperature of the mineralizing fluids but the lack of any evidence of boiling (Solomon et al 1994).

In the areas of intense mineralization the organic matter in the sediments has been hydrothermally altered, but nearby, in the Glyde River Sub-basin, it is well enough preserved to still yield biomarker hydrocarbons. C_{20}, C_{25} and C_{30} acyclic isoprenoids are interpreted as derived from archaea. C_{30} pentacyclic tritelpane occurs in this formation but not other units in the basin; it is presumably derived from the hopane of bacterial cell walls. Low levels of steranes are consistent with palaeontological indications of the presence of algae (Summons et al 1988).

The observations presented above come from several essentially uncorrelated datasets, and it is not certain that a thermophilic microbiota is represented, but it seems inevitable. The ore-forming processes are the subject of much debate and study, but whatever interpretation is accepted it is almost certain that there were hot springs on the basin floor. This example differs from the others described here in that it is not associated with magmatism. It occurs in a sedimentary basin which was undergoing extensional faulting and where continuing sedimentation buried and preserved the hydrothermal system.

Submarine

Shallow seas

Silvermines and Tynagh, Ireland. The first fossil hydrothermal chimney deposit to be recognised was the Carboniferous Silvermines sphalerite, Galena and barite deposit (Larter et al 1981, Boyce et al 1983). This is in a shallow-marine, fault-controlled basin in a continental margin setting. The venting fluids were highly saline brines with temperatures lower than 150 °C. Strong isotopic depletion in the sulfur of the associated pyrite indicates bacterial sulfate reduction. The deposits have up to 8.6% organic carbon by weight. Mounds of laminated pyrite, up to 3 m wide, are suggested by Russell & Hall (1996) to be sulfide stromatolites that formed over thermal vents. Similar mounds occur in the nearby Lisheen deposit.

The nearby Tynagh Pb-Zn-Cu-Ag-$BaSO_4$ deposit is postulated to have formed when acidic hydrothermal solutions (modified from seawater) vented into seawater, depositing minerals as a result of cooling and an increase of pH. Additional sulfide resulted from bacterial reduction of seawater sulfate. Pyritic fossil worm tubes

0.8 mm wide are abundant: they are of two types, distinguished by the spacing of annulations. There may be a third type of worm-like fossil and possibly also a cephalopod (Banks 1985, 1986).

Red Dog, Alaska. This Carboniferous–Permian giant stratiform Pb/Zn sulfide deposit is in an area of very complex structure but seems to have formed in a relatively deep, but not bathyal, marine environment in which turbidites and cherts were being deposited. Tubular fossils occur in chert and barite at sites of inferred hydrothermal discharge (as indicated by the distribution of sulfides and chert). They are sinuous, 3–9 mm wide and up to 10 cm long. Chert pellets 12 mm wide are closely associated with the tubes and have been interpreted as faecal pellets, algal debris or bacterial clumps. The pellets and the tube walls contain spheroidal carbonaceous microfossils 50–200 μm wide. Temperatures of 80–330 °C are interpreted from fluid inclusion and oxygen isotope studies (Moore et al 1986, Young 1989).

Big Cove formation, Newfoundland. Richly fossiliferous mound-shaped masses of carbonate with pyrite, Galena and sphalerite occur in the early Carboniferous Big Cove formation. Fluid inclusion and conodont colour alteration data indicate that the sulfides were deposited by saline hydrothermal fluids with temperatures between 60–110 °C, in a water depth of at least 100 m. The mounds are up to 16 m wide and contain an unusual low-diversity fauna. As well as 1.5–3.0 cm-wide worm tubes, encrusted by 'microbial colonies', there are thickets of bryozoans and accumulations of brachiopods, conularids, selpulid worms and crustaceans. Abundant encrusting and infilling laminated and peloidal carbonates are interpreted as evidence of a rich microbiota (von Bitter et al 1990).

The deep ocean

Zambales ophiolite, Phillipines. A massive copper and zinc sulfide deposit of Eocene age at Barlo contains half-centimetre-wide tubes composed of and encrusted by pyrite. The tube walls are 150 μm thick and consist of framboidal pyrite (Boirat & Fouquet 1986).

Ophiolite, New Caledonia. Millimetric pyrite tubes are set in a matrix of barite and quartz in copper sulfide deposits in late Cretaceous to early Tertiary sea-floor basalts. Normal marine fossils occur in associated sediments (Oudin et al 1985).

Samail ophiolite, Oman. The Bayda ore deposit is considered to have formed in a sea-floor graben on or near a spreading ridge, during the Cretaceous. Three different types of worm tubes are present, all 15 mm wide and randomly oriented, preserved as sinuous molds and casts in a matrix of zinc- and iron-sulfides (sphalerite and pyrite) and minor quartz (Haymon et al 1984, Haymon & Koski 1985). No other fossils are reported.

Troodos ophiolite, Cyprus. Here again fossil worm tubes occur in a Cretaceous massive sulfide deposit, and no other fossils are reported (Oudin & Constantinou 1984).

Urals, Russia. Fossil worm tubes are reported from middle Devonian sulfide ores in the Uralian ophiolite belt (Kuznetsov et al 1988, 1991).

Mount Windsor, Queensland. A volcanic belt related to Late Cambrian to Early Ordovician subduction in northeastern Australia contains the Thalanga Pb-Zn-Cu volcanogenic massive sulfide (VMS) deposit. Stratiform lenses of silica-magnetite-hematite-chert ironstones extend for up to 70 km away from the sulfide deposit. This is a relationship that is common in VMS deposits of all ages back to Archaean. The Mount Windsor ironstones are exceptionally well preserved and contain three types of filamentous microfossils in abundance, preserved as hematitic structures in a matrix of chert: 10 μm-wide septate branching and anastomosing forms, 5 μm-wide aseptate branching and anastomosing forms, and 20–30 μm-wide branching forms. The two smaller forms resemble chemoautotrophic iron-oxidizing bacteria. Very little pyrite is present in the ironstones, but where present it has an isotopic composition indicating an origin through bacterial sulfate reduction. The geochemistry of the Thalanga deposit indicates deposition in more than 1000 m of water and the ironstones have the light rare earth element pattern characteristic of vent precipitates on the modern sea floor. The ironstones are distant from the main vent and are interpreted to have formed at temperatures of 10–40 °C (Duhig et al 1992a,b). No fossils are reported from the VMS deposit itself, where a megafauna might be expected.

Morley, Ontario. 'Algoma-type' iron formations reported from many Archaean successions are analogous to the metal-rich sediments associated with some modern deep-sea hydrothermal environments such as the Red Sea deeps, and with the Mount Windsor deposit described above. The late Archaean (2.7 Ga) example at Morley is interbedded laminated pyrite, chert and carbonaceous cherty mudstone associated with volcanic rocks and turbidites; it formed on the flanks of a submarine volcano in an arc environment. The lamination in a 1 m-thick unit of pyrite includes numerous linked domical structures a few centimetres wide, and is interpreted by Fralick et al (1989) as stromatolitic. One this basis and because of the abundance of kerogen they consider that there were microbial mats present. However, the lamination appears very regular and may be abiogenic, and more work is required to establish the biogenicity of these structures.

Similarly, there are possible stromatolites vaguely associated with hydrothermal features in the 3.5 Ga Barberton Greenstone Belt of South Africa (de Wit et al 1982, de Ronde et al 1994), but there are a great many uncertainties in the interpretation of these.

Strelley Belt, Western Australia. The Sulphur Springs and Kangaroo Caves VMS deposits are unusually little altered for rocks of this age (Early Archaean, 3.26 Ga) and contain exceptionally well preserved sulfide textures apparently indistinguishable from those of modern black smokers. They formed in at least 1000 m of water in a back-arc basin. The geochemistry of the basal parts of the ores indicates fluid temperatures of at least 300 °C (Vearncombe et al 1995). Probable filamentous and coccoid microfossils occur in these deposits (S. Vearncombe, personal communication 1996).

Opportunities and future research

Palaeobiological information is preserved in many forms, including morphological fossils, chemical fossils such as biomarker hydrocarbons, and patterns of distribution of isotopes (particularly in carbon and sulfur). The study of the palaeobiology of hydrothermal ecosystems has barely begun, but already information in all these forms has been won from 13 Phanerozoic deposits. As an indication of the opportunity for finding older hydrothermal deposits, a search of the Georef database revealed 145 references on 'volcanogenic massive sulfide', 'Kuroko-style' or 'epithermal gold' deposits of Archaean age, and 112 of Proterozoic age. And the literature is more extensive than this suggests, as there are over 75 major VMS deposits in Canada alone (Lowe 1985). The Archaean deposits formed in a range of environments from deep maline to subaerial (Lowe 1985, Barley 1992). Small conical stromatolites such as are constructed now by cyanobacteria at temperatures of 30–59 °C in the effluents of subaerial thermal springs (Walter et al 1976) are reported from three late Archaean successions in Australia and Canada (Grey 1981, Hofmann et al 1991, Hofmann & Masson 1994), and as Grey (1981) stated, 'The remarkable similarity between conical stromatolites from Kanowna [Australia] and those from Yellowstone suggests that the possibility of a hot-spring environment should not be overlooked in interpreting these Archaean forms'. Palaeontologists have paid these deposits almost no attention.

Surveys of the distribution of hydrothermal mineral deposits through time (e.g. Lambert et al 1982) show major peaks of abundance in Archaean and Proterozoic successions (Fig. 4). 'Shale-hosted' base-metal sulfide deposits are abundant back to about 1.8 Ga, and although they are generally regarded as epigenetic — that is, as having formed in a subterranean environment — they can be expected to be at least partly exhalative (Solomon et al 1994, Russell & Hall 1996). Another indication comes from the distribution and abundance of banded iron formations. These are subaqueous chemical sediments rich in iron derived from hydrothermal fluids exhaled on to the sea floor, as demonstrated by their tectonic setting and rare earth element composition (Klein & Beukes 1992). They are very abundant, and very thick, in sedimentary rock successions older than 1.8 Ga (Fig. 4). Some are fossiliferous (Walter & Hofmann 1983), but these are remnants of distal, low-temperature communities. None the less, these examples demonstrate that ancient hydrothermal deposits are abundant. This, coupled with the review of fossiliferous hydrothermal deposits above, demonstrates the opportunity for the discovery of a rich palaeobiological record of thermophiles and hyperthermophiles. We now know where and how to look for this record. The known fossil record of thermophilic microorganisms is outlined above, and we may have

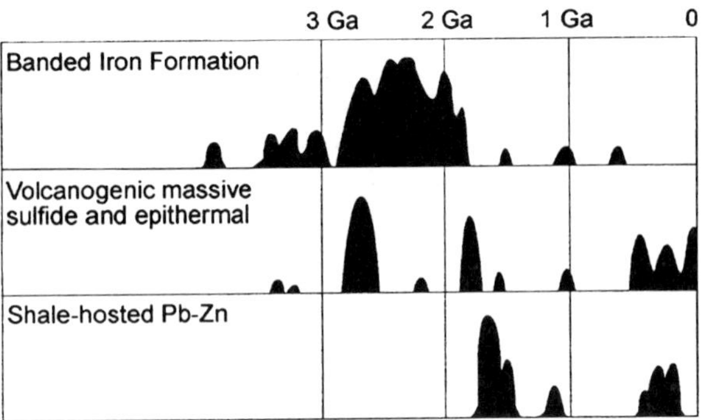

FIG. 4. The distribution through time of some classes of mineral deposits with a hydrothermal component (after Lambert et al 1992). The vertical axis represents a subjective assessment of abundance.

already found some evidence of hyperthermophiles, for instance in the microfossils and biomarkers of the Barney Creek Formation, but cannot yet be sure. It is too early to make any inferences about the evolution of such systems, except to point out the obvious, that pre-Phanerozoic hydrothermal systems lack metazoan fossils (though no 'Terminal' Proterozoic examples seem to have been studied by palaeobiologists).

Modern hydrothermal environments are known to have diverse biological communities, despite the fact that only the surficial parts of these systems have been studied. The third dimension, the much more voluminous subterranean system, remains to be explored. Here palaeobiologists have a special opportunity, as examples of these systems abound in the geological record, and are much more accessible than in modern, active vents. Any strategy for searching for life in ancient hydrothermal systems on either Earth or Mars should include the subterranean environment.

When Baas Becking, Kaplan and Moore wrote their classical paper (Baas Becking et al 1960), they noted that hydrothermal environments were very poorly known. Soon after, Brock (1967) demonstrated that the upper temperature limit for bacterial life is more than 90 °C (and in the same paper he quotes references dating back to 1875 in which it is hypothesized that 'microorganisms of hot springs are relicts of primordial forms of life'). Now the study of such environments has led to the upper temperature limit of life being found to be at least 110 °C, and maybe nearer to 150 °C, and to the discovery of archaea that grow at a pH of 0.8 (Segerer et al 1993). Despite some progress in similarly extending the palaeobiological record, there has not been a concerted effort to explore what are often (wrongly) thought of as 'unusual' palaeoenvironments.

Exploration geologists and miners have long known the importance of hydrothermal systems, as they are sources for much of our gold, silver, copper, lead and zinc. Since the discovery of 'black smokers' only 20 years ago, such systems have been shown to have a major role in chemical cycling within the oceans, and in the heat balance of the Earth. They have proved to be a bonanza for biologists, perhaps even leading to the very root of the tree of extant life. Rather than being just curiosities, they are influencing our thinking on many issues (Humphris 1995, Parson et al 1995). It is demonstrated here that a significant palaeobiological record is preserved in hydrothermal mineral deposits, and that palaeobiologists now have an opportunity to take a new approach to elucidating the earliest history of life on Earth, and perhaps on Mars (Walter & Des Marais 1993), by focusing attention on hydrothermal mineral deposits.

Acknowledgements

The research on which this paper is based was supported by NASA's Exobiology Program. Dick Henley participated in many discussions on this topic and he, Andrew Knoll and David Des Marais tried hard to improve the manuscript. John Bauld helped with literature on biofilms.

References

Baas Becking LGM, Kaplan IR, Moore D 1960 Limits of the natural environment in terms of pH and oxidation-reduction potentials. J Geol 68:243–284

Banks DA 1985 A fossil hydrothermal worm assemblage from the Tynagh lead–zinc deposit in Ireland. Nature 313:128–131

Banks DA 1986 Hydrothermal chimneys and fossil worms from the Tynagh Pb–Zn deposit, Ireland. In: Andrew CJ, Crowe RWA, Finlay S, Pennell WM, Pyne JF (eds) Geology and genesis of mineral deposits in Ireland. Crawe Shaffalitzky, p 441–447

Barley ME 1992 A review of Archean volcanic-hosted massive sulfide and sulfate mineralization in Western Australia. Econ Geol 87:855–872

Baross JA, Hoffman SE 1985 Submarine hydrothermal vents and associated gradient environments as sites for the origin and evolution of life. Origins Life 15:327–345

Berger BR, Henley RW 1989 Advances in understanding of epithermal gold–silver deposits, with special reference to the western United States. Econ Geol 6:405–423

Boirat J-M, Fouquet Y 1986 Découverte de tubes de vers hydrothermaux fossiles dans un amas sulfuré de l'Éocène supérieur (Barlo, ophiolite de Zambalès, Phillipines). C R Acad Sci Paris 302:941–946

Boyce AJ, Coleman ML, Russell MJ 1983 Formation of fossil hydrothermal chimneys and mounds from Silvermines, Ireland. Nature 306:545–550

Brock TD 1967 Life at high temperatures. Science 158:1012–1019

Brown DA, Kamineni DC, Sawicki JA, Beveridge TJ 1994 Minerals associated with biofilms occurring on exposed rock in a granitic underground research laboratory. Appl Env Microbiol 60:3182–3191

Crick IH, Boreham CJ, Cook AC, Powell TG 1988 Petroleum geology and geochemistry of the Middle Proterozoic McArthur Basin, Northern Australia. II. Assessment of source rocks. Am Assoc Petrol Geol Bull 72:1495–1514

Croft WN, George EA 1959 Blue-green algae from the Middle Devonian of Rhynie, Aberdeenshire. Bull Br Mus Nat Hist (Geol) 3:341–353

Cunneen R, Sillitoe RH 1989 Paleozoic hot spring sinter in the Drummond Basin, Queensland, Australia. Econ Geol 84:135–142

de Ronde CEJ, de Wit MJ, Spooner ETC 1994 Early Archean (> 3.2 Ga) Fe-oxide-rich, hydrothermal discharge vents in the Barberton greenstone belt, South Africa. Geol Soc Am Bull 106:86–104

De Wet CCB, Hubert JF 1989 The Scots Bay formation, Nova Scotia, Canada, a Jurassic carbonate lake with silica-rich hydrothermal springs. Sedimentology 36:857–873

de Wit MJ, Hart R, Martin A, Abbott P 1982 Archean abiogenic and probable biogenic structures associated with mineralized hydrothermal vent systems and regional metasomatism, with implications for greenstone belt studies. Econ Geol 77:1783–1802

Deming JW, Baross JA 1993 Deep-sea smokers: windows to a subsurface biosphere? Geochim Cosmochim Acta 57:3219–3230

Doig F, Lollar BS, Ferris FG 1995 Microbial communities in deep Canadian Shield groundwaters: an *in situ* biofilm experiment. Geomicrobiol J 13:91–102

Duhig NC, Davidson GJ, Stolz J 1992a Microbial involvement in the formation of Cambrian sea-floor silica-iron oxide deposits, Australia. Geology 20:511–514

Duhig NC, Stolz J, Davidson GJ, Large RR 1992b Cambrian microbial and silica gel textures preserved in silica-iron exhalites of the Mt Windsor Volcanic Belt, Australia: their petrography, geochemistry and origin. Econ Geol 87:764–784

Edwards DS, Lyon AG 1983 Algae from the Rhynie Chert. Bot J Linn Soc 86:37–55

Fralick PW, Barrett TJ, Jarvis KE, Jarvis I, Schnieders BR, Vande Kemp R 1989 Sulfide–facies iron formation at the Archean Morley occurrence, northwestern Ontario: contrasts with oceanic hydrothermal deposits. Can Mineral 27:601–616

Grey K 1981 Small conical stromatolites from the Archaean near Kanowna, Western Australia. Western Australia Geological Survey Annual Report 1980. Geological Survey, p 90–94

Haymon RA, Koski RA 1985 Evidence of an ancient hydrothermal vent community: fossil worm tubes in Cretaceous sulfide deposits of the Samail Ophiolite, Oman. Biol Soc Wash Bull 6:57–65

Haymon RA, Koski RA, Sinclair C 1984 Fossils of hydrothermal vent worms from Cretaceous sulfide ores of the Samail Ophiolite, Oman. Science 223:1407–1409

Heald P, Foley NK, Hayba DO 1987 Comparative anatomy of volcanic-hosted epithermal deposits: acid-sulfate and adularia-sericite types. Econ Geol 82:1–26

Henley RW, Ellis AJ 1983 Geothermal systems ancient and modern. Earth-Sci Rev 19:1–50

Henley RW 1996 Chemical and physical context for life in terrestrial hydrothermal systems: chemical reactors for the early development of life and hydrothermal ecosystems. In: Hydrothermal ecosystems on Earth (and Mars?). Wiley, Chichester (Ciba Found Symp 202) p 61–82

Hofmann HJ, Masson M 1994 Archean stromatolites from Abitibi greenstone belt, Quebec, Canada. Geol Soc Am Bull 106:424–429

Hofmann HJ, Sage RP, Berdusco EN 1991 Archean stromatolites in Michipicoten Group siderite ore at Wawa, Ontario. Econ Geol 86:1023–1030

Humphris SE (ed) 1995 Seafloor hydrothermal systems: physical, chemical, biological, and geological interactions. American Geophysical Union, Washington DC, p 273–296

Jannasch HW 1995 Microbial interactions with hydrothermal fluids. In: Humphris SE (ed) Seafloor hydrothermal systems: physical, chemical, biological, and geological interactions. American Geophysical Union, Washington DC, p 273–296

Klein C, Beukes NJ 1992 Time distribution, stratigraphy, and sedimentological setting, and geochemistry of Precambrian iron formations. In: Schopf JW, Klein C (eds) The Proterozoic biosphere: a multidisciplinary study. Oxford University Press, Oxford, p 335–338

Kuznetsov AP, Maslennikov VV, Zaykov BB, Sobetskiy VA 1988 Fauna sul'fidnikh gidrotermal'nikh kholmov Ural'skogo paleookeana (Sredniy Devon). Dokl Akad Nauk SSSR 303:1477–1481

Kuznetsov AP, Maslennikov VV, Zaykov BB, Zonenshain LP 1991 Fossil hydrothermal vent fauna in Devonian massive sulphide deposits of the Uralian ophiolites. Deep Sea Newslett 17:9–11

Krupp RE, Seward TM 1987 The Rotokawa geothermal system, New Zealand: an active epithermal gold-depositing environment. Econ Geol 82:1109–1129

Lambert IB, Drexel JF, Donnely TH, Knutson J 1982 Origin of breccias in the Mount Painter area, South Australia. J Geol Soc Aust 29:115–125

Larter RCL, Boyce AJ, Russell MJ 1981 Hydrothermal pyrite chimneys from the Ballynoe Baryte deposit, Silvermines, County Tipperary, Ireland. Mineral Deposita 16:309–318

Logan RG, Murray WJ, Williams N 1990 HYC silver-lead-zinc deposit in McArthur River. In: Hughes FE (ed) Geology of the mineral deposits of Australia. Australian Institute of Mining and Metallurgy, Melbourne, p 907–911

Lowe DR 1985 Sedimentary environment as a control on the formation and preservation of Archean volcanogenic massive sulfide deposits. In: Ayres LD, Thurston PC, Card KD, Weber W (eds) Evolution of Archean supracrustal sequences. Geol Assoc Canada Spec Pap 28:193–201

Lowell RP, Rona PA, Von Herzen RP 1995 Sea-floor hydrothermal systems. J Geophys Res 100:327–352

Lutz RA, Kennish MJ 1993 Ecology of deep-sea hydrothermal vent communities: a review. Rev Geophys 31:211–242

Marshall KC 1976 Interfaces in microbial ecology. Harvard University Press, Cambridge, MA

Moore DW, Young LE, Modene JS, Plahuta JT 1986 Geologic setting and genesis of the Red Dog zinc-lead-silver deposit, western Brooks Range, Alaska. Econ Geol 81:1696–1727

Nisbet EG 1986 RNA, hydrothermal systems, zeolites and the origin of life. Episodes 9:83–89

Oehler JH 1977 Microflora of the H.Y.C. Pyritic Shale Member of the Barney Creek Formation (McArthur Group), middle Proterozoic of northern Australia. Alcheringa 1:314–349

Oehler JH, Logan RG 1977 Microfossils, cherts, and associated mineralization in the Proterozoic McArthur (H.Y.C.) lead-zinc-silver deposit. Econ Geol 72:1393–1409

Oudin E, Constantinou G 1984 Black smoker chimney fragments in Cyprus sulphide deposits. Nature 308:349–353

Oudin E, Bouladon J, Paris J-P 1985 Vers hydrothermaux dans une minéralisation sulfurée des ophiolites de Nouvelle-Calédonie. C R Acad Sci Paris 301:157–162

Packer BM 1991 Sedimentology, paleontology, and stable-isotope geochemistry of selected formations in the 2.7-billion-year-old Fortescue Group, Western Australia. PhD thesis, University of California, Los Angeles, USA

Packer BM, Walter MR 1986 Late Archean hot spring deposits, Pilbara Block, Western Australia. 12th International Sedimentology Congress, Canberra, p 232

Parson LM, Walker CL, Dixon DR 1995 Hydrothermal vents and processes. Geological Society, Bath (Geological Society Special Series 87)

Pederson K 1993 The deep subterranean biosphere. Earth-Sci Rev 34:243–260

Parkes RJ, Cragg BA, Bale SJ et al 1994 Deep bacterial biosphere in Pacific Ocean sediments. Nature 371:410–413

Rice CM, Ashcroft WA, Batten DJ et al 1995 A Devonian auriferous hot spring system, Rhynie, Scotland. J Geol Soc 152:229–250

Rolfe WDI, Durant GP, Baird WJ, Chaplin C, Paton RL, Reekie RJ 1994 The East Kirkton Limestone, Visean, of West Lothian, Scotland: introduction and stratigraphy. Trans R Soc Edinb Earth Sci 84:177–188

Rona PA, Scott SD 1993 A special issue on sea-floor hydrothermal mineralization: new perspectives — preface. Econ Geol 88:1935–1975

Russell MJ, Hall AJ 1996 The emergence of life from iron monosulphide bubbles at a submarine hydrothermal redox and pH front. J Geol Soc Lond, in press

Russell MJ, Daniel RM, Hall AJ, Sherringham JA 1994 A hydrothermally precipitated catalytic iron sulphide membrane as a first step toward life. J Mol Evol 39:231–243

Segerer AH, Burggraf S, Fiala G et al 1993 Life in hot springs and hydrothermal vents. Origins Life Evol Biosphere 23:77–90

Smith RE 1979 Interpretation of volcanic relations and low grade metamorphic alteration, Maddina volcanics, Western Australia. CSIRO Mineral Res Lab Div Mineralogy Report FP21

Solomon M, Groves DI, Jaques AL 1994 The geology and origin of Australia's mineral deposits. Clarendon Press, Oxford

Stetter KO 1994 The lesson of Archaebacteria, In: Bengtson S (ed) Early life on Earth. Columbia, New York (Nobel Symp 84) p 143–160

Stetter KO, Huber R, Blöchl E et al 1993 Hyperthermophilic archaea are thriving in deep North Sea and Alaskan oil reservoirs. Nature 365:743–745

Summons RE, Powell TG, Boreham CJ 1988 Petroleum geology and geochemistry of the Middle Proterozoic McArthur Basin, northern Australia. III. Composition of extractable hydrocarbons. Geochim Cosmochim Acta 52:1747–1763

Trewin NH 1994 Depositional environment and preservation of biota in the Lower Devonian hot-springs of Rhynie, Aberdeenshire, Scotland. Trans R Soc Edinb 84:433–442

Tunnicliffe V 1991 The biology of hydrothermal vents: ecology and evolution. Ocean Mar Biol 29:319–407

Vearncombe S, Barley ME, Groves DI, McNaughton NJ, Mikucki EJ, Vearcombe JR 1995 3.26 Ga blacksmoker-type mineralization in the Strelley Belt, Pilbara Craton, Western Australia. J Geol Soc 152:587–590

von Bitter PH, Scott SD, Schenk PE 1990 Early Carboniferous low-temperature hydrothermal vent communities from Newfoundland. Nature 344:145–148

Walter MR, Des Marais DJ 1993 Preservation of biological information in thermal spring deposits: developing a strategy for the search for fossil life on Mars. Icarus 10:129–143

Walter MR, Hofmann HJ 1983 The palaeontology and palaeoecology of Precambrian iron formations. In: Trendall AF, Morris RC (eds) Proterozoic iron formations: facts and problems. Elsevier, Amsterdam, p 373–400

Walter MR, Bauld J, Brock TD 1976 Microbiology and morphogenesis of columnar stromatolites (*Conophyton, Vacerrilla*) from hot springs in Yellowstone National Park. In: Walter MR (ed) Stromatolites. Elsevier, Amsterdam, p 273–310

Walter MR, Farmer JD, Des Marais DJ, Hinman N 1996 Palaeobiology of Palaeozoic thermal spring deposits in the Drummond Basin. Palaios, in press

Ward DM, Bauld J, Castenholz RW, Pierson BK 1992 Modern phototrophic microbial mats: anoxygenic, intermittently oxygenic/anoxygenic, thermal, eukaryotic and terrestrial. In: Schopf JW, Klein C (eds)The Proterozoic biosphere. Cambridge University Press, New York, p 309–324

White NC, Wood DG, Lee MC 1989 Epithermal sinters of Paleozoic age in north Queensland, Australia. Geology 17:718–722

Williams N 1980 Precambrian mineralisation in the McArthur-Cloncurry region, with special reference to stratiform lead–zinc deposits. In: Henderson RA, Stephenson PJ (eds) The geology and geophysics of northeastern Australia. Geological Society of Australia, p 89–107

Woese CR 1987 Bacterial evolution. Microbiol Rev 51:221–271

Young LE 1989 Geology and genesis of the Red Dog deposit, western Brooks Range, Alaska. CIM Bull, Sept 1989, p 57–67

DISCUSSION

Henley: The deposit at Mount Painter is a subaerial sinter, isn't it?

Walter: I don't think so. There are numerous geodes with radial arrangements of large quartz crystals. They were clearly open spaces, presumably in a subterranean system.

Henley: But they sit within what is basically a sinter, don't they?

Walter: I went there originally because this deposit was described as having sinters 200 m thick. When I got there I found massive quartz reefs and just a suggestion of sinter here and there right at the top of Mount Painter.

Henley: Is that not the growth of a big sinter mound that progressively accreted?

Walter: No; I looked very carefully and saw very little that I would interpret as a subaerial sinter.

This place is also of interest because it is frequently used to interpret Olympic Dam, the Proterozoic hydrothermal system that Jack Farmer has compared with a potential hydrothermal system on Mars.

Knoll: We know from younger Proterozoic deposits (which are very unlikely to have been deposited under the kind of conditions you mentioned) that you can get very thin pyritized filaments, at least some of which started out life as much fatter cyanobacteria. What we see is something that has degraded and has then been diagenetically replaced by pyrite. My comment is simply a general one: one has to interpret all of these Precambrian fossils through a degradational taphonomic veil, and try to reconstruct a morphology that might have been somewhat different from what is actually seen in the fossil.

Stetter: These ultrathin filaments are familiar to me, because they are widespread in modern-day hot springs. We have cultivated some of them and found that they are all members of the genus *Thermofilum*, an anaerobic sulfur-respiring archaeum which grows at around 100 °C. It can be as thin as 0.15 μm; almost at the limit of the resolution of light microscope.

Knoll: Are the organically preserved cyanobacteria-like fossils and the pyritized fossils found in physically separate laminae or are they mixed together in single thin sections?

Walter: Now we're going back to John Oehler's work of nineteen years ago. It is not clear from his papers. I think they occur together.

Knoll: In places such as the Draken Formation in Spitsbergen, which is a good old-fashioned tidal flat, there are still horizons that are dominated by pyritized fossils.

Walter: One of the things that really intrigues me about the HYC deposit is that we are able to do biomarker geochemistry and isotope geochemistry here as well. We can also apply organic petrography to the morphology of the fossils. Ian Crick, who has looked at this deposit, sees wavy laminated mats which he calls 'lamalginite'. He interprets this as indicating the former presence of benthic microbial mats. So there are a number of converging lines of evidence from different palaeobiological techniques. This is what we have to find and take back to the Archaean; this might be a place to hone our skills.

Barns: How well is morphology conserved through evolution? That is, do we have reason to think that if we see what looks like a thermophile preserved in the rock that it corresponds to a thermophile today, or is morphology a character that changes through time?

Walter: In the case of the cyanobacteria, the morphologies appear to have been conserved for a long time.

Barns: Even within modern cyanobacteria there is a lot of phenotypic plasticity, depending on growth conditions.

Knoll: You are raising an important point. If you grow a strain of cyanobacteria in the lab and put it under all sorts of growth conditions that it rarely sees in nature, it can assume a great variety of morphologies. But under the given set of growth conditions where it's actually found in nature, plasticity is often limited. Geologists can characterize the range of depositional conditions in which fossil remains occur and have found repeatedly that cyanobacteria both from ancient sites and from what are considered to be their modern environmental analogues, are morphologically very similar.

Stetter: In cyanobacteria there's a lot morphological plasticity, but we see hyperthermophiles with identical morphologies in both the field and the lab. I have no experience of fossils, but what I see here excites me. For example, one of your fossils looks exactly like *Thermosipho*, which has unique C32 unbranched ether lipids. This belongs to the Thermotogales, which is a very deep branch in the phylogenetic tree. Its lipids could be a biomarker.

Nisbet: We are trying to extract from the fossil records some sense of evolutionary history. As a field geologist, my own response when I'm mapping along and I get something horribly complicated like this, is to mark it down as 'horribly complicated' and go on. There are very few areas in the Archaean where the degree of structural preservation is good enough (the metamorphic grade is low enough) to yield useful information. Most field geologists like myself don't really know what to look for. One of our priorities ought to be some sort of atlas of things to look for in thin section. The only hard information, of a sort of rigour that is now coming from molecular biology, is isotopic — the morphology is not going to give us much. If we are to find any really hard information, it will have to come from approaches such as very fine detailed isotopical work.

Walter: I think you are being too sceptical. You should remember what has been learned about the fossil history of cyanobacteria simply from the morphology of organisms.

Nisbet: But I'm talking Archaean — not much morphology left! We might find better material, I hope.

Walter: The biomarkers that Roger Summons and others work on are far more specific in what they indicate about the characteristics of organisms than the isotope studies.

Summons: Even more specific than that is the chemical structure of the biomarker combined with the individual isotopic composition of that molecule. Isotopic analyses at the molecular level are highly diagnostic.

Cowan: How far back in the record that you have followed might one expect to find reasonable biomarkers? At what point are the different biomarkers likely to drop out?

Walter: I would like to answer that anecdotally. 20–30 years ago people tried this sort of approach and they found all sorts of things, including amino acids, in rocks. These turned out to be contaminants and the field fell into disrepute for some time. Then, 10 years ago, Roger Summons and others got involved, and the general view was that new analytical techniques such as gas chromatography–mass spectrometry made fresh approaches possible. People were pretty sceptical then about the possibility of finding any well preserved organic matter in the Precambrian. Now, as well as all the elegant work Roger has done on late and middle Proterozoic sediments where there are abundant biomarkers, we know of biomarkers dating back to 2.5 Ga. From 10 years ago, when it was all impossible, we have progressed to a situation where the record has been pushed back to 2.5 Ga and I see no reason why it should not extend to the Archaean. The sulfide textures in the deposits from the Pilbara indicate extraordinarily low metamorphic grades.

Nisbet: I accept that places such as North Pole in the Pilbara, Western Australia, have had a very gentle thermal history after the original event. In the Belingwe belt, in Zimbabwe, we have a few rocks with very gentle thermal history. But this is extremely rare. Perhaps we should focus on those places with the very gentlest history.

Trewin: I tend to agree with Malcolm Walter that we have to look at biomarkers, but also we need to know a lot more about the taphonomy of these organisms and how they are going to be preserved. One approach is cultivating them and observing experimental preservation. For example, we often see something which looks like a filament preserved in pyrite: but we have to know whether it is the internal diameter or the external case of that filament preserved, or whether the pyrite has grown on something very tiny which we can no longer directly observe.

The Rhynie cherts: an early Devonian ecosystem preserved by hydrothermal activity

Nigel H. Trewin

Department of Geology and Petroleum Geology, Meston Building, Kings College, University of Aberdeen, Aberdeen AB24 3UE, UK

Abstract. The Rhynie cherts contain a remarkable early Devonian terrestrial to freshwater biota preserved in siliceous sinter by the action of a precious-metal-bearing hot spring system. Arthropods, vascular and non-vascular plants, algae, fungi and cyanobacteria are present. Preservation ranges from perfect 3D cellular permineralization to compacted coalified films, and can be related to both silicification processes and stages of biological and physical degradation of the plants at the time of silicification. Plants occasionally have original subaerial vertical axes preserved in growth position, and rhizomes bearing rhizoids. The plant litter of the substrate is also partly silicified. Silicification of organic material took place in hot spring pools, by surface flooding of areas with growing plants, and by permeation of the substrate. Sinters recognized include botryoidal geyserite typical of vent margins, and laminated sinter comparable with that of modern sinter terraces. Massive, vuggy, brecciated and nodular sinter textures are also present. At the microscopic level, textures associated with filamentous elements of the biota, and with the preservation of plants, closely match those present in modern sinters. Oxygen isotope and organic geochemical data from the Rhynie cherts indicate a temperature of 90–120 °C. This is apparently greater than the temperature at which elements of the biota were preserved and represents subsequent shallow burial in the hot spring system. The range of temperature and chemistry present at the surface provided high local environmental gradients. Current work attempts to identify thermophilic elements of the biota and document environmental zonation of biota relative to hot spring vents.

1996 Evolution of hydrothermal ecosystems on Earth (and Mars?). Wiley, Chichester (Ciba Foundation Symposium 202) p 131–149

The Rhynie chert of Rhynie, Aberdeenshire (Grampian Region), NE Scotland, is an early Devonian hot spring deposit which contains the superbly silicified remains of early terrestrial plants and arthropods. The cherts were originally deposited as highly porous amorphous silica sinters comparable to modern examples in hot spring areas, and were cemented by silica and transformed to chert in the shallow subsurface of the hydrothermal system. The cherts represent the oldest well-preserved terrestrial ecosystem in the world and the biota has been extensively studied, with many new elements of the biota being recognized in recent years. The superb preservation makes the site unique for the study of early terrestrial plants, and the cherts are also the oldest subaerial surface expression of a gold-bearing hydrothermal system in the world.

The cherts were discovered by Mackie (1913) as loose blocks, and subsequently material was obtained by surface trenching in weathered material. The classic work on the plants by Kidston & Lang (1917, 1920a,b, 1921a,b) has been followed by numerous papers detailing the palaeontology. Early suggestions by Mackie (1913) and Kidston & Lang (1917) that the chert was the deposit of a hot spring system (which they compared with Yellowstone) remained untested until Rice & Trewin (1988) demonstrated that the chert and associated silicified lavas and sandstones have anomalously high gold and arsenic concentrations, indicating the action of a precious-metal-bearing hot spring system.

Subsequent work based on a mineral exploration programme in the area demonstrated that the cherts occur within a sequence of alluvial plain and ephemeral lacustrine deposits (Trewin & Rice 1992). A new fossiliferous chert (Windyfield chert) was also located 700 m from the original locality, and a large fragment of geyserite, indicative of the presence of a hot spring vent, was reported.

In 1988 a short (35 m) inclined borehole produced the first cores from the plant-bearing chert and associated sediments. General features of this core were reported by Trewin & Rice (1992) and Trewin (1994), and details of the succession and palaeontology are contained in the thesis by Powell (1994). An overview of knowledge of the hydrothermal system (Rice et al 1995) provided a radiometric age of 396 ± 12 Ma for the cherts, confirming a general Pragian age based on palynology. Oxygen isotopic analysis of the chert gave interpreted temperatures in the range 90–120 °C. This is too high for surface depositional temperatures and probably represents the temperature at which the originally highly porous opaline sinter was converted to solid chert in the shallow subsurface.

The biota of the Rhynie and Windyfield cherts

The main elements of the biota described from the Rhynie and Windyfield cherts are listed below with appropriate references for botanical and zoological detail. The affinities of several plants are under review. A more detailed review is contained in Cleal & Thomas (1995).

Tacheophyte plants

Rhynia gwynne-vaughanii (Rhyniaceae) (Kidston & Lang 1920a, Edwards 1986).
Horneophyton lignieri (Rhyniaceae) (Kidston & Lang 1920a, El-Saadawy & Lacey 1979a).
Asteroxylon mackiei (Lycopsida) (Kidston & Lang 1920b).
Trichopherophyton teuchansii (Zosterophyllophytales) (Lyon & Edwards 1991).
New Zosterophyll plant awaiting formal description (Powell 1994).

Plants, Incertae sedis

Aglaophyton major (Kidston & Lang 1920a, Edwards 1986).
Nothia aphylla (Lyon 1964, El-Saadawy & Lacey 1979b).

Several gametophytes of the plants have also been described (e.g. Remy & Remy 1980).

Algae

Representatives of Charophyta and Chlorophyta are present (see Croft & George 1959, Edwards & Lyon 1983, Kidston & Lang 1921b). The problematic Nematophytales have been discussed by Lyon (1962) and are represented by *Nematophyton taiti* (Kidston & Lang 1921b) and *Nematoplexus rhyniensis* (Lyon 1962).

Fungi

Several types of resting spores and fungal hyphae have been described by Kidston & Lang (1921b) and Boullard & Lemoigne (1971). Taylor et al (1992a,b) have described parasitic relationships between aquatic fungi and the alga *Palaeonitella*. Mycoparasitism has been described by Hass et al (1994).

Cyanobacteria

Six cyanobacteria have been described from the chert (Croft & George 1959, Kidston & Lang 1921b, Edwards & Lyon 1983). Small (max. few mm) domal structures and mats associated with sinter surfaces and coating silicified plant axes are also probably cyanobacterial in origin and possibly thermophilic. Filaments 2–3 μm in diameter occur in laminae of about 0.2 mm which comprise filaments alternately vertically and horizontally oriented, typical of phototactic control as seen in stromatolites.

Arthropods

Lepidocaris rhyniensis (Branchiopoda) (Scourfield 1925).
Palaeocharinus rhyniensis (trigonotarbid arachnid). Hirst (1923) described four species of *Palaeocharinus* and one of *Palaeocharinoides*. Shear et al (1987) suggest the two genera are synonymous.
Protacarus crani (Pachygnathideae; mite) (Hirst 1923).
Rhyniella praecursor (Collembola) (Hirst & Maulik 1926).
Other arthropods are present, based on fragmentary material, and are not listed here.

Chert textures

The macrotextures of the Rhynie cherts have been summarized by Powell et al (1991), Powell (1994) and Trewin (1994). This brief summary follows the earlier work with additional observations derived from chert blocks from the Windyfield area. Some features are illustrated in Fig. 1.

(1) Vuggy to massive cherts (Fig. 1C), often with well preserved plants: these are sometimes partly in growth position with upright stems up to 15 cm long. Geopetal chalcedonic infills of cavities and of plant stems are present and these sometimes show changes in levels due to tilting during the period of silicification. Open vugs that remain are lined by small quartz crystals. Deposited by flooding of area of plant growth by hot-spring water.

(2) Lenticular cherts form composite beds up to 75 cm thick. Lenses of dark organic-rich chert have partings of carbonaceous cherty sandstone. The angular relations of these partings indicate that the chert lenses are up to 15 cm thick and can thin laterally to nothing in 20 cm. Centres of the lenticular patches are essentially similar to the massive cherts, but are darker with more organic material. Possibly formed in small pools or hollows flooded with hot spring water.

(3) Laminated cherts (Fig. 1A) have irregular lamination on a submillimetre to centimetre scale and are not always plant-bearing. Plants are frequently confined to specific bands within the chert. Fine sandstone partings may be present which separate chert laminae of chalcedony or equal-sized microcrystalline quartz. Typical texture of laminated sinter terraces, deposited by shallow hot flowing water.

(4) Nodular cherts are thin (under 5 cm) nodular cherts which occur in cherty sandstones. They are generally plant-bearing, but plants are often poorly preserved. Sandstones are compacted around the chert nodules. Formed by silicification around plant debris by permeation of fluids in the substrate.

(5) Brecciated cherts (Fig. 1B) are extensively fractured and the fractures are sealed with chert or quartz. The brecciation affects fully silicified plants, and fractures contain chert-cemented sediment. Calcite and baryte also occur in fractures in cherts caused by later faulting. Surface brecciation may be due to shock, silica transformations, desiccation, and collapse of delicate sinter textures. Deeper brecciation is due to compaction. Faulting can cause brecciation of all textures, but fault fractures are filled with quartz, baryte or calcite rather than chert.

(6) Botryoidal texture (Fig. 1D) is seen in a single block $20 \times 30 \times 30$ cm which consists of oriented lobes 10–15 cm long with a botryoidal surface texture. Botryoids are 1–2 cm long and have a pustulose surface. Very fine concentric laminae occur in sections of the botryoids. The texture resembles that of geyserite from the splash zone of a vent.

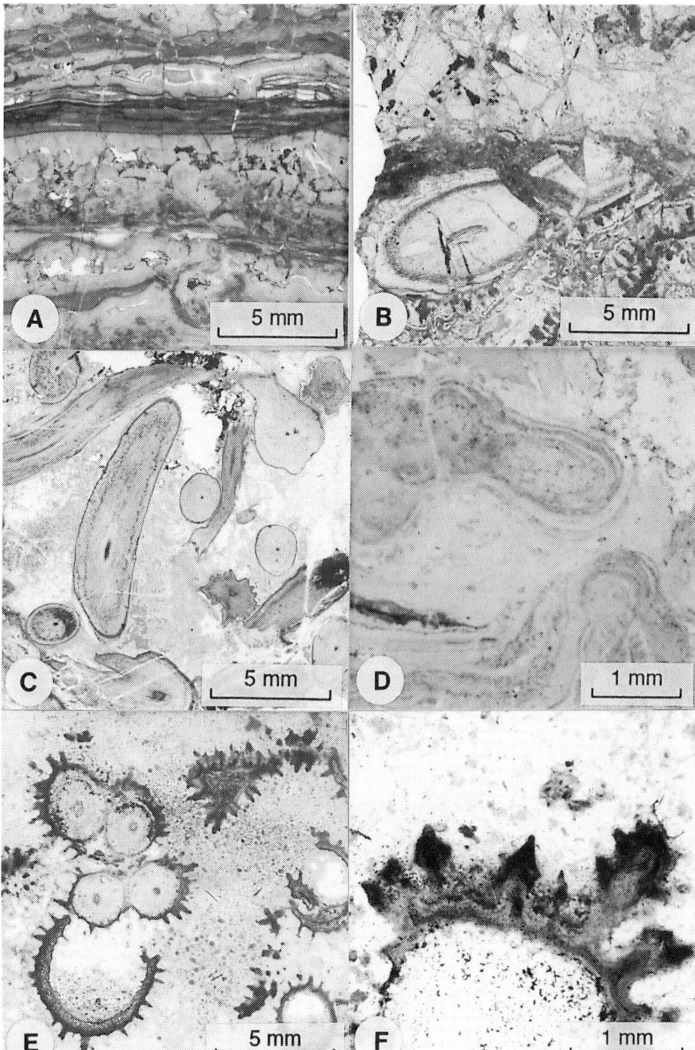

FIG. 1. Chert textures in thin section. (A) Laminated sinter with vuggy bands separating microlaminated chert. (B) Brecciated chert. Note brecciation of previously silicified plant and subsequent resealing by chert. (C) Massive chert with excellent full preservation of plant stems as in Fig. 2G. (D) Texture of botryoidal geyserite. (E) Massive to vuggy Windyfield chert with both degraded and complete plant axes. Note geopetal fills to plant fragments and in 'straws'. A digitate cyanobacterial mat covers parts of the plant stems. (F) Detail of laminated cyanobacterial mat growing on plant epidermis. Dark colour due to pyrite.

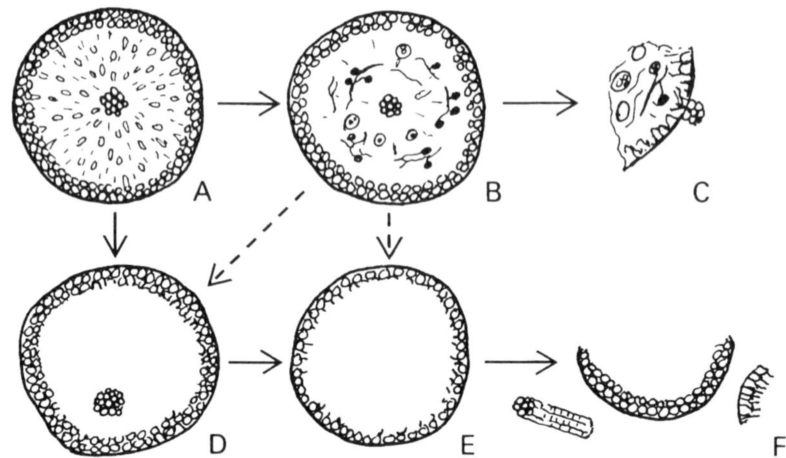
1. PLANT DEGRADATION
A
B
C
D
E
F

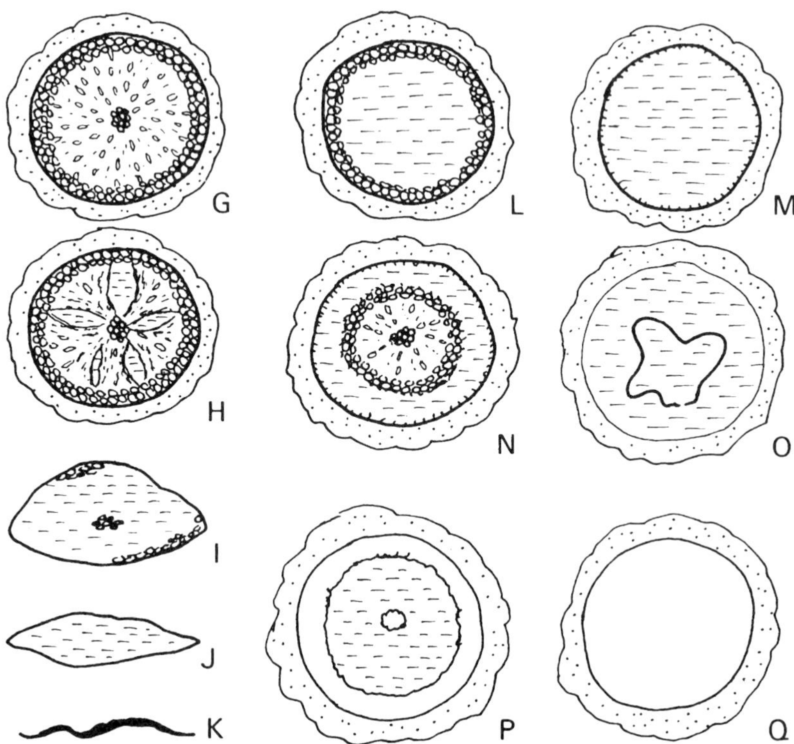
2. PRESERVATION EFFECTS IN CHERT
G
L
M
H
N
O
I
J
K
P
Q

Degradation of plants of prior to silicification

The 35 m core of the Rhynie cherts (Trewin 1994, Powell 1994) displays plant-bearing cherts interbedded with carbonaceous sandstones and laminated and desiccation-cracked shales of alluvial plain and lacustrine origin. This implies that the cored sequence lay at the cool margin of a hot-spring eruption centre. The presence of subaerial plants and arthropods reinforces this view, which conforms with zonation described in Upper Devonian sinters in Queensland (Walter et al 1996). However, at the times of influx of hot fluids there is clearly an opportunity for thermophilic species to flourish, even at the same time that the plants are undergoing degradation and/or preservation. Plant degradation features clearly show the silicification events to be episodic and short lived.

Plants within some cherts show perfect cellular preservation of 3D subaerial axes which in some cases are preserved in growth position to a height of 15 cm. The subaerial axes are connected to rhizomes and the undersides of rhizomes bear delicate rhizoids. Features of the growing tips of subaerial axes, sporangia in different stages of development and spores preserved in the process of germination show that silicification was sudden and rapid, and took place by hot-spring waters invading areas where plant communities were growing. Most plant-bearing cherts overlie carbonaceous sandstones; the original sand formed the substrate for plant growth.

Many beds contain plant material which was clearly physically degraded prior to, and during, silicification. The outer cortex and xylem strands of the plant are most resistant to decay, and thus isolated hollow straws and fragments of plant axes can be found with a partial sediment fill (Fig. 1E and Fig. 2). Some cherts represent a mix of detrital sediment, plant debris and spores, and rare arthropod fragments.

Biological degradation of the plants is commonly seen, and fungi appear to have played a major role. Plant axes show different degrees of internal and surface infestation of tissue with fungal hyphae and cysts, the silicification process providing a snapshot of the breakdown process (Fig. 2).

FIG. 2. Semi-diagrammatic illustration of features of (1) plant degradation prior to silicification, and (2) effects associated with plant preservation in the Rhynie and Windyfield cherts. (1) Plant degradation. (A) Unaltered plant axis with central xylem strand and thicker walled marginal ring of cells. (B) Decay by fungal and probably bacterial attack. (C) Resulting fragments of fungal -infected plants. (D) Decay of weaker cells allows central xylem strand to fall to floor of hollow 'straws'. (E) Hollow straw. (F) Fragments of xylem strand and outer cuticle with marginal cell ring. (2) Preservation effects in the chert. Stipple, initial silica coating of plant; horizontal dashes, later fill or replacement; blank, porosity. (G–K) Range from perfect 3D cellular preservation (G) to coalified compression (K). Intermediate stage (H) shows shrinkage of plant tissue to produce radial chert-filled gashes. (I) Compaction with loss of cellular structure. (J) Chert lens with no internal structure. (I), (J) and (K) occur in the transition from chert to cherty sandstone. (L) Silicification only permeates outer ring of cells, centre later decays and is filled with chert; this appears similar to the silicification of (E) above. (M) Only the outer cuticle is preserved. (N) Here the plant tissue has shrunk away from the outer cuticle and the annulus so formed has filled with chert. (O) The cuticle has shrunk away from the enclosing chert and forms a curl in the void which is subsequently filled. (P) The xylem strand and outer ring occur as pore space. (Q) Only a pore remains to show previous presence of plant stem.

Thus the plant bearing cherts resulted from specific, very short-lived silicification events resulting from the invasion of areas of plant growth and decay by hot-spring waters. Chert beds do not exceed 90 cm in thickness, and such beds are composite, containing several surfaces where plants colonized the old sinter surface prior to the next silicification event. Many of the chert beds only represent a single hot spring flood event.

Preservation of biota by silicification

The variable degree of silicification of the flora imparts another set of preservational features to the deposit (Fig. 2). It is not always possible to determine whether a preservation texture or feature is due to pre-silicification degradation, or the silicification process itself. Furthermore, it must be stressed that these processes overlap. Silicification is frequently incomplete and non-mineralized plant tissue may remain and continue to decay after an initial silicification event. The textural features of plant preservation seen in the Rhynie chert can be closely matched with features in modern sinters (Figs 3, 4), despite the obvious differences in the vegetation.

There are three main situations in which silicification of biota has taken place in the Rhynie cherts: (1) surface flooding; (2) hot spring pool; and (3) subsurface permeation. These three situations are not mutually exclusive. Surface flooding may result in the creation of small pools on an irregular surface, and subsurface permeation may continue processes commenced at the surface.

Surface flooding from hot springs takes many forms, ranging from the violence of geyser eruptions to gentle overflow from an orifice producing a terrace of laminated sinter, over which outflow streams migrate as the sinter builds up, so producing episodic deposition at any one spot. At Rhynie, laminated sinter has been figured from loose blocks (Trewin 1994) and the episodic nature of sinter deposition is demonstrated by the presence of several horizons with macroplant colonization within individual chert beds (Fig. 5). Some specimens show semi-prostrate oriented axes which could have been aligned by flow. Surface flow also appears to have produced beds of massive to vuggy plant-bearing chert through the invasion by hot spring waters of areas with plant growth. The lack of lamination in the cherts implicates a single extended flooding event.

Hot spring pools occur in many modern hot spring areas where they display an amazing variety of form, temperature and chemistry, and a corresponding wide range of biota. Pools with widely variable conditions occur in close proximity: those associated with active discharge and which are too hot for any biota to survive can exist less than a metre away from pools cool enough to support plant life and aquatic arthropods. Abandoned vents on sinter terraces are left as small pools when activity or drainage moves to another area of the terrace. Dry, abandoned pools are rapidly colonized by subaerial plants. Flooding of such abandoned pools by hot waters can result in preservation of either a subaerial or a subaqueous biota. In the Rhynie chert, lenticular chert pods up to 20 cm thick and containing upright plant axes may represent flooding of such hollows. The presence of the aquatic arthropod *Lepidocaris* implies the presence of cool water in pools prior to silicification. Blocks of the Windyfield chert contain plant

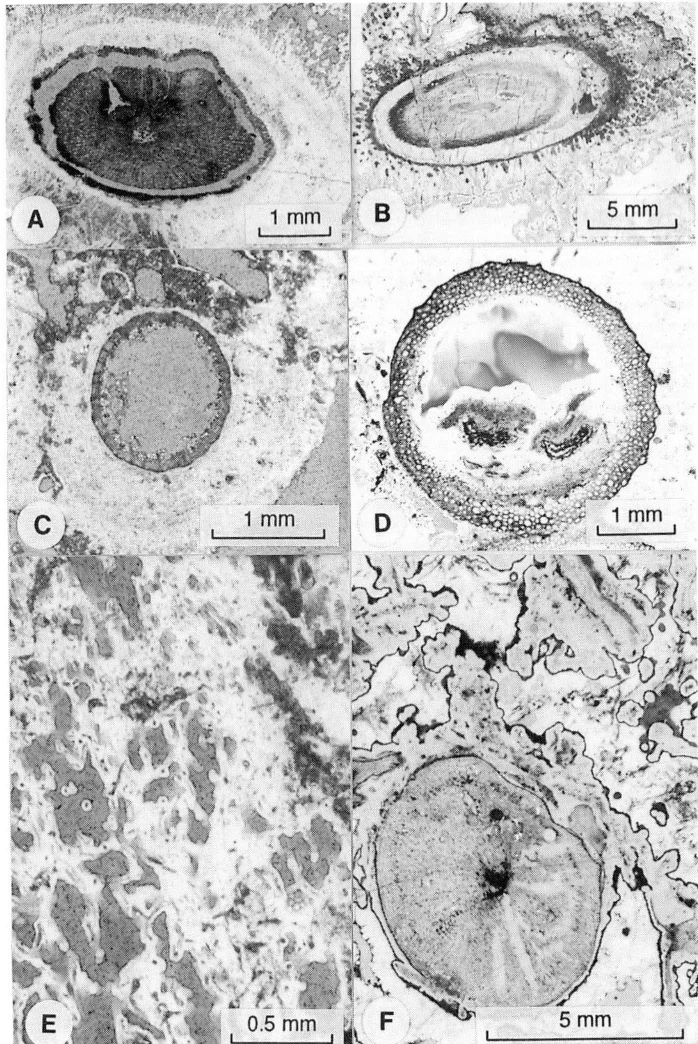

FIG. 3. Textural comparisons between Lower Devonian and recent sinters. (A) Plant stem in sinter showing shrinkage to leave ring of porosity. Whakarewarewa, Rotorua, NZ. (B) Plant axis in Windyfield chert showing shrinkage of tissue from epidermis (See Fig. 2N). (C) Plant axis with only the outer cells preserved, leaving a pore. Orakei Korako, Taupo, NZ. (D) Stem in Windyfield chert with outer ring of cells and later geopetal fill of the original cavity. (E) Deposition of sinter on surfaces of network of cyanobacterial filaments. Orakei Korako, Taupo, NZ. (F) Plant axis in Rhynie chert showing tissue shrinkage (see Fig. 2H) and surrounded by chert network of filaments as in (E) above.

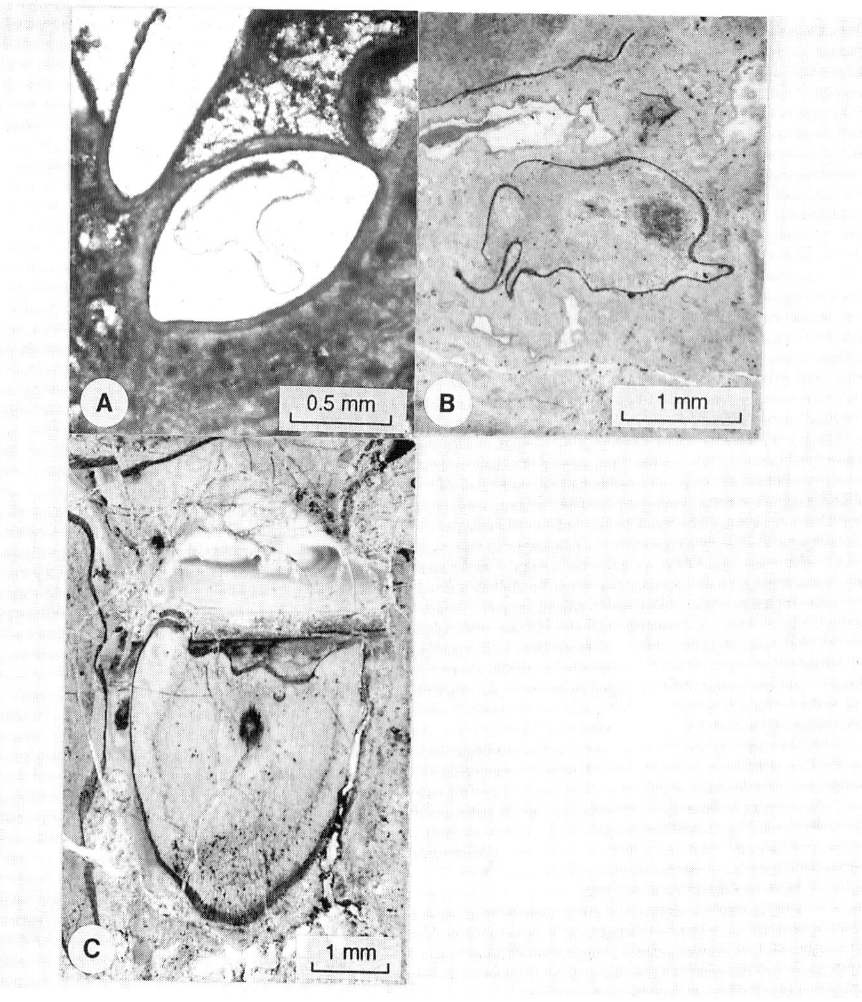

FIG. 4. Textural comparisons between Lower Devonian and recent sinters. (A) Section of sinter deposited on small leaves. Only the outer leaf cuticle remained, and has shrunk from the wall into the pore previously occupied by the leaf. Whakarewarewa, Rotorua, NZ. (B) Shrunken cuticle in Rhynie chert. (C) Rhynie chert showing cuticle collapse from top of preserved plant axis, and later geopetal silica fill in the void. Lower part of axis is well preserved.

stems that appear to have been floating in a vertical inverted orientation. These stems are encrusted with what appears to be a cyanobacterial mat with tiny domal outgrowths accentuating features on the epidermis of the plant (Fig. 1E, F). The same chert bed also contains *Lepidocaris* and numerous tiny coprolites, possibly from this animal. In a hot spring environment it only requires a minor change in the hydrothermal plumbing

FIG. 5. Cut surface of block of Rhynie chert showing levels with plant growth separated by laminated sinter. Brecciation on a small scale is associated with the laminated sinter. The darker subvertical feature is an early fracture, partly filled with sediment and silicified.

for a pool with a flourishing biota to be invaded with hot water, killing and preserving the biota.

Subsurface permeation of fluids to produce silicification could result in two ways. The first is downward percolation of hot water from the surface when hot spring waters invade an area. The degree to which this could occur would depend on the permeability and water saturation of the substrate. The fact that rhizomes bearing rhizoids are preserved indicates that minor permeation into the surface litter took place, but there is seldom evidence of downward permeation into sands and none into muds.

The second and major effect is upward or lateral permeation of sediment and sinters. Small nodular chert patches frequently developed around rhizomes prior to compaction, and as sinters were buried the porous texture was progressively filled with silica.

Brecciated textures with resealing by chert are common, and show disoriented geopetal features. Fracturing and resealing took place within individual beds, and sometimes at the surface, since cyanobacterial filaments occur in the fracture-filling chert.

Zonation of biota in the Rhynie system

Despite the wealth of palaeontological work that has been published on the Rhynie cherts, virtually nothing is known about the lateral and vertical variation within the chert deposits, either in terms of sinter textures, or biotic zonation. The simple reason for this is that the cherts are not exposed, and prior to 1988 all work was done on blocks of chert either obtained loose in fields, or by shallow trenching. Trenching only sampled the zone of surface weathering, and it was not until the borehole was drilled in 1988 that anything was known of the strata between the chert beds and the general sedimentology (Trewin & Rice 1992). Whilst this borehole gave valuable information, neither the top nor base of the chert-bearing sequence was recognized.

The general character of the cherts in the borehole is indicative of the cooler margin of a hot spring system with only periodic invasion of the area with hydrothermal waters depositing sinter. The general presence of plants and arthropods confirms this view and conforms with zonation of the Drummond Basin sinters as described by Walter et al (1996).

However, the discovery of the Windyfield chert as loose blocks some 700 m from the original locality reveals that deposits of the hotter parts of the system are probably preserved. The botryoidal sinter block (Fig. 1D, and Fig. 2 in Trewin 1994) interpreted as geyserite, compares closely with splash textures around modern geyser vents, and probably formed at a temperature >73°C and no more than 5–10 m from a vent (Walter 1976a). Scarce blocks of laminated sinter show lamination typical of sinter terraces deposited by shallow flowing water, and contain evidence of cyanobacterial mats and small domal growths. The typical 'streamer' structures produced by flow-oriented cyanobacterial filaments (White et al 1989, Walter et al 1996) have not yet been observed, but the few blocks found do not split along the lamination to show this feature. Such blocks do show the typical brecciation and recementation of the finely laminated sinter.

Attempts to estimate depositional temperatures of the Rhynie chert by isotopic and organic geochemical means (Rice et al 1995) give temperatures in the range 90–120 °C. This temperature range seems too hot for the cooler part of the system, but is reasonable for geyserite deposited in a vent. It is probable that the results obtained reflect a hotter shallow burial temperature, perhaps that at which the originally highly porous amorphous opaline sinter was converted to chert. It seems unlikely that the sinter would become an isotopically closed system at the time of initial surface deposition.

In fossil hydrothermal ecosystems it is frequently difficult to estimate the degree to which elements of the biota were thermophilic. Observations of modern sinter depositional textures controlled by growing organisms (e.g. Walter 1976b, Cady & Farmer 1996, this volume) provide the best evidence for the interpretation of fossil examples.

Walter et al (1996) have demonstrated the value of ancient–modern comparisons in the case of Upper Devonian deposits, and some of these textures can be recognized in the Rhynie cherts.

Problems arise in the common situation where non-thermophilic species are introduced to hot spring waters (e.g. plants falling, or fronds dipping from a live plant into a hot pool), or where hot waters invade a pool with a cooler-water biota. In these cases a thermophilic biota may encrust the non-thermophile. Whilst this situation is obvious in recent deposits, ancient deposits pose more of a problem.

At Rhynie there is no evidence that the macroplants were adapted to a hot spring environment. In most cases they grew on a sandy substrate prior to inundation by hot-spring waters. In cases where plants grew on sinter they appear to have colonized the surface between pulses of sinter deposition. Four species of cyanobacteria with both prostrate and upright filaments have been described (*Archaeothrix oscillatoriformis* and *A. contexta* [Kidston & Lang 1921b] and *Langiella scourfieldi* and *Kidstoniella fritschi* [Croft & George 1959]). These forms were possibly thermophilic and impart texture to laminated chert where poorly developed palisade fabric is developed.

It is clear that great potential exists for an investigation of biotic zonation and sinter variation in the Rhynie and Windyfield area. It is hoped that detailed study of micro-textures in the chert and comparison with modern sinters will provide the best evidence for the presence and nature of thermophilic elements of the biota of this ancient hot spring.

Conclusions

The deposits of the Rhynie–Windyfield hot spring system studied to date contain a varied biota which is generally representative of the cooler parts of the hydrothermal system. Evidence exists for the preservation of hot spring vents and sinters of the hotter parts of the system, but drilling is required to obtain in *situ* cores before biotas and their distribution can be meaningfully assessed. The varied biota of the Rhynie records an abundance of life in the immediate area of this 400 Ma hot-spring system. Textures in the cherts and taphonomic features of the biota compare closely with modern examples.

Acknowledgements

It is a pleasure to acknowledge the contribution of all involved in recent interpretations of Rhynie geology, particularly by Clare Powell in her thesis work, during the receipt of a NERC Research Studentship. The Royal Society is thanked for support to visit hot spring areas in New Zealand in 1992. This contribution arises from a wider study of early terrestrial environments supported by the Carnegie Trust for the Universities of Scotland and the Leverhulme Trust. Walter Ritchie is thanked for photographic work, Barry Fulton for drafting and Karen Chalmers for processing the text.

References

Boullard B, Lemoigne Y 1971 Les champignons endophytes du Rhynia gwynne-vaughanii K& L. Étude morphologique et déductions sur leur biologie. Botaniste 54:49–89

Cady SL, Farmer JD 1996 Fossilization processes in siliceous thermal springs: trends in preservation along the thermal gradient. In: Evolution of hydrothermal ecosystems on Earth (and Mars?). Wiley, Chichester (Ciba Found Symp 202) p 150–173

Cleal CJ, Thomas BA 1995 Palaeozoic palaeobotany of Great Britain. Geological Conservation Review Series 9. Joint Nature Conservation Committee. Chapman & Hall, London

Croft WN, George EA 1959 Blue–green algae from the Middle Devonian of Rhynie, Aberdeenshire. Bull Br Mus Nat Hist (Geol) 3:341–353

Edwards DS 1986 *Aglaophyton major*, a non-vascular land plant from the Devonian Rhynie Chert. Bot J Linn Soc 93:173–204

Edwards DS, Lyon AG 1983 Algae from the Rhynie Chert. Bot J Linn Soc 86:37–55

El-Saadawy WE, Lacey WS 1979a The sporangia of *Horneophyton lignieri* (Kidston & Lang) Barghoorn & Darrah. Rev Palaeobot Palynol 28:137–144

El-Saadawy WE, Lacey WS 1979b Observations on *Nothia aphylla* Lyon ex Hoeg. Rev Palaeobot Palynol 27:119–147

Hass H, Taylor TN, Remy W 1994 Fungi from the Lower Devonian Rhynie chert: mycoparasitism. Am J Bot 81:29–37

Hirst S 1923 On some arachnid remains from the Old Red Sandstone (Rhynie Chert Bed, Aberdeenshire). Ann Mag Nat Hist 9:455–474

Hirst S, Maulik S 1926 On some arthropod remains from the Rhynie Chert (Old Red Sandstone). Geol Mag 63:69–71

Kidston R, Lang WH 1917 On Old Red Sandstone plants showing structure from the Rhynie chert bed, Aberdeenshire. I. *Rhynia gwynne-vaughani* Kidston & Lang. Trans R Soc Edinb 51:761–784

Kidston R, Lang WH 1920a On Old Red Sandstone plants showing structure from the Rhynie chert bed, Aberdeenshire. II. Additional notes on *Rhynia gwynne-vaughani* Kidston & Lang, descriptions of *Rhynia major* n. sp. and *Hornea lignieri* n., g., n. sp. Trans R Soc Edinb 52:603–627

Kidston R, Lang WH 1920b On Old Red Sandstone plants showing structure from the Rhynie Chert bed, Aberdeenshire. III. *Asteroxylon mackiei*, Kidston & Lang. Trans R Soc Edinb 52:643–680

Kidston R, Lang WH 1921a On Old Red Sandstone plants showing structures from the Rhynie Chert bed, Aberdeenshire. IV. Restorations of vascular cryptograms, and discussion of their bearing on the general morphology of the Pteridophyta, and the origin of the organisation of land-plants. Trans R Soc Edinb 52:831–854

Kidston R, Lang WH 1921b On Old Red Sandstone plants showing structures from the Rhynie chert bed, Aberdeenshire. V. The Thallophyta occurring in the peat-bed; the succession of the plants throughout a vertical section of the bed, and the conditions of accumulation and preservation of the deposit. Trans R Soc Edinb 52:855–902

Lyon AG 1962 On the fragmentary remains of an organism referable to the Nematophytales, from the Rhynie chert, '*Nematoplexus rhyniensis*' gen. et sp. nov. Trans R Soc Edinb 65:79–87

Lyon AG 1964 The probable fertile region of *Asteroxylon mackiei* K and L. Nature 203:1082–1083

Lyon AG, Edwards D 1991 The first zosterophyll fom the Lower Devonian Rhynie Chert, Aberdeenshire. Trans R Soc Edinb Earth Sci 82:323–332

Mackie W 1913 The rock series of Craigbeg and Ord Hill, Rhynie, Aberdeenshire. Trans Edinb Geol Soc 10:205–236

Powell CL 1994 The palaeoenvironments of the Rhynie cherts. PhD thesis, University of Aberdeen

Powell CL, Trewin NH, Edwards D 1991 Sinter textures and plant preservation in the Rhynie Chert. Br Sedimentological Res Group Abstracts, Posters, Edinburgh

Remy W, Remy W 1980 Devonian gametophytes with anatomically preserved gametangia. Science 208:295–296

Rice CM, Trewin NH 1988 A Lower Devonian gold-bearing hot-spring system, Rhynie, Scotland. Trans Inst Mining Metallurgy B 97:141–144

Rice CM, Ashcroft WA, Batten DJ et al 1995 A Devonian auriferous hot spring system, Rhynie, Scotland. J Geol Soc 152:229–250

Scourfield DJ 1925 On a new type of crustacean from the Old Red Sandstone (Rhynie Chert Bed, Aberdeenshire) — *Lepidocaris rhyniensis* gen. and sp. nov. Phil Trans R Soc Lond 214:153–187

Shear WA, Selden PA, Rolfe WDI, Bonamo PM, Grierson JD 1987 New terrestrial arachnids from the Devonian of Gilboa, New York (Arachnida, Trigonotarbida). Novitates 290: 1B–74B

Taylor TN, Remy W, Haas H 1992a Parasitism in a 400-million-year-old green alga. Nature 357:493–494

Taylor TN, Haass H, Remy W 1992b Devonian fungi: interactions with the green alga *Palaeonitella*. Mycologia 84:901–910

Trewin NH 1994 Depositional environment and preservation of biota in the Lower Devonian hot-springs of Rhynie, Aberdeenshire, Scotland. Trans R Soc Edinb 84:433–442

Trewin NH, Rice CM 1992 Stratigraphy and sedimentology of the Devonian Rhynie chert locality. Scott J Geol 28:37–47

Walter MR 1976a Geyserites of Yellowstone National Park: an example of abiogenic stromatolites. In: Walter MR (ed) Stromatolites. Elsevier, Amsterdam, p 87–112

Walter MR 1976b Hot-spring sediments in Yellowstone National Park. In: Walter MR (ed) Stromatolites. Elsevier, Amsterdam, p 489–498

Walter MR, Des Marais D, Farmer JD, Hinman NW 1996 Lithofacies and biofacies of mid-Palaeozoic thermal spring deposits in the Drummond Basin, Queensland, Australia. Palaios, in press

White NC, Wood DG, Lee MC 1989 Epithermal sinters of Palaeozoic age in north Queensland, Australia. Geology 17:718–722

DISCUSSION

Cady: In some of the Rhynie Chert thin sections you have shown us there are regions where the morphological fidelity of the organic remains is better than adjacent regions. Is this due to post-burial alteration?

Trewin: The chert will have been buried in the hot spring system. Hot liquid penetrates fractures in the chert, causing localized heating. The organic material is simply burnt off in the adjacent porous chert.

Shock: I would say that is partially right. In terms of what is preserved, I think the openness of the system is the crucial factor, not the temperature. If the system becomes open as the fluid is coming through the fracture then reactions can take place, but the parts of the chert that are not opened up again can preserve organic matter even at quite high temperatures.

Knoll: In support of that, one sees those same types of micro-gradients, going from exquisite to terrible preservation, in tidal flat silicified deposits where hydrothermal fluids aren't involved.

Farmer: I wanted to comment about the importance of permeability as a major control on preservation of organic matter. There could be an indirect correlation with temperature, because the silica precipitation is driven by decreasing temperature. It's my impression that the completeness and the rate of infilling gets better as you move down the temperature gradient. You tend to get very dense sinters at or near vents, but in terms of infilling, sinter frameworks tend to be very porous until you get down to the lower temperature end.

Pentecost: Do you find any fluid inclusions, or evidence that they might have once been present in the Rhynie chert?

Trewin: Not in the Rhynie chert itself. Fluid inclusions analysed to date come from quartz veins, which are of deeper origin. Noble gases obtained by crushing the chert give an estimate of isotopic argon ratios in the Devonian atmosphere which differ from today's values (Rice et al 1995).

Pentecost: Fluid inclusions might give an idea of the composition of the hydrothermal waters.

Trewin: Yes, we're hoping that in a future investigation we will be able to find fluid inclusions in more deeply buried chert. However, this awaits funding for drilling!

Walter: You made it sound very complicated, with the constant re-working by fluids at different temperatures and the juxtaposition of disparate facies.

Trewin: I think the initial burial stage is very complicated. I'm looking at the cooler end of the system at Rhynie with the plants and the arthropods preserved. At the edges of the system, where the silica-rich waters were cooling down, they were invading areas of plant growth. So the drainage and the plumbing of the system is changing and fluids are invading areas of plant growth and killing them off. Sinter deposition then takes place for a very short time, and clastic sedimentation covers the sinter. In between the cherts in some places there are shales with desiccation cracks and also finely laminated muds from the lake deposits. The silica-rich sinter is only preserved where eruptions were subaerial.

Walter: If you had the advantage of surface exposures you might be able to see through some of those complexities, in the way that we can with the Drummond Basin deposits. We think we can see the original thermal gradients, no doubt with all those kinds of complications imposed during diagenesis. However, overall there's a structure that can be related to the initial thermal gradients.

Trewin: We have plans for a drilling programme which we hope will define the vent area, and then we'll be able to say something about zonation in the Rhynie hydrothermal system.

Cowan: Can one assume that the deposition process is rather less complex at the higher temperature end of the deposition spectrum? For example, the formation of geyserite, where you've got a relatively regular deposition rate.

Trewin: Yes, that's reasonable.

Cowan: So the microbial fossil record in high temperature deposition regions might be easier to interpret?

Trewin: Yes, but when you have macroplants and arthropods at the generally cool end of the system which occasionally becomes hot, then it is quite difficult to say whether any filamentous organism I find is a thermophile or not.

Cady: We have observed that the best morphological preservation of sheathed cyanobacteria occurs in mats that grow in the subaqueous parts of the system. There is actually a range of preservational states at the low temperature end of the system, presumably because of the dynamic balance between rates of silicifcation and decomposition. Although we observe this variation in the modern environment, the important question to ask concerns whether the variation in preservation affects our ability to identify these remains as particular species in older sinters.

Henley: I think you were suggesting that the complexity and the difficulty of interpreting these cherts comes about because of their burial: it makes your job very difficult to recognize conditions at the original surface. To me, however, it seems that the burial has actually given you an opportunity, because the cross-cutting fractures may carry the deeper subterranean (possibly hyperthermophile) environments. Have you looked at the cross-cutting opaline veins?

Trewin: Yes we have, but we have not yet found any evidence of life. The only place we see this is in sinter I interpret as near-surface, where there is a brecciated sinter and the actual fill, or breccia cement, contains filaments in growth position. Again, the sinter terrace may have broken up (as they usually do), filamentous organisms might have grown in the cracks and then the surface was re-flooded with silica-rich water. The problem is that we don't know the temperature at which the organisms were living or the temperature at which they were preserved.

Cady: That is a good point. We've observed in modern environments that the direction, rate, and temperature of fluid flow changes during an eruption cycle. There is also the problem of overprinting of lower temperature lithofacies by higher temperature lithofacies, and vice versa, if the hydrodynamics of the system changes for longer periods of time.

Parkes: If you are getting organic preservation, and if there are some unique phospholipids indicative of hyperthermophiles, then these biomarkers might indicate the presence of hyperthermophiles irrespective of morphological preservation. The bacterial membrane also changes with temperature. Therefore, if you get very good preservation, you could even use the biomarkers to tell you the gradation of temperature with time.

Summons: If the samples have been to temperatures of 150–200 °C for significant geological periods, the chemical structures within the organic matter may have been altered or destroyed. A possibility for circumventing this is to look at hydrocarbons trapped in fluid inclusions, which are less at risk of destruction. Hydrocarbon-bearing fluid inclusions would be one of the best places to look for biomarkers.

Walter: Haven't hydrocarbons already been reported from the Rhynie chert?

Trewin: These were taken from the shales associated with the Rhynie chert, which were much cooler. If we can find a nice piece of chert that was at a low enough temperature right on the margin of the system, and which wasn't involved in the general heating as the whole system was buried, then we might be able to obtain some useful information from biomarkers.

Summons: Another possibility is to isolate microscopically or dissect out pieces of plant material. This can then be pyrolysed in the inlet system of a mass spectrometer and the products analysed for diagnostic carbon skeletons.

Over the last couple of years we have done some work on the HYC, a possible hydrothermal marine deposit. We were looking at organic compounds as a way to measure temperature gradients. We found that the heating alteration has been so severe that all diagnostic information has been lost. Another problem is the contamination effect of hydrocarbons being produced from the host or country rock. It is very difficult to imagine that you could identify the biomarkers from a tiny amount of thermophile biomass if it has been mixed with a much larger amount of hydrocarbon generated from the host rock by the heat of the hydrothermal activity. What you are almost certain to see is the geochemical signature of the host rock.

Walter: But if you think about those sorts of systems, any subaqueous hydrothermal system has a plume of hot water that goes into the ambient water body. Those plumes often contain archaea. So we're not talking just about the immediate environment around the hot spring, we're also talking about a mechanism for dispersing the information from that environment. So in the shales that are deposited distally from these springs, we could well see the signature of the organisms that lived in the springs.

Parkes: Has anybody actually conducted a biomarker transect in an existing hydrothermal system to see whether lipid biomarkers are preserved? If so, at what temperature do we get maximum preservation?

Stetter: David White's group at Talahassee, FL, are doing things like that, but I don't know how successful they have been.

Parkes: They were looking at the food web relationships between the free-living high temperature organisms and animals like the shrimps using $\delta^{13}C$ to indicate whether they're predominantly getting their energy from the autotrophic activity (Rieley et al 1995). I don't think they were actually targeting the thermophilic phytanyl ether lipids.

Farmer: I wanted to draw attention to some additional observations from two sources: the modern systems that Sherry Cady and I have been looking at, and the Drummond Basin sinters, which I have been looking at with Malcolm Walter. We find fairly well preserved plant materials all the way through the system. Even in some of the detrital layers in the higher temperature lithofacies you can get fairly well preserved, albeit fragmental, plant material. But, as we move down the system we observe an increase in the abundance of permineralized whole plant fragments with good structural preservation. In thermal spring outflows the diversity of metazoan grazers increases dramatically at about 15–20 °C. For example, I've noted in thermal spring outflows at Yellowstone that small beetles which graze on the *Calothrix* mats appear at this temperature range. We don't observe organically preserved (permineralized) microbial fossils above temperatures of 20 °C or so. And even at lower temperatures it is only occasionally that you will have a permineralized filament. By and large only the cyanobacterial sheaths are permineralized, particularly the thicker sheaths of

Calothrix. As you move up the temperature gradient, preservation is dominated by filament moulds, and all the organics are stripped away during early diagenesis.

You showed an example of a mat of cyanobacterial filaments sitting on a fossilized plant stem (Fig. 1E, F). In the microbial part, is the preservation organic or are those actually filament moulds?

Trewin: It's very difficult to tell. The general brown colour of preservation is the same as the brown colour that is generally organic. However, there is iron present in the rocks and at present I would not rule out the possibility that the colour is due to iron.

The specimen shown in Fig. 1E, F comes from the Windyfield chert, near where the geyserite (Fig. 1D) was found. The plant fragments appear to have been floating in a small pool in an inverted position because some of the stems branch downwards in the block of chert. In some sections there are coprolites indicating the presence of arthropods. I imagine the hot-spring plumbing system changed and the pool was invaded by hot siliceous waters, killing and preserving the whole ecosystem. You can see this process in action on the shore of Lake Rotorua (NZ) in the active area near Sulphur Point. With careful observation you can find two pools just a foot or so apart, one of which has animals living it, and the other is boiling. It only requires a very tiny change in the plumbing to invade the cool pool with hot water and preserve the fauna and flora.

Farmer: In Yellowstone, vents and outflow channels have moved many times during the period of our studies.

References

Rice CM, Ashcroft WA, Batten DJ et al 1995 A Devonian auriferous hot spring system, Rhynie, Scotland. J Geol Soc 152:229–250

Rieley G, Van Dover CL, Hedrick DB, White DC, Eglinton G 1995 Lipid characteristics of hydrothermal vent organisms from 9 °N, East Pacific Rise. J Geol Soc Spec Publ 87:329–342

Fossilization processes in siliceous thermal springs: trends in preservation along thermal gradients

S. L. Cady and J. D. Farmer

NASA Ames Research Center, MS 239–4, Moffett Field, CA 94035–1000, USA

Abstract. To enhance our ability to extract palaeobiological and palaeoenvironmental information from ancient thermal spring deposits, we have studied the processes responsible for the development and preservation of stromatolites in modern subaerial thermal spring systems in Yellowstone National Park (USA). We investigated specimens collected from silica-depositing thermal springs along the thermal gradient using petrographic techniques and scanning electron microscopy. Although it is known that thermophilic cyanobacteria control the morphogenesis of thermal spring stromatolites below 73 °C, we have found that biofilms which contain filamentous thermophiles contribute to the microstructural development of subaerial geyserites that occur along the inner rims of thermal spring pools and geyser effluents. Biofilms intermittently colonize the surfaces of subaerial geyserites and provide a favoured substrate for opaline silica precipitation. We have also found that the preservation of biotically produced microfabrics of thermal spring sinters reflects dynamic balances between rates of population growth, decomposition of organic matter, silica deposition and early diagenesis. Major trends in preservation of thermophilic organisms along the thermal gradient are defined by differences in the mode of fossilization, including replacement, encrustation and permineralization.

1996 Evolution of hydrothermal ecosystems on Earth (and Mars?). Wiley, Chichester (Ciba Foundation Symposium 202) p 150–173

Given the significance of hydrothermal systems in the early evolution of the biosphere (this volume: Barns et al 1996, Stetter 1996, Walter 1996), we look to the fossil record for evidence regarding the palaeobiology and palaeoenvironments of these ecosystems. Subaerial thermal springs are of interest because they contain many genera of thermophilic microorganisms that are likely to have existed early in Earth's history (Ward et al 1989). Recent studies have shown that fossiliferous analogues of modern subaerial siliceous thermal spring sinters (e.g. Drummond Basin, Australia [Walter et al 1996] and Rhynie Chert, Scotland [Trewin 1994, 1996 this volume, Rice et al 1995]) retain many of their primary macro-scale textural characteristics, even after diagenetic overprinting. However, the accurate reconstruction of ancient communities from

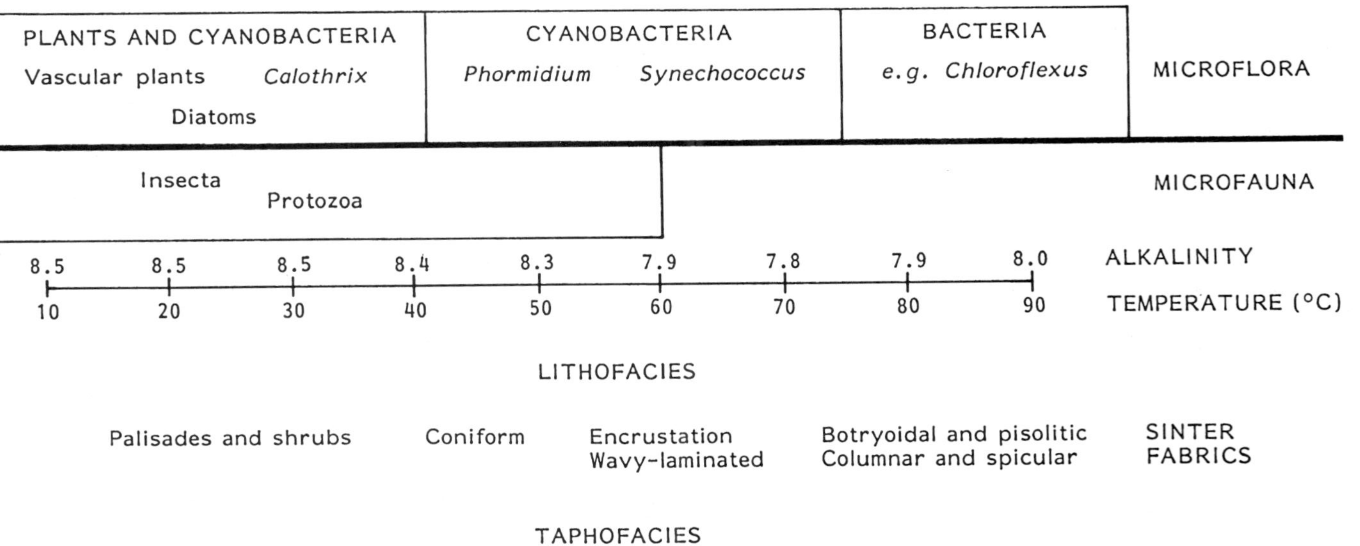

FIG. 1. Biofacies and lithofacies model for siliceous sinters developed by Walter (1976a) based on his studies of silica-depositing thermal springs in Yellowstone National Park (USA). Generalized taphofacies model summarizes observations of authors regarding preservational modes along thermal/pH gradient in silica-depositing thermal springs in Yellowstone.

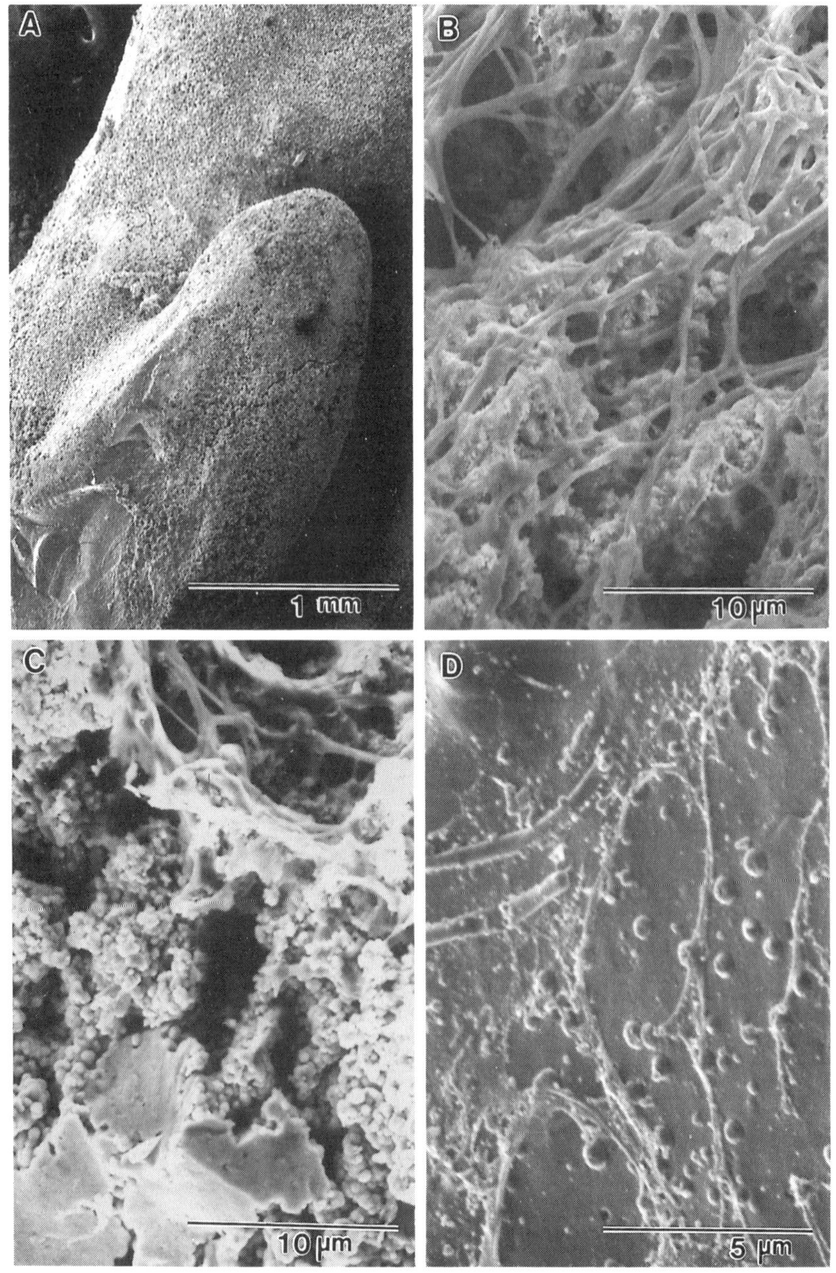
A
1 mm
B
10 µm
C
10 µm
D
5 µm

fossilized analogues is complicated by preservational biases and by the loss of fine-scale microstructural characteristics during diagenesis. A complementary approach is to study how modern hydrothermal ecosystems are converted to their fossil counterparts. Here we discuss examples from our studies of modern siliceous sinters in Yellowstone National Park (USA) to demonstrate how biotic and abiotic factors contribute to the morphogenesis and microstructural development of geyserites and thermal spring stromatolites. The apparent rarity of organically preserved microfossils in ancient siliceous sinters emphasizes the importance of understanding how microorganisms contribute to sinter morphogenesis.

Thermal spring stromatolites form in association with distinct microbial communities whose distribution along thermal spring outflows is controlled by steep thermal gradients and the composition of the hydrothermal fluids and dissolved gases (e.g. Brock 1978, Castenholtz 1984, Ward et al 1989). Walter (1976a), on the basis of studies in Yellowstone, developed a general framework for silica-depositing thermal springs that illustrates the association between the principle types of siliceous sinter and the dominant mat-forming bacteria in the overlying microbial mat community (Fig. 1). We include here a preliminary taphofacies model that summarizes major trends in early preservation we have observed in thermal spring geyserites and siliceous sinters from Yellowstone. Differences in modes of preservation are discussed in terms of dynamic balances between rates of population growth, organic matter decomposition, mineral deposition and early diagenesis. A main objective for future studies is to elucidate the interrelationships between these parameters and incorporate them into a more refined taphofacies model for understanding patterns of preservation in ancient sinter deposits.

Methods

To preserve microorganisms for laboratory investigation using low-voltage scanning electron microscopy (LVSEM), we fixed samples of sinter in the field using a 2.5% glutaraldehyde solution prepared from filtered spring water taken from sites where samples were collected. Samples were post-fixed in the laboratory using a 1M osmium tetroxide solution in cacodylate buffer, dehydrated in a graded ethanol series, critical point dried, and coated with a thin layer of amorphous carbon or gold to increase surface conductivity.

FIG. 2. LVSEM micrographs of spicular geyserite from an unnamed spring in Shoshone Geyser Basin, Yellowstone National Park (USA). (A) Spicules are characterized by an irregular porous surface texture that results from (B) the favoured deposition of opaline silica on discontinuous biofilms that contain filamentous thermophilic bacteria. (C) Cross-sectional view of fractured spicule showing the surface biofilm, a porous rind of opaline silica and an inner dense matrix of opaline silica that formed as a result of secondary infilling. (D) Fractured surface of another spicule shows a non-porous layer of opaline silica that contains well-preserved morphological remains of filamentous and coccoidal(?) thermophiles.

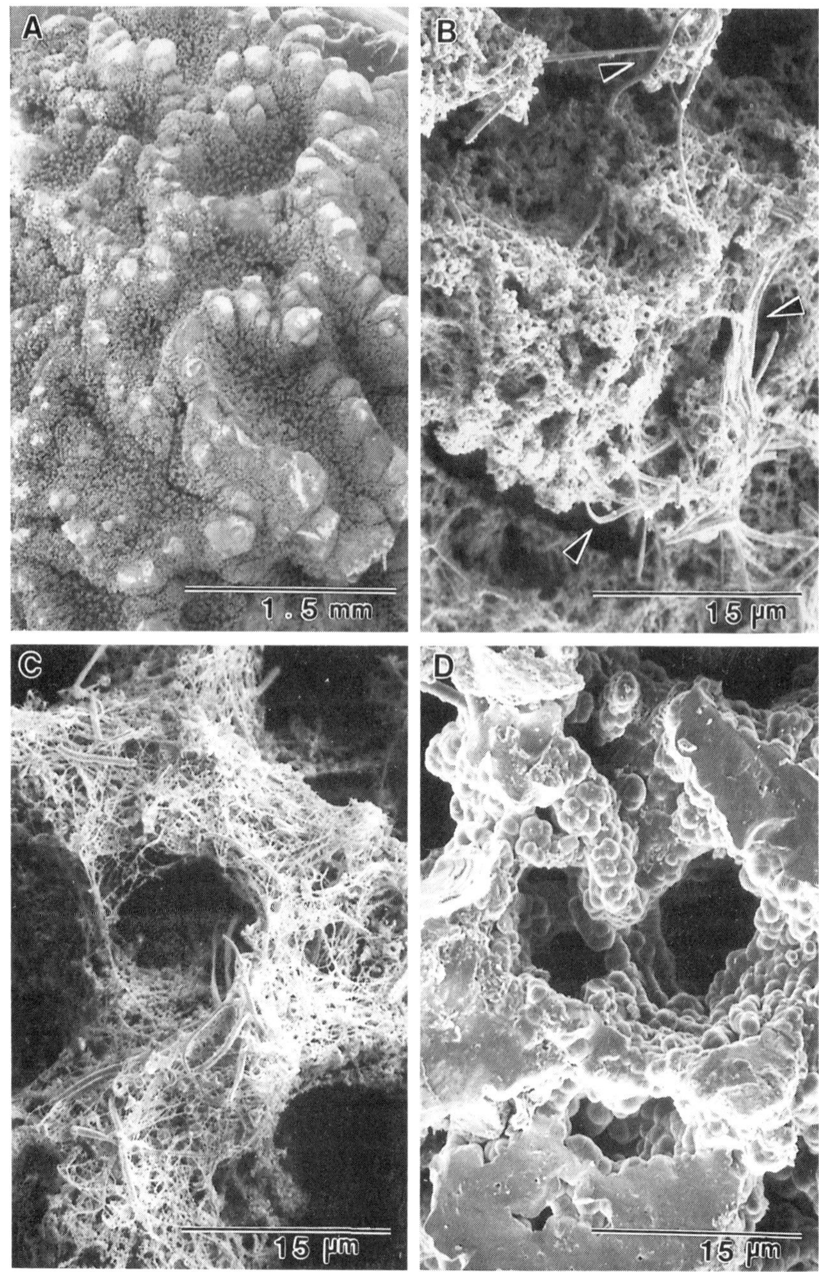
A
B
C
D
1.5 mm
15 µm
15 µm
15 µm

High temperature, near-vent environments

The term geyserite is used (e.g. White et al 1964) to describe siliceous sinter that precipitates within or immediately adjacent to thermal springs and geyser effluents from hydrothermal fluids ejected at or above surface boiling temperatures—waters once thought to be 'sterile' (Allen 1934). Geyserites lack petrographically identifiable microfossils and commonly display micro cross-laminated fabrics characterized by banded laminae microns to submicrons thick (Walter 1976b). The lower temperature limit for geyserite precipitation was defined by Walter (1976b) as the 73 °C isotherm, which coincides with the upper temperature limit for photosynthetic bacteria (e.g. Brock 1978). Although non-photosynthetic bacteria were known to inhabit the spring waters at Yellowstone, Walter (1976b) was unable to prove that microorganisms contributed to geyserite morphogenesis and concluded, therefore, that geyserites formed abiogenically as a result of the rapid cooling and evaporation of thermal spring and geyser waters (e.g. White et al 1956). Walter (1976b) also noted that even at temperatures lower than 73 °C there was no clear evidence that microorganisms played a role in silica precipitation. We are not aware of any studies which have demonstrated that silica precipitation can be induced by the metabolic activity of bacteria. We have found, however, that organisms do contribute to the microstructural development of geyserites by providing a favoured substrate for opaline silica precipitation. Observations made using LVSEM on field-fixed specimens demonstrate that the morphogenesis of spicular and columnar geyserites reflects the coupled interaction between biotic and abiotic sedimentary processes.

Spicular geyserites, less than a millimetre or two in diameter and up to several millimetres in length (Fig. 2A), form along the innermost margins of pool rims. Petrographic examination shows that the microfabric of spicules consists of steeply convex to parabolic laminae that vary in thickness from tens to hundreds of microns. Submicron-thick laminae are visible with scanning electron microscopy, and submicroscopic laminae on the order of tens of nanometres thick have been identified using transmission electron microscopy (unpublished observation of S. L. Cady). LVSEM micrographs reveal that laterally discontinuous, micron-thick biofilms that contain submicron-size filaments intermittently colonize the surfaces of spicular geyserites (Fig. 2B). The divergence of adjacent filaments creates a ridged network that extends across the surfaces of the individual spicules. The highest concentrations of microbial

FIG. 3. LVSEM micrographs of columnar geyserite from unnamed spouter located near Octopus Pool, Yellowstone National Park (USA). (A) Top portion of knobby-shaped column is characterized by an irregular surface texture and protrusions of fine-grained opaline silica. (B) Viable filaments (arrows) occur along the outermost surface of substratum composed of network of silicified filment moulds. (B, C) Diverse microbial network that forms between the protrusions contains filament moulds that formed by the encrustation of filaments by opaline silica, and the apparent loss of microbial remains. (D) Silicified equivalent of (C) near base of knobby-shaped column demonstrates influence of microbial network in the development of abiotic fabric.

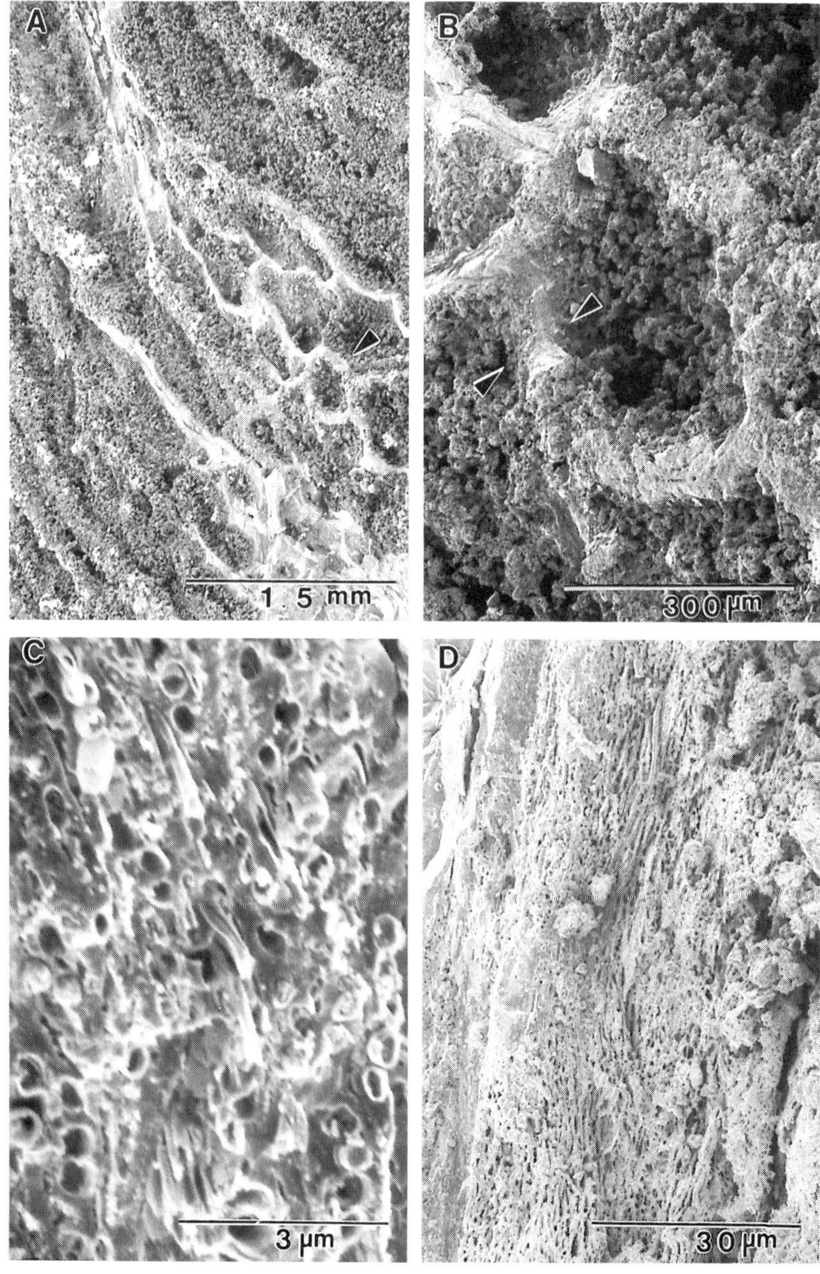
A
1.5 mm
B
300 µm
C
3 µm
D
30 µm

filaments were identified at the tips of spicules that formed along the inside rim of thermal springs at the air–water interface. These observations indicate that the directed accretion and growth of spicules results from the localized precipitation of opaline silica where filaments are concentrated. Additional evidence that the biofilms serve as loci for opaline silica deposition is seen in cross-sectional views of fractured spicules.

The internal microstructure of spicules consists of porous and non-porous submicroscopic laminae. A porous rind develops beneath the biofilms as a result of the favoured deposition of opaline silica on the surfaces of older biofilms and/or their degraded remains (Fig. 2C). The subsequent growth and coalescence of opaline silica grains on the organic substrate imparts a botryoidal texture on exposed surfaces within the porous rind. Porous laminae develop because the rate of opaline silica deposition at the outermost surface of the spicule outpaces the rate of secondary infilling of the porous rind. Dense, non-porous laminae may form as a result of the concomitant infiltration of the porous rind with opaline silica, or, in the absence of a biofilm, as a result of abiotic opaline silica deposition. The occasional presence of well-preserved morphological remains of microorganisms in the non-porous laminae (Fig. 2D) indicates that a considerable amount of secondary infilling may occur within a relatively short period. Such a condition could exist, for instance, if the inner pool rim became temporarily submerged due to fluctuations in the water level of the thermal spring. The mechanisms by which biofilms in the porous laminae and the morphological remains in the non-porous laminae become preserved in opaline silica are currently being investigated.

Microorganisms also influence the development of the microstructure of columnar geyserites. Prior to collection of the knobby-shaped column shown in Fig. 3A, we observed that it was periodically submerged by geyser eruptions (for approximately 3 min) every 15–18 min (92 °C effluent water temperature). LVSEM micrographs reveal that between the knobby-shaped protrusions lies a diverse consortium of coccoids and filaments that form an elaborate microbial network (Fig. 3B). Silicification of the microbial network occurs as a result of the encrustation of microorganisms by opaline silica. Viable filaments (see arrows) occur primarily on the outer surfaces of the silicified substrate, which consists of filament moulds devoid of any morphologically identifiable ultrastructural organic remains. A comparison of the surface texture at the top of the periodically submerged column (Fig. 3C) with that at the base of the column,

---

FIG. 4. LVSEM micrographs of cornices on side wall of columnar geyserite from unnamed spouter located near Octopus Pool, Yellowstone National Park (USA). (A) Parallel and anastomosing cornices along side wall of columnar geyserite. (B) LVSEM investigation of fractured surface reveals that cornices consist of non-porous and (C) porous laminae that contain a high concentration of filament moulds with silicified extracellular and/or intracellular organic remains. (D) Laminae with high concentration of filament moulds also occur in the regions between the cornices. LVSEM micrographs shown in Figs 4C and 4D are rotated approximately 30° clockwise from images shown in Figs 4A and 4B.

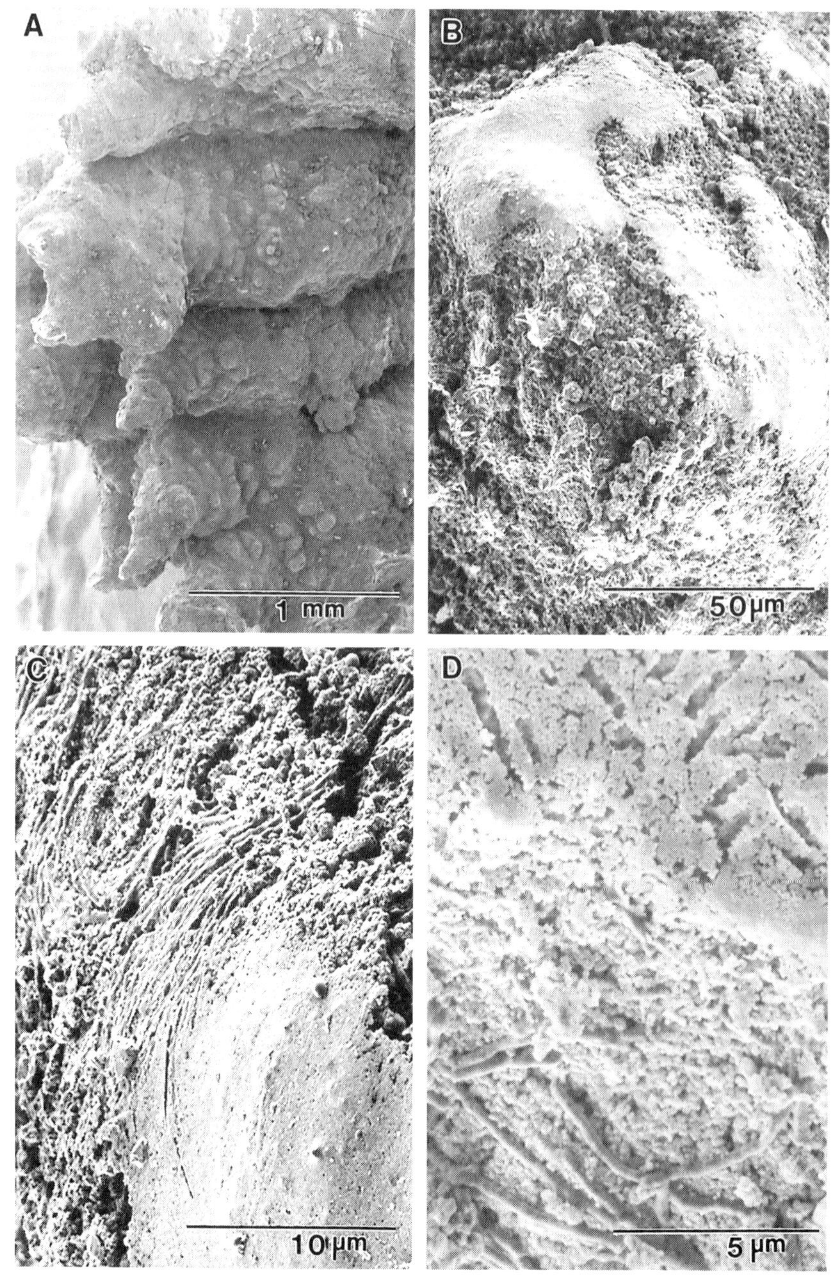
A
B
C
D
1 mm
50 µm
10 µm
5 µm

which is submerged most of the time (Fig. 3D), suggests that in the latter, a silicified network of filament moulds served as a substrate for opaline silica precipitation. Although preservation of this biotically directed fabric is unlikely without differences in the mineralogy of primary and secondary infilling phases, it is clear that the presence of the microbes controlled the localization of opaline silica deposition.

Regularly spaced cornices are one of the distinguishing microstructural characteristics of columnar geyserite (see Walter 1976b, Fig. 33). Walter (1976b) determined from field measurements that cornices, like the ones shown in Fig. 4A, form approximately once each year. An LVSEM micrograph of a fractured cornice surface (arrow, Fig. 4B) reveals the presence of porous laminae that contain high concentrations of filament moulds (Fig. 4C). Some of the filament moulds contain silicified microbial remains in various stages of degradation; intracellular remains protrude from some of the filament moulds whereas extracellular remains protrude from others. The approximately circular shape of many of the moulds indicates that the microorganisms must have been rapidly encased in opaline silica prior to degradation. LVSEM also reveals that laminae containing high concentrations of filament moulds (Fig. 4D) are intercalated throughout the columnar geyserite with non-porous, abiotic laminae. The aperiodic intercalation of biotic and abiotic laminae suggest that local variations in the microenvironments ultimately control whether or not the microbes colonize the geyserites and, consequently, serve as a substratum for opaline silica deposition. Local variations in microenvironments include perturbations in the hydrodynamics or fluid chemistry of the system, as well as seasonal changes that occur because of the geographic location of the hydrothemmal system.

The preferred nucleation of silica on biofilms also produces discontinuous submicroscopic laminae on the surfaces of microspicules that can develop on spicules (Fig. 5A) and columns. As shown in Fig. 5B, the discontinuous laminae consist of dense patches of opaline silica that form on the sides (and top, not shown) of microspicules. The patches presumably accrete as a result of the evaporation and resultant precipitation of opaline silica from fluid droplets held in place by surface tension (e.g. Walter 1976b). These patches of extremely fine-grained opaline silica are almost always associated with relatively high concentrations of filaments. Filaments within the thicker parts of the laminae are completely entombed within the mineral matrix. Those filaments located around the edges of the opaline silica patches become partially

FIG. 5. LVSEM micrographs of microspicules on surface of spicular geyserite from an unnamed spring in Shoshone Geyser Basin, Yellowstone National Park (USA). LVSEM micrograph shown in Fig. 5A rotated approximately 90° counterclockwise from vertical. (A) Microspicules that occur on the tops of actively accreting spicules are oriented toward the direction of the thermal spring pool. (B) Top view of microspicule shows discontinuous patches of opaline silica associated with (C) high concentration of silica-encrusted filaments. The patches form discontinuous submicroscopic laminae. (D) Comparison of viable filaments (bottom) and partial filament moulds (top) at the edge of a discontinuous lamina.

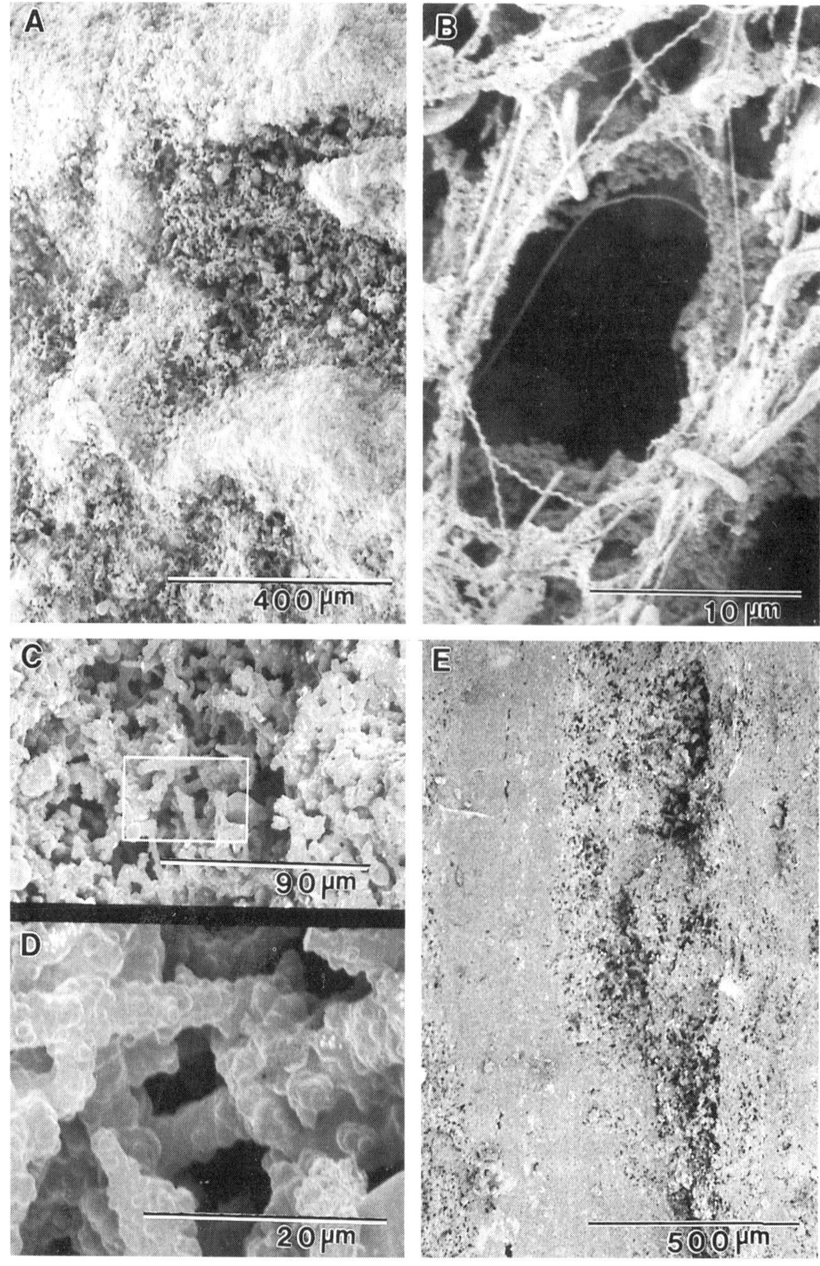
A
400 µm
B
10 µm
C
90 µm
D
20 µm
E
500 µm

encrusted in the dense matrix (Fig. 5C) and rapidly decompose to form filament moulds (Fig. 5D).

These examples illustrate how the potential to fossilize and preserve microorganisms within the subaerial geyserite regime depends upon the relative rates of population growth, microbial decomposition, primary silica deposition and secondary infilling. Petrographic observation by the authors of 2.5 Ma sub-Recent sinters from Steamboat Springs, Nevada (USA) indicate that the diagenetic recrystallization of opaline silica occurs earlier in porous rather than non-porous laminae of spicular and columnar geyserites. It appears that the loss of submicroscopic detail in the porous laminae during early diagenesis masks evidence of their biotic character. Sinters at Steamboat Springs have been described by White et al (1964).

Moderately high temperature pools and channels

At temperatures of ~60–73 °C, thin yellow-to-orange mats dominated by the cyanobacterium *Synechococcus* and the gliding, filamentous, photosynthetic bacterium *Chloroflexus* cover the sinter surfaces of pond floors and outflows. At the transition zone (72–82 °C) between stratiform (flat-laminated) geyserite and moderately high temperature sinter, LVSEM micrographs reveal that microbial communities occur within irregular depressions bounded by ridges of dense, mat-free sinter (Fig. 6A). The microbial assemblage consists of rod-shaped *Synechococcus* and several species of long, thin filaments, including a *Spirulina*-like organism, that are attached to webs and strings of mucilage (Fig. 6B). The ridges appear to be areas of rapid accretion, although they lack any obvious morphological evidence of microorganisms or biofilms. Mats that develop within the depressions display little evidence of early mineralization.

A transition zone between the ridge tops and the valley depressions is indicated by the presence of silicified filament moulds along the outermost regions of the ridge margins.

FIG. 6. LVSEM micrographs of high temperature sinter comprising vent pool floor deposits at Octopus Spring, Lower Geyser Basin (A, B) and an unnamed spring in Shoshone Geyser Basin (C, D), Yellowstone National Park (USA). LVSEM micrograph shown in Fig. 6A rotated approximately 90° counterclockwise from vertical. (A) Sinter surface consists of elongate ridges of relatively non-porous opaline silica, separated by shallow depressions with active mat communities. Alternating spring flow and convective circulation generates bi-directional flow over surface with temperatures varying from about 72–82 °C. Unsilicified active mat community occurs in shallow depressions whereas mat organisms along ridge margins have undergone various degrees of silicification by encrustation. No evidence for microorganisms or biofilms were observed on ridges. (B) Mat community that exists near the upper temperature limit of *Synechococcus* mats. Mat assemblage consists of *Synechococcus* rods and several species of long thin filaments, including a coiled *Spirulina*-like organism, attached to webs and strings of dried mucilage. (D) Close-up view of porous area in (C) indicates that filamentous mat communities were eventually silicified by encrustation as the sinter surface accreted. (E) LVSEM of cross-section of laminated sinter showing alternation of porous patches that contain silicified filaments and finely laminated sinter with no apparent biotic component.

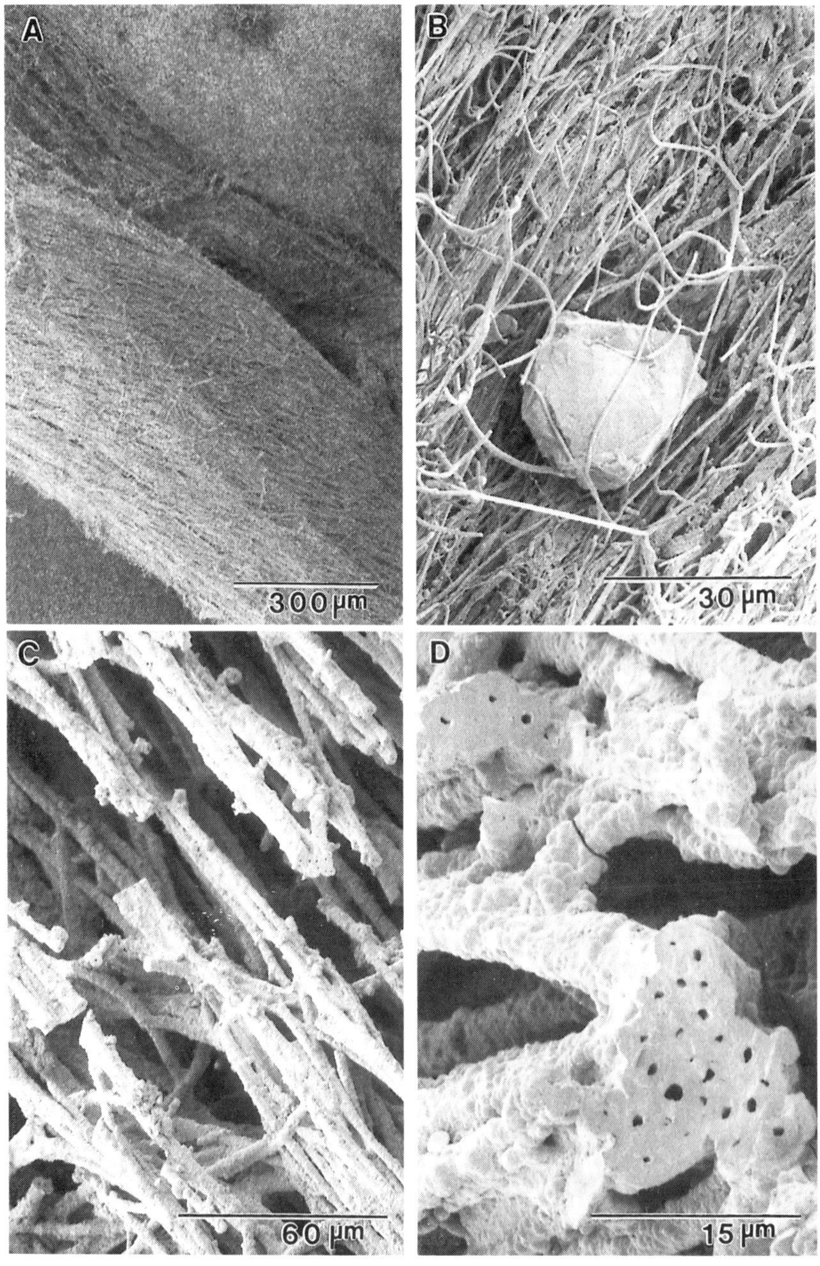
A
B
C
D
300 μm
30 μm
60 μm
15 μm

The vent pool floor at Octopus Spring from which the sample was collected is subjected to alternating spring flow and convective circulation that generates bi-directional flow over the sinter surface. Comparison with laminated sinters from an unnamed pool in Shoshone Geyser Basin (Fig. 6E) indicate that during accretion of the sinter surface, porous areas harboured filamentous mat communities that were eventually silicified by encrustation (Fig. 6C, D) and overstepped by fine laminae of opaline silica.

Mid-temperature pools and channels

At mid-temperatures ($\sim$35–59 °C), mat growth rates are high relative to rates of opaline silica deposition. Thick, gelatinous, green-to-orange-coloured mats dominated by the filamentous cyanobacterium *Phormidium*, are observed in outflow channels and terrace ponds (see Walter et al 1976). Within ponds, microbial phototaxis leads to clumping of cells and the development of preferred filament orientations. Such processes lead to the development of higher-order composite fabrics, including network, pinnacle and coniform morphotypes.

The microfabric of coniform stromatolites formed by *Phormidium* are dominated by finely fibrous palisades of silicified filaments (Fig. 7A). Phototaxis produces a strong subvertical alignment of filaments (Fig. 7B). An LVSEM micrograph of a Recent *Phormidium*-type sinter reveals that the continued accretion of opaline silica onto the external surfaces of individual filaments eventually cemented them together to form palisades with cores of multiple filament moulds (Fig. 7C, D). The micron-size filament moulds are difficult to resolve with light microscopy, and the microfabric appears finely fibrous. Although higher-order tufted mat structures are preserved, organically preserved filaments have not yet been identified in sub-Recent *Phormidium*-type sinters ($<$0.07 Ma) from Artist Point in Yellowstone National Park (Nancy Hinman, personal communication).

FIG. 7. LVSEM micrograph of *Phormidium* filaments within a tuft that developed on the floor of a mid-temperature pond at Fountain Paint Pots, Lower Geyser Basin (A, B), and silicified equivalent of Recent siliceous sinter from Queen's Laundry Terrace, Sentinel Meadows (C, D), Yellowstone National Park (USA). LVSEM micrograph shown in Fig. 7A rotated approximately 60° counterclockwise from upward growth direction. (A) Strong alignment of *Phormidium* filaments reflects construction of tuft by phototaxis. (B) Close-up view of *Phormidium* tuft in Fig. 7A shows entrapment of sinter detritus by cyanobacterial filaments that are mostly vertically oriented in the growing mat. (C) Vertically aligned palisades of *Phormidium*-type silica-encrusted filaments. (D) Close up of sinter shown in Fig. 7C shows cross sections of palisades which have filament moulds in their cores, formed by the encrustation of a *Phormidium*-sized organism. Each palisade is a composite structure that formed by coalescence of adjacent encrusted filaments. The interior of filament moulds sometimes has a thin lining of botryoidal silica but cellular materials appear to have been removed by decay or autolysis following encrustation.

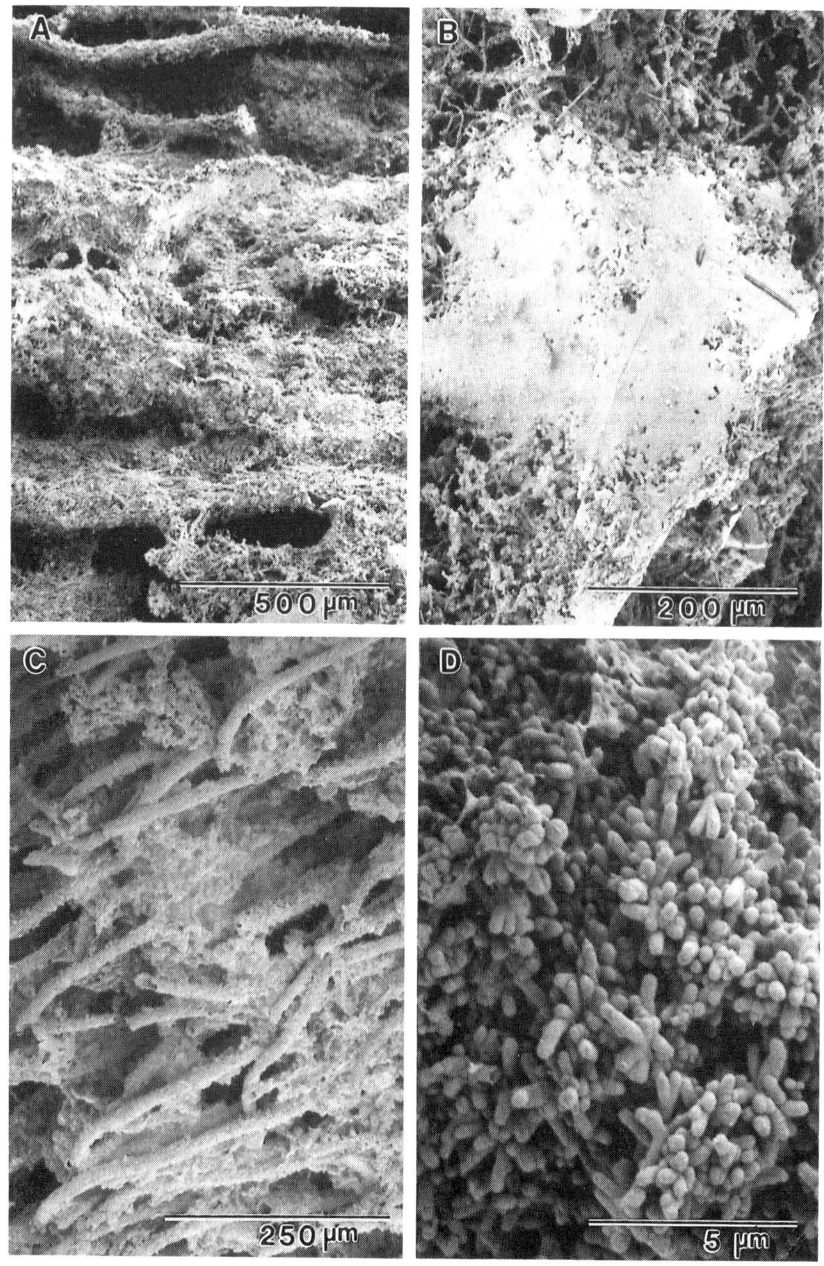
A
500 µm
B
200 µm
C
250 µm
D
5 µm

Low temperature terraces

At low temperatures (<35 °C), dark brownish-green *Calothrix* mats cover shallow terracette pools, forming flat carpets of vertically oriented filaments. Radially oriented *Calothrix* filaments also occur as overgrowths on mat-derived detritus and form small spherulites that are transported down broad terrace slopes and accumulate behind terracette dams. In deeper, low temperature ponds, *Calothrix* growth produces pustular mat surfaces. Shrub-like *Calothrix* mats form overgrowths on coniform *Phormidium*-type sinters in ponds where temperatures drop below 35 °C (e.g. because of lateral channel shifting).

The microfabric of *Calothrix* mats consists of vertical palisades which anastomose and branch (Fig. 8A). Palisades are covered by a diverse community of epibionts that include small filamentous and coccoid forms, diatoms and even sessile protozoans. Each palisade typically has a core region that contains one or more filament moulds surrounded by porous patches of smaller, randomly oriented filaments and coccoids (Fig. 8B, 8C). As the various epibionts become encrusted with silica, they are incorporated into the palisade and form a porous rind around the palisade core. Cross-bridging of the palisades by growth of the epibiont community contributes to the complexity of the porous framework, which is eventually infilled by opaline silica. High concentrations of heterotrophic bacteria coat the surfaces of unsilicified *Calothrix* sheaths (Fig. 8D). Silicified equivalents of these bacteria probably account for the presence of a porous outer rim that surrounds some of the permineralized *Calothrix*-sized sheaths (e.g. Fig. 9C).

Petrographic study by the authors of Recent low temperature sinters from Excelsior Geyser Crater and Queen's Laundry Terrace (Yellowstone National Park) indicates that although organically preserved cyanobacterial sheaths are common, the *Calothrix*-type filaments display a range of degradational variants interpreted to reflect differences in the rates of organic decomposition and silica deposition. The recalcitrant nature of extracellular sheaths and capsules of cyanobacteria is well known (e.g. Golubic & Hofmann 1976). Taphonomic studies of cyanobacteria from siliceous

FIG. 8. LVSEM micrographs of Recent palisade sinter collected from the wall of Excelsior Geyser Crater, Midway Geyser Basin (A, B), and an active, unnamed spring in Shoshone Geyser Basin (C, D), Yellowstone National Park (USA). LVSEM micrographs shown in Figs 8A and C rotated approximately 90° counterclockwise from upward growth direction. (A) Palisade fabric formed by *Calothrix*-type filaments, a vertically growing cyanobacterium that dominates mat communities developed below 35 °C. The ragged appearance of palisade surfaces results from the presence of a silicified community of small filamentous microbes that colonized palisade surfaces. (B) Cross section of palisade showing filament mould in core area formed by encrustation and overgrowth of *Calothrix*-sized filament, followed by accretion and encrustation of epibiont community. (C) *Calothrix* mat from floor of low temperature terrace pond consists of palisades formed by coalescence of silicified sheaths of a vertically growing cyanobacterium. (D) Close up of an individual *Calothrix* sheath showing a surface population of small, rod-shaped coccoids and associated mucilage.

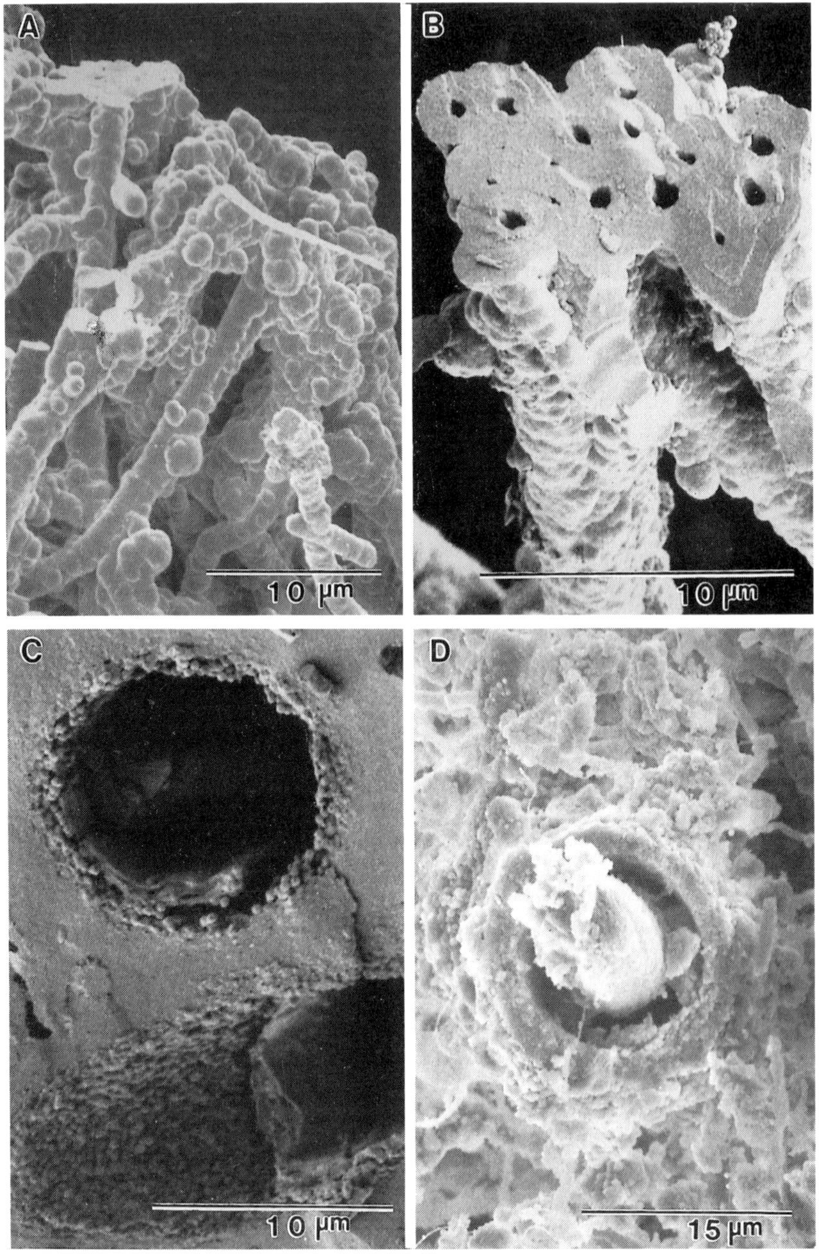
A
10 μm
B
10 μm
C
10 μm
D
15 μm

subaerial thermal spring deposits in Kamchatka have shown that extracellular sheaths and capsules of cyanobacteria are selectively preserved during silicification (e.g. Gerasimenko & Krylov 1983). Organic preservation of cellular remains in the distal parts of the system, however, may also be enhanced by additional factors that include significant amounts of secondary infilling, changes in pH and microenvironments, and occurrence of these mats in a predominantly subaqueous environment. Although organically preserved sheath materials were not observed petrographically by the authors in sub-Recent *Calothrix*-type sinters from McGinness Hills (approximately 3.0 Ma, Nevada, USA), the palisade frameworks contain abundant filament moulds that have been infilled by secondary silica.

Implications for preservation in ancient deposits

As the LVSEM micrographs shown in Fig. 9 illustrate, thermophiles in subaerial silica-depositing thermal springs are preserved by several modes of fossilization, which include replacement, encrustation and permineralization. We are currently investigating how the different mechanisms of microbial preservation in hydrothermal environments affect the morphological and ultrastructural fidelity of microbes in progressively older sinters. Knoll (1985) suggested that the model proposed by Leo & Barghoorn (1976) for the silicification of wood may be applicable to silica permineralization of microbial remains. Leo & Barghoorn (1976) proposed that the vascular tissues of plants, as well as their degradational products, provide a template for silica deposition via hydrogen bonding of exposed hydroxyl and functional groups with monosilicic and polysilicic acid. As discussed by Knoll (1985), a limited amount of microbial cellular decomposition under anaerobic conditions, accompanied by early mineralization, may also enhance the expectancy for preservation of cellular material. Recent studies indicate that silica deposition may be promoted by mechanisms in addition to partial cellular degradation. Schultze-Lam et al (1993) reported that silicification of thermophilic cyanobacteria in Icelandic hot springs is predominantly a surface-mediated process in which sheaths concentrate silica and initiate the mineralization process. Fossilization experiments (Westall 1994) indicate that silicification

---

FIG. 9. LVSEM micrographs illustrating the different modes of fossilization of thermophilic bacteria identified in (A) subaerial geyserites and (B–D) thermal spring stromatolites from Yellowstone National Park (USA). (A) Encrusted/replaced filamentous microorganisms preserved in a horizontal fenestral cavity in modern subaerial geyserite, unnamed spring, Sentinel Meadows. (B) *Phormidium*-type Recent sinter from Queen's Laundry Terrace, Sentinel Meadow, showing fossilization by filament encrustation and the subsequent removal of cellular materials to form a filament mould. (C) *Calothrix*-type sinter from Recent Excelsior Geyser Crater, Midway Geyser Basin, showing preservation of sheath by permineralization. Filament moulds contain an outer porous rind and adjacent dense lamellar material interpreted to be the silica infused sheaths of *Calothrix*-type filaments. (D) *Calothrix*-dominated sinter from active unnamed spring in Sentinel Meadows showing silicified extracellular sheath and intracellular remains.

mechanisms are species-specific; the character of the cell wall and external layers influences the type of mechanism and rate of fossilization. It has also been shown that the chemical composition of the solute influences the fidelity of microbial preservation. Ferris et al (1988) demonstrated experimentally that a delay in the rate of cellular degradation of iron-laden thermophilic bacteria allowed the silicification of intact cellular structures, presumably because of the inhibitory effect of iron on wall-degrading autolysins.

Our study of geyserites from active systems has allowed us to recognize that both biotic and abiotic processes contribute to geyserite morphogenesis. Filamentous biofilms, when present, contribute to the formation of subaerial geyserites by providing a favoured substrate for the localized precipitation of opaline silica. It will be important for future studies to determine whether geyserites can be produced abiotically and whether the biotic and abiotic microstructural characteristics of geyserites can be distinguished after diagenetic overprinting.

We have also observed general trends in preservation along the thermal gradient. Permineralization of cyanobacterial sheaths in Recent and sub-Recent materials has been petrographically identified by the authors only in low temperature ($<35\,^{\circ}C$) microfacies. We have observed that filament moulds, rather than permineralized remains, are preserved in the mid-temperature to moderately high temperature pools and channels. Although microorganisms cannot be resolved petrographically in high temperature geyserites, the occurrence of high concentrations of filament moulds identified using LVSEM indicates that encrustation and degradation of microorganisms is the dominant mode of fossilization at the high temperature end of the system. To summarize, the potential to preserve organic remains increases with decreasing temperature along the thermal gradient of subaerial silica-depositing thermal springs. These materials, however, must be investigated at the ultrastructural level, where distinctions between preservational modes can be more exact. Exceptions to the general trend in preservation result from variations in the local microenvironment.

We have also found that the persistence of macro-scale biofabrics in sub-Recent and ancient siliceous sinters can be related to diagenetic effects, which tend to accentuate differences between primary and secondary textures in sinter stromatolites. The rate and pathway of the diagenetic recrystallization of opaline silica to quartz may be influenced by differences in the trace composition, crystallinity, microstructure, grain size and habit of primary and secondary phases, as well as by differences in the porosity of laminae in geyserites and thermal spring stromatolites.

Acknowledgements

This research was supported by a National Research Council research associateship to Sherry Cady and by a grant from NASA's Exobiology Program to Jack Farmer. We thank the staff of the National Park Service at Yellowstone National Park for their logistical assistance in the field, especially Robert Lindstrom and Roderick Hutchinson. Laboratory and microscope facilities were kindly provided to the authors by David Blake at NASA Ames Research Center, and to Sherry Cady at the National Center for Electron Microscopy, Lawrence Berkeley National

Laboratory (United States Department of Energy Contract #DE-AC03-76SF00098). Improvements to the manuscript were made by Malcolm Walter and David Des Marais.

References

Allen ET 1934 The agency of algae in the deposition of travertine and silica from thermal waters. Am J Sci 28:373–389

Barns SM, Delwiche CF, Palmer JD, Dawson SC, Hershberger KL, Pace NR 1996 Phylogenetic perspectives on microbial life in hydrothermal ecosystems, past and present. In: Evolution of hydrothermal ecosystems on Earth (and Mars?). Wiley, Chichester (Ciba Found Symp 202) p 24–39

Brock TD 1978 Thermophilic microorganisms and life at high temperatures. Springer-Verlag, New York

Castenholtz RW 1984 Composition of hot spring microbial mats: a summary. In: Cohen Y, Castenholtz RW, Halvorson HO (eds) Microbial mats: stromatolites. Alan R. Liss, New York, p 107–119

Ferris FG, Fyfe WS, Beveridge TJ 1988 Metallic ion binding by *Bacillus subtilis*: implications for the fossilization of microorganisms. Geology 16:149–152

Gerasimenko LM, Krylov IN 1983 Postmortem alterations of cyanobacteria in the algal–bacterial films in the hot springs of Kamchatka. Dokl Akad Nauk 272:215–218

Golubic S, Hofmann HJ 1976 Comparison of modern and mid-Precambrian Entophysalidaceae (Cyanophyta) in stromatolitic algal mats: cell division and degradation. J Paleontol 50:1074–1082

Knoll AH 1985 Exceptional preservation of photosynthetic organisms in silicified carbonates and silicified peats. Philos Trans R Soc Lond Ser B Biol Sci 311:111–122

Leo RF, Barghoorn ES 1976 Silicification of wood. Botanical Museum Leaflets, 25-1. Botanical Museum, Harvard University, Cambridge, MA

Rice CM, Ashcroft WA, Batten DJ et al 1995 A Devonian auriferous hot spring system, Rhynie, Scotland. J Geol Soc 152:229–250

Schultze-Lam S, Ferris G, Wiese R 1993 Silicification of cyanobacteria in an Icelandic hotspring microbial mat. Geological Society of America, Abstracts with Programs, 26:192A

Stetter KO 1996 Hyperthermophiles in the history of life. In: Evolution of hydrothermal ecosystems on Earth (and Mars?). Wiley, Chichester (Ciba Found Symp 202) p 1–18

Trewin NH 1994 Depositional environment and preservation of biota on the Lower Devonian hot-springs of Rhynie, Aberdeenshire, Scotland. Trans R Soc Edinb 84:433–442

Trewin NH 1996 The Rhynie cherts: an early Devonian ecosystem preserved by hydrothermal activity. In: Evolution of hydrothermal ecosystems on Earth (and Mars?). Wiley, Chichester (Ciba Found Symp 202) p 131–149

Walter MR 1976a Hot-spring sediments in Yellowstone National Park. In: Walter MR (ed) Stromatolites. Elsevier, Amsterdam, p 489–498

Walter MR 1976b Geyserites of Yellowstone National Park: an example of abiogenic stromatolites. In: Walter MR (ed) Stromatolites. Elsevier, Amsterdam, p 87–112

Walter MR 1996 Ancient hydrothermal ecosystems on Earth: a new palaeobiological frontier. In: Evolution of hydrothermal ecosystems on Earth (and Mars?). Wiley, Chichester (Ciba Found Symp 202) p 112–130

Walter MR, Bauld J, Brock TD 1976 Microbiology and morphogenesis of columnar stromatolites (*Conophyton, Vacerrilla*) from hot springs in Yellowstone National Park. In: Walter MR (ed) Stromatolites. Elsevier, Amsterdam, p 273–310

Walter MR, Des Marais DJ, Farmer JD, Hinman N 1996 Palaeobiology of mid-Palaeozoic thermal spring deposits in the Drummond Basin, Queensland, Australia. Palaios, in press

Ward DM, Weller R, Shiea J, Castenholtz RW, Cohen Y 1989 Hot spring microbial mats: anoxygenic and oxygenic mats of possible evolutionary significance. In: Cohen Y, Rosenberg E (eds) Microbial mats, physiological ecology of benthic microbial communities. American Society for Microbiology, Washington DC

Westall F 1994 How bacteria fossilize: experimental observations and some theoretical considerations. In: Awramik SM (ed) Death Valley International Stromatolite Symposium (Abstracts), Laughlin, Nevada, October 15–17, 1994, p 88–89

White DE, Brannock WW, Murata KJ 1956 Silica in hot-spring waters. Geochim Cosmochim Acta 10:27–59

White DE, Thompson GA, Sandberg GH 1964 Rocks, structure, and geologic history of Steamboat Springs thermal area, Washoe County, Nevada. Geol Surv Prof Paper 458B

DISCUSSION

Walter: Have you made a positive identification of any of these filamentous organisms as archaea?

Cady: No, but we are currently using transmission electron microscopy to investigate how they become silicified. This work necessitates collaboration with microbiologists familiar with the identity and ultrastructural characteristics of the organisms; we hope to have that information in the near future.

Stetter: Did you say that silicate begins to crystallize at the tips of these filamentous organisms?

Cady: No, silica is preferentially deposited at the tips of the spicules, which are a particular morphotype of geyserite that forms along the inner rim of thermal spring pools and around the mouths of some geyser effluents. We have observed the highest concentrations of viable filaments at the tips of the spicules that form near the air–water interface. The preferred deposition of opaline silica on the organisms at the tips of the spicules is presumably what causes the geyserite to accrete with the elongated spicular form.

Stetter: Sometimes we encounter problems in our attempts to clone hyperthermophiles in glass capillaries under the laser microscope, since they rapidly attach onto the glass surface. Filaments, especially, appear to be very sticky, mainly at their ends. In their natural environments, these organisms are faced with a dilemma: on the one hand they have to attach to the surface or they will be washed away to colder areas, and on the other, according to your results, there is silica precipitation, which also kills them. They have a difficult life.

Cowan: With respect to the filamentous organisms that you see in the geyserites, I would expect there to be a substantial population of the bacterial genus *Thermus*. This genus is a significant colonizer of boiling terrestrial hydrothermal environments, even though it will only grow up to about 80 °C in the laboratory. It is very well known that if you dip a slide into a boiling hydrothermal pool, you get massive colonization of the glass surfaces, and *Thermus* is one of the predominant colonizers.

Walter: The temperature point is interesting: you don't really know the temperature of the water that is actually hitting the tip of those spicules.

Cady: That is true, but we have observed that the highest concentration of filaments occur on spicules that form at the air–water interface, where they are not only splashed with spring water, but often become temporarily submerged due to fluctuations in the water level of the pool.

Walter: At the elevation of Yellowstone, water boils at about 96 °C, and you are somewhat below this because you're right at the air–water interface. It would be interesting to know what the temperature actually is.

Farmer: We've measured the temperature of the vents: the problem is that the erupted hot water travels through the air before it splashes onto sinter surfaces, and we don't know its temperature by the time it reaches the surface. The change could be significant, so vent temperature is probably not an accurate parameter.

Cady: It is worth emphasizing that although Brock (1978) demonstrated that extreme thermophiles live in boiling pools by submerging glass slides which became colonized, we have shown for the first time that biofilms influence the microstructural development of geyserite. The preferential deposition of silica at the tips, where filaments are concentrated, indicates that the biofilms also have a significant influence on the morphological development of spicular geyserite.

Cowan: I agree entirely. In hydrothermal systems the total cell populations are orders of magnitude higher in the sediment fractions than in the aqueous phase. The consensus is that cells are strongly adhered to sediment particles and I would expect the same to apply in these deposition areas.

Walter: Sherry is being polite in not pointing out that what she's doing is completely revising work I did 25 years ago when I concluded that these structures were entirely abiogenic.

Pentecost: The cyanobacteria found in travertines appear to be much the same as those you have found. They are motile and phototactic, and often they will creep under the travertine, if it is porous enough, to form endolithic communities. Do you have any evidence for endolithic communities in your sinters?

Cady: We have observed what appear to be stratified endolithic communities on columnar geyserite located several centimetres away from the edges of the pools. Jack Farmer may have something more to say about the presence of endoliths in the mid-to-low temperature sinters.

Farmer: We do find endoliths in dried areas around inactive vents. Even in columnar geyserites near vents, you can take small geyserite columns, break them in half, and see the bands of colour a millimetre or so below the surface. Whether these are endoliths or entombed thermophilic cyanobacteria is unclear, and we don't know yet whether they are still viable. Certainly the pigments are still there!

Cowan: There are some interesting issues relating to the local conditions required to preserve the biological material. The mechanisms causing biological structures to dissipate before silica deposition certainly include enzymic and chemical lysis. Anything that inhibits these processes is likely to promote the preservation of the biological structure, irrespective of what happens later. Such effects may be critical within the first 12 h of the deposition process. Silica may play a key role, possibly by 'immobilizing' the cells. You suggested that metal ion inactivation may also play a role: we might speculate that inactivation of degradative enzymes is also important.

Knoll: A lot of the beautiful textures that you see with the scanning electron microscope are really present in part because there's a lot of pore space that allows you to see the way the silica has been deposited. Have you done any vertical studies going down into the sinters to see what happens to these textures as diagenesis occludes that pore space? In other words, once you have something that instead of 50% porosity is 100% silica, are the textures still as apparent as they are at the surface?

Cady: Yes, the relatively rapid encrustation of the microorganisms enhances preservation of the microbial mat architecture. Even if organic remains degrade prior to permineralization, differences in the mechanism, timing and temperature of formation of primary and diagenetic silica phases accentuate textural differences between the primary biofabric and the authigenic matrix.

Trewin: Many of the textures Sherry Cady has illustrated are identical to those in the Rhynie chert, both on a macro (e.g. plant stem) or micro (cyanobacterial filament) scale. In the Rhynie chert all opaline silica has been converted to microquartz. However, the biogenic structures are present, so it does seem that the textures of value for identification of biota can remain even after the whole deposit has been converted into a mosaic of chert.

Cady: The presence of degraded remains in the biotic laminae increases the number of potential nucleation sites for opaline silica deposition. These laminae are characterized by an extremely fine-grained matrix of opaline silica that can be recognized petrographically, even after the opaline silica transforms to quartz.

Knoll: In the Gunflint Chert, which might be one of the best Precambrian analogues of this system, differences in crystal size between the organic-rich and organic-free areas are still seen. This can be preserved for a long time.

Farmer: We have also been looking at these systems in places other than Yellowstone. In most cases, organic preservation is poor, consistent with what we have seen in Yellowstone. Tracking deposits back in the rock record a few million years, and into the 350 Ma old Drummond Basin material that Malcolm Walter showed in his paper (Walter 1996, this volume), there isn't much organic matter preserved there either. Most biological information consists of biofabrics. Dave Des Marais has completed total organic carbon analyses on the sinters from the Drummond Basin. I wonder if he could share some of those results?

Des Marais: They are consistent with what was said earlier about poor preservation at higher temperatures and better preservation as you approach ambient conditions. The most organic-rich material is found at the low-temperature end of the system, and it contains as much as a few tenths of a percent of organic carbon. In sinter formed above 70 °C, the organic carbon content is below the detection limit. We've made isotope measurements of this, and even at the high temperature end we see values identical to the isotopic values we see at the low temperature end, which is dominated by plant fragment material. It seems likely that small plant fragments are being incorporated in high temperature sinters, and even those small bits are dominating the organic carbon present. This implies that bacterial preservation is pretty poor relative to plant preservation. We're still working out the proper way of analysing more recent sinters, keeping in mind that that the organic matter is quite sensitive to the procedures that we would use, for example, to remove carbonates from the travertines.

Stetter: Could the reduced preservation at high temperatures also be due to differences in cellular structure? Hyperthermophiles seldom have the rigid peptidoglycan wall found in most bacteria; instead, they usually have a rather loose surface layer made up of glycoproteins which may be destroyed more easily.

Cady: Yes, it is likely that differences in cellular structure would result in the preferential preservation of some species over others. Don't the archaea have quite a diverse range of cell envelopes and extracellular structures?

Stetter: There are only a few examples I know of. Rod-shaped methanogens such as *Methanobacterium*, *Methanothermus* and *Methanopyrus* have a pseudomurein cell wall and grow at temperatures between 60 and 110 °C depending on the genus (Zeikus & Wolfe 1972, Lauerer et al 1986, Huber et al 1989). *Methanobacterium thermoautotrophicum* and *Methanothermus* sp. occur in terrestrial hot springs while *Methanopyrus* sp. thrives within submarine black-smoker walls.

Cady: Another point worth making is that petrographic and morphological descriptions of these materials should be supplemented by biomarker and isotopic studies similar to those of our colleagues Roger Summons and David Des Marais. To correctly interpret the preservational biases in the fossil record, it is important for us to include all of these techniques when evaluating the biogenicity of ancient hydrothermal materials.

References

Brock TD 1978 Thermophilic organisms and life at high temperatures. Springer-Verlag, New York

Huber R, Kurr M, Jannasch HW, Stetter KO 1989 A novel group of abyssal methanogenic archaebacteria (*Methanopyrus*) growing at 110 °C. Nature 342:833–834

Lauerer G, Kristjansson JK, Langworthy TA, König H, Stetter KO 1986 *Methanothermus sociabilis* sp. nov., a second species within the *Methanothermaceae* growing at 97 °C. Syst Appl Microbiol 8:100–105

Walter MR 1996 Ancient hydrothermal ecosystems on Earth: a new palaeobiological frontier. In: Evolution of hydrothermal ecosystems on Earth (and Mars?). Wiley, Chichester (Ciba Found Symp 202) p 112–130

Zeikus JG, Wolfe RS 1972 *Methanobacterium thermoautotrophicus* sp. n., an anaerobic, autotrophic, extreme thermophile. J Bacteriol 109:707–713

Lipid biomarkers for bacterial ecosystems: studies of cultured organisms, hydrothermal environments and ancient sediments

Roger E. Summons*, Linda L. Jahnke† and Bernd R.T. Simoneit‡

**Australian Geological Survey Organisation, GPO Box 378, Canberra, ACT 2601, Australia, †Exobiology Branch, NASA Ames Research Center, Moffett Field, CA 94035, USA and ‡College of Oceanic and Atmospheric Sciences, Oregon State University, Corvallis, OR 97331–5503, USA*

Abstract. This paper forms part of our long-term goal of using molecular structure and carbon isotopic signals preserved as hydrocarbons in ancient sediments to improve understanding of the early evolution of Earth's surface environment. We are particularly concerned with biomarkers which are informative about aerobiosis. Here, we combine bacterial biochemistry with the organic geochemistry of contemporary and ancient hydrothermal ecosystems to construct models for the nature, behaviour and preservation potential of primitive microbial communities. We use a combined molecular and isotopic approach to characterize lipids produced by cultured bacteria and test a variety of culture conditions which affect their biosynthesis. This information is then compared with lipid mixtures isolated from contemporary hot springs and evaluated for the kinds of chemical change that would accompany burial and incorporation into the sedimentary record. In this study we have shown that growth temperature does not appear to alter isotopic fractionation within the lipid classes produced by a methanotropic bacterium. We also found that cultured cyanobacteria biosynthesize diagnostic methylalkanes and dimethylalkanes with the latter only made when growing under low pCO_2. In an examination of a microbial mat sample from Octopus Spring, Yellowstone National Park (USA), we could readily identify chemical structures with ^{13}C contents which were diagnostic for the phototrophic organisms such as cyanobacteria and *Chloroflexus*. We could not, however, find molecular evidence for operation of a methane cycle in the particular mat samples we studied.

1996 Evolution of hydrothermal ecosystems on Earth (and Mars?). Wiley, Chichester (Ciba Foundation Symposium 202) p 174–194

Our interest in the organic geochemistry of hydrothermal ecosystems stems from the commonly held hypothesis that such environments could have spawned Earth's original inhabitants. Even if this was not the case, extant hydrothermal environments are now the domain of bacterial consortia thought to resemble closely those which dominated and shaped the Earth's surface and subsurface for most of geological time.

Chemical fossils, in the form of diagnostic lipids and isotopic signatures, constitute important tools for making reliable connections between modern bacterially dominated (e.g. hydrothermal) ecosystems, their ancient counterparts, and Earth's earliest life. Chemical fossils may also be the only recognizable remnant of many groups of bacteria, and particularly the archaea, which leave no textural or other visible remains and cannot be detected by conventional palaeontological methods. Of special interest in the present work are chemical fossils which inform us about the development and evolution of oxygenic photosynthesis and aerobiosis, since these processes are at the heart of the modern carbon cycle and sustain complex life. This paper outlines our approach to the study of patterns of isotopic dispersion in the lipids of cultured cyanobacteria and methanotrophs, how these patterns compare with those measured in an extant hydrothermal mat system and then, how these observations might facilitate interpretation of hydrocarbon signatures preserved in Archaean and Proterozoic sediments.

Compound-specific isotope analysis

While the lipid chemistry of bacteria is dominated by relatively simple chemical structures, often having limited phylogenetic significance, the metabolic diversity for carbon assimilation and energy harvesting is well known. This suggests that there will be major isotopic differences between lipids produced by say, cyanobacteria, photosynthetic bacteria, eubacterial heterotrophs and the archaea. Combined molecular and isotopic data, therefore, offer scope to be much more specific about the origins of simple *n*-alkyl and isoprenoid lipids.

Isotope ratio monitoring gas chromatography–mass spectrometry (GC-MS), or compound-specific isotope analysis (CSIA), is an analytical tool that uses a stable isotope mass spectrometer as the real-time detector for a gas chromatograph. The interface is a combustion and/or micro reaction device which quantitatively converts individual compounds in the chromatographic effluent to gases, carbon dioxide or nitrogen, which are amenable to continuous flow isotopic analysis. The method can be sensitive to 0.5–5 nmole of carbon or nitrogen and, for compounds at or near natural isotopic abundances, measurements to within 0.3‰ (1‰ for nitrogen) are possible. The feasibility of this tool for routine natural abundance isotope measurements on individual compounds has now seen it applied to a wide range of geochemical problems (e.g. Freeman et al 1990, Hayes et al 1990, Logan et al 1995). This opens the way for us to measure and contrast the isotopic compositions of a wide variety of simple lipids from bacteria grown in pure culture under various conditions. These data then help us to understand comparable patterns measured for bacteria growing naturally in microbial mat and water column ecosystems and in their fossil counterparts.

Molecular and isotopic fossils from the Precambrian: a record ripe for 'remastering'

Organic geochemical studies of ancient sediments have long been conducted for their perceived potential to unlock information about the early evolution of the terrestrial surface environment. To date, these have been inclined toward isotopic analyses of preserved carbon in its bulk forms (i.e. carbonates and kerogens) and have yielded valuable insights into the early carbon cycle and a working chemostratigraphy for the Terminal Proterozoic. Recognizing that hydrocarbons (*n*-alkanes, *iso*-alkanes and acyclic isoprenoids) occur abundantly in certain Proterozoic and Archaean formations, and may be indigenous to those rocks, we have been pursuing ways to extract meaningful information about the circumstances of their synthesis.

One often encounters the negative view that these hydrocarbons have either migrated in from younger rocks, have been altered to the point that all useful information has been obliterated or that they have been formed by abiological, i.e. 'random', chemical processes deep in the earth and, consequently, do not reflect surface biology at all. Hayes et al (1983, 1992) have reviewed these issues and concluded that reliable molecular fossils do exist in the Precambrian and that they may encode useful facts about ancient biochemistries and environments. Although there are some complex and phylogenetically significant lipids such as steroids and hopanoids preserved in old rocks, the most abundant compounds by far are simple straight and branched-chain hydrocarbons. These are often viewed as lacking diagnostic chemical structures and hence, devoid of significant information. However, Logan et al (1995) demonstrated that even these simple molecules can be used to make a strong case for a redox-stratified global ocean throughout most of the Proterozoic. Whereas our understanding of the progress in oxygenation of Earth's surface has formerly depended upon occurrence of diagnostic minerals such as detrital uraninites, banded iron formations and palaeosols, organic geochemistry now offers a renascent and independent tool for studying Precambrian environments.

Calvin-Benson cycle carbon assimilation and methanogenesis–methanotrophy are the two most obvious carbon cycle processes that can be identified through the bulk isotopic character of preserved organic matter. They are recognized, albeit with different degrees of certainty, in the Archaean and Proterozoic (e.g. Hayes 1983, Hayes et al 1983, 1992, Schidlowski 1988). This work suggests that a more insightful record may exist if, for example, we could combine the isotopic information with diagnostic molecular structures. While there is generally a strong appreciation for how carbon isotopes are fractionated at the organismic level, further advances in applying this approach are hampered by a lack of fundamental knowledge about isotopic dispersion at the molecular level. This is particularly the case for prokaryotes, which are the engine of the global carbon cycle.

Preservable chemical fossils from hydrothermal environments

Contemporary hydrothermal ecosystems

The nature of the sedimentary regimes leads to sharply different geochemical features and preservation potentials for organic matter in modern terrestrial and submarine

TABLE 1 A summary of known or possible Octopus Spring mat isolates and diagnostic lipids that may be derived from them

		Polar lipid alkylation			
Metabolic category	*Organism(s)*	*Fatty acids*	*Ethers*	*Pigments*	*Other free lipids*
Phototrophy	*Synechococcus lividus*	C_{16}, C_{18}, $C_{16:1}$, $C_{18:1}$		Chl-a-O-phytol carotenoids	*n*-C_{17-19}, isoalkanes?
Phototrophy/heterotrophy[a]	*Chloroflexus aurantiacus*	C_{16}–C_{18}, $C_{16:1}$–$C_{18:1}$		Bchl-c-O-octadecanol carotenoids	C_{28}–C_{38} wax esters verrucosanol[b]
Aerobic and facultative heterotrophy	*Isosphaera pallida* *Thermomicrobium roseum* *Thermus* sp. *Bacillus stearothermophilus*	C_{18}, $C_{18:1}$, C_{18}, MeC_{18}, iC_{15}–iC_{17}, iC_{15}–iC_{17}, aiC_{15}, aiC_{17}		carotenoids	
Anaerobic fermentation	*Clostridium* sp.	iC_{15}, iC_{17}			
Sulfate reduction	*Thermodesulfobacterium commune*		aiC_{17}, aiC_{18}, iC_{16}–iC_{18}, C_{16}, C_{18}		
Methanogenesis	*Methanobacterium thermoautotrophicum*		phytanyl and biphytanyl		

Adapted from Ward et al (1989) and other data subsequently published by Zeng et al (1992a,b).
[a]Although *C. aurantiacus* is capable of autotrophic growth, it is thought to be operating as a photoheterotroph in this environment.
[b]Hefter et al (1994).

hydrothermal systems. Biogeochemical investigations of contemporary terrestrial deposits, such as those in Yellowstone National Park (Wyoming), New Zealand and Iceland, have followed the theme that they serve as models for 'primitive,' microbial (e.g. stromatolitic) ecosystems from the past. Consequently, our interest is in characterizing biomarker lipids from specific types of biota.

As stated above, quite simple molecules dominate bitumens entrained in the oldest organic-rich sediments. Fortuitously, or otherwise, bacterial mats from modern hot springs have lipid distributions with striking similarities to preserved hydrocarbons from the Proterozoic and Archaean. Points of resemblance include abundant, low molecular weight, linear and simple branched alkyl chains, acyclic isoprenoids such as phytenes, phytol and carotenoids (e.g. Dobson et al 1988, Robinson & Eglinton 1990, Shiea et al 1990, Ward et al 1989, Zeng 1992a,b). Polycyclic isoprenoids such as steroids and hopanoids are often undetectable or occur at very low levels in these mats. Points of difference stem mainly from the fact that modern systems reflect the living microbiota and are rich in functionalized lipids such as glycerol esters and ethers, fatty acids, alcohols and wax esters whereas fossil assemblages contain mostly saturated hydrocarbon skeletons. A prime example to illustrate these similarities is the *Synechococcus–Chloroflexus* mat community from Octopus Spring in Yellowstone National Park. Table 1, edited from Ward et al (1989) and Zeng et al (1992a,b), summarizes the microbiota and their diagnostic lipids as found in this system. Of particular interest, in the context of the present work, is the potential to isotopically analyse, in a single sample, individual biomarkers which are likely to be diagnostic for cyanobacteria (*n*-C_{17} and C_{18}MA), *Chloroflexus aurantiacus* (wax esters and verrucosanol), methanogens (phytanyl ethers) and possibly methanotrophs (C_{29}–C_{31} hopanoids). Besides the well-known lipid chemical work, there has been one seminal study of the carbon and deuterium isotopic compositions of these Yellowstone mats. Estep (1984) measured $\delta^{13}C$ for total organic carbon in a variety of hot springs and found a strong positive correlation between ^{13}C-depletion and the availability of dissolved inorganic carbon (DIC).

In contrast to terrestrial hydrothermal systems, submarine systems have a better potential for long-term preservation from the perspectives of the permanence of water cover, organic matter alteration and preservation processes under thermal conditions, i.e. proneness for incorporation into the sediment pile. Because of this, we know about the twin aspects of biota–biomarker relationships and the long-term fate of the organic matter. Hydrothermal alteration of immature organic detritus, generally under strongly reducing conditions, occurs in these high temperature and rapid fluid flow regimes. Pressure in the marine systems (1 to >3 km water depth) maintains the fluid state. The agent of thermal alteration and mass transfer, hot circulating water (temperature range warm to >400 °C) is responsible for molecular alterations which are primarily reductive. The 'reduced' nutrients, such as hydrogen, hydrogen sulfide, methane and higher hydrocarbons, emanating with the fluids from hydrothermal vent systems, support an ecosystem with biota carrying out chemosynthesis for biomass production and respiration (e.g. Childress 1988, Jones 1985, Laubier 1988, Simoneit 1990). The lipid and biomarker compositions of these organisms are just beginning to be

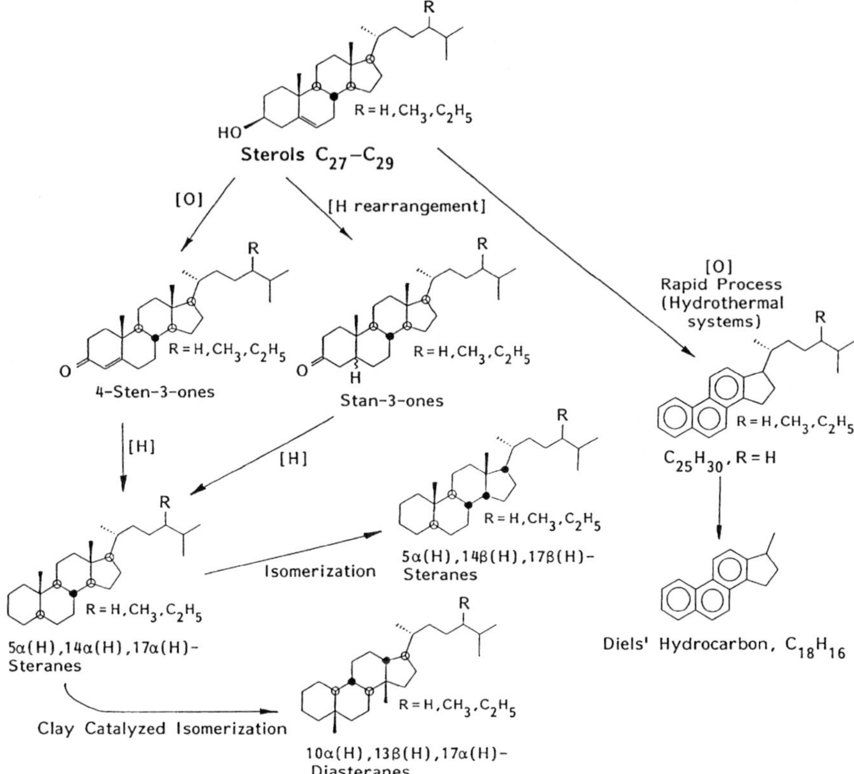

FIG. 1. Hydrothermal alteration scheme for sterols by reduction to the geo-steranes and oxidation to aromatic hydrocarbon derivatives.

elucidated (e.g. Brault et al 1989, Comita et al 1984, Rieley et al 1995). Extrapolation of biomarker evolution in hydrothermal environments to ancient ecosystems will become more definitive as more molecular data are reported.

Biomarkers in hydrothermally fossilized biota

An example of a biomarker product-precursor compound group, the steroids, is illustrated in Fig. 1, showing both the reductive and oxidative diagenesis–preservation pathways observed in the geosphere, especially in hydrothermal systems (e.g. Simoneit 1995). It is obvious that one precursor natural product can yield many derivatives as fossils in a geological record, and that all these specific products can be correlated back to their precursor. Thus, the detailed characterization of biomarker mixtures in terms of sources and degree of alteration permits the assessment of: (a) extant life and major contributing species, and (b) extinct life, major contributing species and geo/hydrothermal alteration. The preservation of biomarkers is illustrated with the lipid signatures from tube worms and *Beggiatoa*, both major biomass contributors at hydrothermal vents. The hydrothermal mounds in Guaymas Basin are composed of mixed minerals (mainly sulfates, carbonates, silicates, sulfides and silica). Samples of *Beggiatoa* mat and a tube worm fossil from Guaymas Basin (Alvin dives 1975 and 1976) were selected for biomarker analysis. The GC-MS data for the total extracts are summarized in Fig. 2. The major resolved components for the lipids of the bacterial mat are *n*-alkanes ranging from C_{17} to C_{27} with no carbon number predominance and C_{max} at 17 and 22, elemental sulfur, diploptenes (17α and 17β), and cholest-2-ene (Fig. 2a). Traces of polynuclear aromatic hydrocarbons (PAHs) are also present. The heptadecane and diploptene indicate a primary residue from bacteria while the other alkanes, PAHs and sterane biomarkers are a hydrothermal petroleum component derived from other organic detritus in the system. The major resolved compounds for the tube worm fossil are diasteranes, steranes, cholesterol and PAH, consisting of benzo(a&e)-pyrenes, benzo(ghi)perylene, coronene and heavy PAH with M_r 326 and 328 (Fig. 2b). Only steroid biomarkers are detectable and consist of C_{27} diacholestenes with lesser amounts of C_{27} and C_{28} steranes. C_{29} homologues, Diels' hydrocarbon, triterpanes and *n*-alkanes as found in typical hydrothermal petroleums from the Guaymas Basin are not detectable. Thus, this biomarker mixture indicates an origin primarily from organic matter of macroorganisms (i.e. tube worms) with subsequent addition of PAH from high temperature fluids.

Biomarkers in ancient hydrothermal systems

The previous examples demonstrated that biomarkers from specific submarine hydrothermal vent organisms can be preserved in association with their fossils. The hydrothermal alteration process of organic matter in contact with hot fluids progresses from reductive to more oxidative reactions as the temperature increases. Reduction is strongly mediated by metal sulfides and other catalytic surfaces, with the hydrogen being derived from both water and organic matter (Leif 1993). Oxidation of organic matter in the system is enhanced by the presence of sulfur and sulfate, yielding hydrogen sulfide (Leif et al 1992). At very high temperatures, organic matter is only partly destroyed, probably because the thermogenic products are soluble in the ambient fluid which rapidly moves away from the hot zone by convection. Hydrothermal products, including biomarkers, are then trapped and preserved in interstitial fractures and voids or in fluid inclusions (e.g. Peter et al 1990).

Although terrestrial hot springs have fossil analogues recognized from preserved rock fabrics, there is limited potential for long-term organic matter preservation because atmospheric contact leads to rapid degradation. One fossil system that has been studied extensively is the Rhynie chert and here, interest has centred on morphological preservation of local fauna and flora (Trewin 1994). However, low total organic carbon (TOC) content and high maturity suggest that these cherts may not yield molecular fossils.

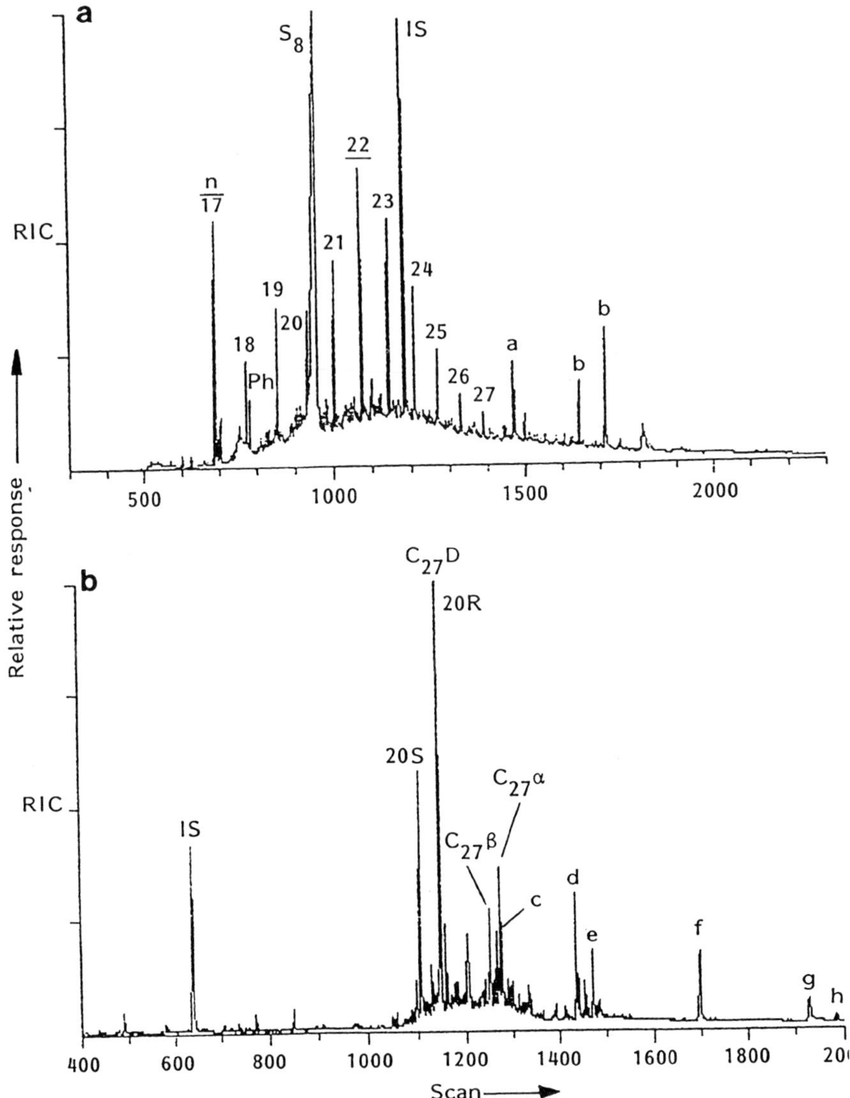

FIG. 2. GC-MS chromatograms for samples GB-1975 and 1976. (a) *Beggiatoa* mat. (b) Tube worm fossils. Numbers represent *n*-alkane chain length. Ph, phytane; S8, sulfur; a, cholest-2-ene; b, diploptene; IS, internal standard; $C_{27}D$, diacholest-13(17)-ene (20R+S); C_{27} α and β, 5α and 5β-cholestanes; c, benzpyrenes; d, cholesterol; e, benzo[ghi]perylene; f, coronene; g, PAH M_r 326; h, PAH M_r 328.

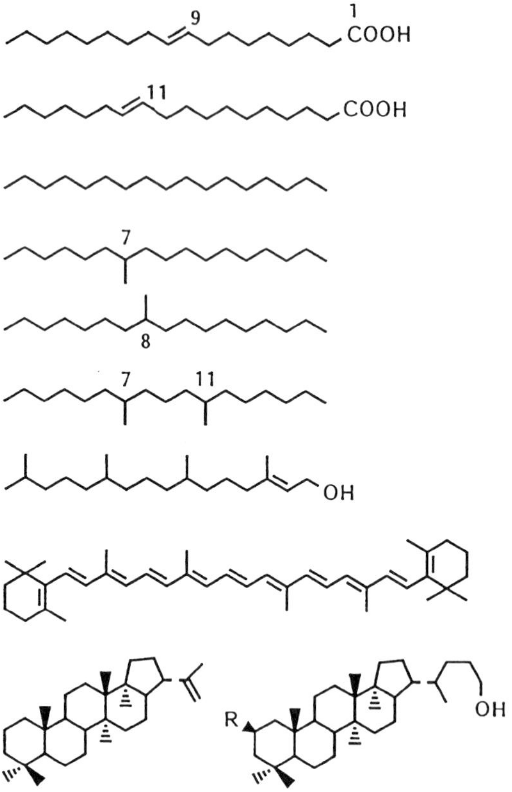

FIG. 3. Structures of *Phormidium luridum* lipids which have been characterized in this study.

Cultured bacteria: lipids and isotopic compositions

In order to obtain more information about lipids produced by extant microbes with affinities to geologically ancient groups, and given our focus on oxygenic photosynthesis and the methane cycle, we selected a number of specific organisms for detailed study. Of these, the cyanobacterium *Phormidium luridum* was chosen for its relationship to conophyton stromatolite-forming cyanobacteria and for its ability to produce a range of hydrocarbon skeletons which are prevalent in old bitumens. These skeletons, illustrated in Fig. 3, include methylalkanes (MAs) and dimethylalkanes (DMAs)—hydrocarbons which are particularly abundant in Proterozoic and Archaean sediments and in recent cyanobacterial mats (e.g. Robinson & Eglinton 1990, Shiea et al 1990, Kenig et al 1995). It is noteworthy that thermophilic *Phormidium* species occur widely in hot spring mat communities and are especially prominent in the modern conophytons found there.

TABLE 2 Fatty acid composition of *P. luridum* from three growth experiments (P1–P3)

	% Composition of fatty acids for various growth conditions		
Fatty acid	*P3: standing/air*	*P1: sparged/CO_2+N_2 inoculum=P3*	*P2: sparged/CO_2+N_2 inoculum=P1*
a-C_{15}	1.9	0.9	0.3
n-C_{16}	28.6	30.5	32.5
n-$C_{16:1}$, Δ9	28.3	23.5	14.8
a-C_{17}	1	0.9	0.3
n-C_{17}	0.3	0.4	0.4
cyc-C_{17}	0.5	0.5	0.3
n-C_{18}	0.5	1.1	4.9
n-$C_{18:1}$, Δ9	1.7	2.7	9.8
n-$C_{18:1}$, Δ11	3.2	1.7	0.5
n-$C_{18:2}$	27.2	13.3	18.5
n-$C_{18:3}$	6.8	24.6	17.5

Details given in footnotes to Table 3.

In our studies of *P. luridum*, we found that lipid contents vary markedly with growth conditions. An observation of particular interest and significance is that abundances of some esterified polyunsaturated ($C_{18:2}$) fatty acids along with MAs and DMAs which are apparently biosynthetically related, are strongly regulated by DIC availability. This is simply illustrated by the combined data in Tables 2 and 3. Cells grown under high DIC, provided by sparging with tank CO_2 in nitrogen (P2–3, 5, 10&11), tank CO_2 in air (P7) or as a Na_2CO_3 supplement (P4) produced lower amounts of MAs and no DMAs. CSIA analysis shows that some MAs and most DMAs in P2 and P3 were inherited from the inoculum. Cells grown under air as standing cultures, where atmospheric CO_2 is limited by diffusion, produced a higher proportion of their hydrocarbons as MAs and DMAs. Moreover, the relative abundance of these compounds increased with the age of the cultures even though the total hydrocarbon content declined. These data constitute the first direct observation of DMA biosynthesis by a cultured cyanobacterium and recognition of specific conditions under which this biosynthesis is enhanced. They also suggest that MAs and DMAs constitute direct indicators of oxygenic photosynthesis and an indirect signal for palaeoenvironmental pCO_2. Further, we conducted analyses of the bacteriohopanepolyol contents of these cells and found that hopanoids could only be detected in long-term growth experiments, also suggesting that production is influenced by conditions or stages of growth. *P. luridum* hopanoids, after oxidative cleavage of the polyfunctionalized side-chain (viz. Summons et al 1994), comprised a partly co-eluting mixture of bishomohopanol and its 2β-methyl analogue.

TABLE 3 Isotopic compositions of total organic carbon, total lipid and individual compounds from *Phormidium luridum* grown for different times and conditions of dissolved inorganic carbon and O_2 availability

								Acyclic hydrocarbons (include % composition)				Methylated fatty acids		
Experiment	*Conditions*	*Biomass*	*Residue*	*Lipid*	*Phytol*	*Hopanol*[a] C_{32}+ 2βMe C_{32}	*Carotene*	*Content* (μg.g^{-1} dry)	*n*-C_{17}	*MA* C_{18}	*DMA* C_{19}	$C_{16:1}$	$C_{16:0}$	$C_{18:\#}$
P1	Sparging 5% CO_2/N_2 (−39.9‰) inoculum = P3	−39.8	NA	−41.2	−44.6	ND	NA	NA	−52.0 (65)	−25.4 (28)	−25.4 (7)	−39.0	−50.3	−47.5
P2	Sparging 5% CO_2/N_2 (−39.9‰) inoculum = P2	−60.1 (ε = 21.5)	NA	−63.1	−65.1	ND	NA	NA	−69.4 (93)	−33.5 (6)	−26.5 (1)	−65.0	−67.6	−66.6
P3	Standing in air (∼−7.8‰) inoculum for P1	−18.4 (ε = 10.5)	NA	−23.8	−22.0	ND	NA	NA	−29.9 (42)	−26.6 (47)	−27.2 (11)	−28.0	−27.2	−31.0
P4	Standing Na_2CO_3[b] (−8‰ but open to air)	−22.2	−21.0	−37.6	−27.2	ND	NA	NA	−34.5 (70)	−30.0 (30)	ND (0)	−31.6	−31.8	−29.7
P5	Standing in air 19 d	−19.0 (ε = 11.0)	−17.7	−24.6	−22.8	TR	NA	NA	−30.0 (39)	−26.7 (47)	(9)	−29.3	−28.1	−29.0
P6	Sparging 1% CO_2/N_2 (−39.9‰), 192 h	−57.1 (ε = 18.5)	−56.7	−61.1	−62.9	ND	NA	NA	−69.0 (84)	−32.8 (16)	ND (0)	−64.8	−66.3	−63.9

P7	Sparging 1% CO_2/air ($\delta^{13}C$ DIC NA)	−38.0	−37.1	NA	−42.6	ND	NA	NA	−51.0 (100)	ND	ND	−46.6	−48.5	−44.4
P10	Sparging with air 192 h	NA	NA	NA	NA	TR	NA	380	−34.6 (62)	−33.5 (19)	−31.0 (19)	NA	NA	NA
P11-1	Standing in air 66 d	−20.7 (ε = 12.5)	NA	NA	NA	−26.4	−22.0	3770	−29.7 (29)	−27.9 (62)	−27.2 (9)	NA	NA	NA
P11-2	Standing in air 75 d	−19.9 (ε = 11.5)	NA	NA	NA	−25.4	−22.1	3480	−28.7 (27)	−27.0 (62)	−26.4 (11)	NA	NA	NA
P11-3	Standing in air 99 d	NA	NA	NA	NA	−24.8	−21.5	2620	−28.0 (20)	−27.1 (66)	−26.0 (13)	NA	NA	NA
P11-4	Standing in air 99 d unbuffered (final pH 11)	NA	NA	NA	NA	−24.6	NA	1420	−24.8 (47)	−27.2 (49)	(4)	NA	NA	NA

All cultures were grown in a slightly modified BG11 (carbon-free) medium plus 5 mM tricine, pH 7.9 buffer at 24 °C and light intensity of 40 μEM^{-2}. Standing cells were grown in 2-1 Eyrlenmeyer flasks containing 1 l of medium and were exposed to air through a cotton wool plug. Sparged cultures were grown under similar conditions except that gas was supplied by bubbling a mixture of CO_2 in N_2 or air (tank), or air (atmospheric) through 2.5 l of medium in a 3 l water-jacketed growth flask. As elsewhere in this paper, carbon isotopic compositions are reported as delta values (parts per mil) where $\delta^{13}C$ (‰) = $[R_{sample} - R_{standard}/R_{standard}] \times 10^3$.

[a]Hopanol fractions were analysed as acetate derivatives and data corrected for the isotope effect of derivatization.

[b]Dissolved inorganic carbon concentration increased by addition of Na_2CO_3.

NA, not analysed; ND, not detected; TR, trace amount present.

A feature of the *P. luridum* isotope data (Table 3) is that overall isotopic fractionation (ε) is positively correlated with DIC availability. This is consistent with expectations and also accords with the observations on hydrothermal cyanobacterial mat communities by Estep (1984). On the other hand, isotopic dispersion between biomass and lipid and within different classes of lipids is much greater than expected (see review in Hayes 1993), with up to 12‰ difference between biomass and *n*-C_{17} and 8‰ between phytol and *n*-C_{17}.

We have also grown a number of methanotrophs such as *Methylococcus capsulatus* and *Methylomonas methanica* (M1923) because of their potential as indicators of aerobiosis and their known ability to produce distinctive triterpenoid lipids such as hopanoids. Moreover, methanotroph hopanoids may have characteristic substitution patterns in both the ring and side-chain which render their fossil derivatives identifiable as saturated hydrocarbons. We have previously reported (Summons et al 1994) the carbon isotopic fractionation patterns of these methanotrophs and their lipids and have made the following observations:

1. Two different methane monooxygenase (MMO) enzymes exert the primary control on isotope fractionation in biomass (pMMO $\varepsilon \sim 30$‰; sMMO $\varepsilon \sim 16$‰)
2. Polyisoprenoid lipids are strongly ^{13}C-depleted compared with biomass and more depleted than polymethylenic lipids.

At this stage it is not known whether or not the latter finding is a specific trait of certain classes of bacteria (where biochemical pathways for the synthesis of acetate precursors to both lipid classes may be different) or whether it is due to the mode of carbon assimilation, for example, growth on reduced carbon species as opposed to autotrophy. The *n*-alkyl-isoprenoid isotopic relationship seen in methanotrophic bacteria is contrary to that observed for a range of eukaryote autotrophs and to predictions made on the basis of what is currently known about biochemical pathways leading to their formation (Hayes 1993). Our experiments with *P. luridum* provide some clues. In the experiments P3–5, for example, the phytol derived from chlorophyll is isotopically 'normal' in the sense that it is heavier than *n*-alkyl lipid as exemplified by C_{16} and C_{18} fatty acids and *n*-C_{17}. However, the isotopic difference between isoprenoid and *n*-alkyl lipid is much larger than expected. Carotene in experiments P11(1–4) behaves similarly in that it is heavier than *n*-C_{17} by up to 7‰. The hopanoid data from these experiments are complex to evaluate because the hopanoids comprise a mixture of C_{32} (bishomohopanol) with approximately 25% of its of 2β-methyl analogue. Additionally, two of the 32 carbons are non-isoprenoid in origin being from the ribose moiety of the bacteriohopanepolyol. Further work is planned to investigate these issues since the ^{13}C differentials between simple isoprenoids (e.g. Pr and Ph) and *n*-alkanes are now easily measured and are diagnostic characteristics of ancient bitumens (e.g. Logan et al 1995).

Since some methanotrophs grow over a range of temperatures we conducted a study of a psychrotolerant strain of *M. methanica* (M1923) grown from 10–35 °C in order to determine whether temperature caused any significant alteration to the isotopic

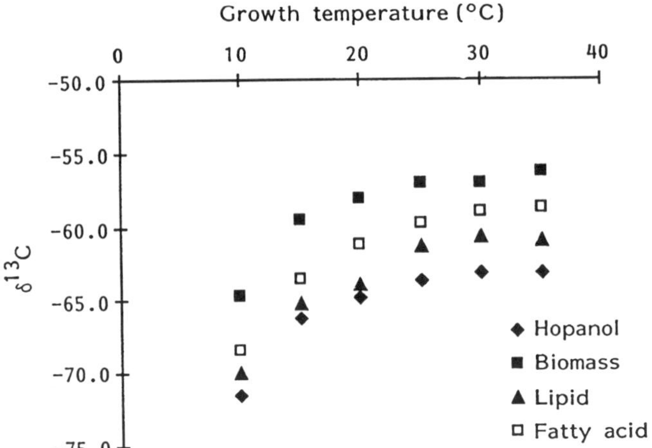

FIG. 4. Carbon isotopic compositions of biomass and individual lipids in M1923 as a function of growth over a range of temperatures.

relationships between different lipid classes. These results are shown in Fig. 4 and illustrate that, while the overall isotopic fractionation changes with temperature (possibly due to differentials in the solubility of substrate and proportions of sMMO and pMMO), there is no change to the observation that polyisoprenoids (exemplified by hopanoids) are considerably more ^{13}C-depleted than polymethylenic lipids (exemplified by fatty acids). The M1923 data also suggest that isotopic dispersions in different lipid classes of other bacteria growing at elevated temperatures will be comparable to those of organisms grown at ambient laboratory temperatures.

Lipid composition and CSIA analysis of an Octopus Spring mat

Table 4 summarizes the bulk characteristics and lipid contents of a *Synechococcus–Chloroflexus* mat from Octopus Spring, and our isotopic analyses of different layers and individual components are given in Table 5. There is a systematic decrease in δ^{13}C TOC down through the layers, possibly reflecting selective preservation of isotopically depleted (e.g. lipid) components in the decaying matter as compared with the actively growing surface layers. This is in marked contrast to the behaviour of some individual lipids such as the *Chloroflexus*-derived wax esters which show a general trend toward ^{13}C enrichment away from the surface layers and may reflect a variety of processes, including fractionation associated with the partial degradation of these species or the evolution of pCO_2 in surface waters over time. *Chloroflexus aurantiacus* in these mats may be growing photoheterotrophically (Bauld & Brock 1974), photoautotrophically or by a combination of both. Isotopic fractionation (expressed only as $\Delta\delta$) is known for autotrophic growth and has been measured at -13.7‰ (Holo & Sirevag 1986). If we assume ~5‰ difference between wax ester and intact biomass in *C. aurantiacus* the latter

TABLE 4 Total organic carbon and lipid contents of isolated layers of a *Synechococcus–Chloroflexus* mat from Octopus Spring, Yellowstone National Park

				Total lipid (TL)		*Esterified fatty acid (FA)*		*Bacteriohopanepolyol (BHP)*	
Mat layer[a]	*Total dry wt. (mg)*	*Total carbon (TC) (%, w/w)*	*Total nitrogen (%, w/w)*	*Total (mg)*	*TL/TC ($\mu g.mg^{-1}$)*	*Total (μg)*	*FA/TL ($\mu g.mg^{-1}$)*	*Total (μg)*	*BHP/TL ($ng.mg^{-1}$)*
0–1 mm	313	45.4	7.1	27	187	5430	204	3.1	118
1–2 mm	237	40.1	5.6	25	261	5740	231	4.3	172
3–4 mm	253	32.5	3.3	21	251	2930	142	5.3	256
5–7 mm	260	20.8	2.0	16	300	1800	111	6.3	391
8–11 mm	436	17.0	1.8	17	231	1940	114	6.8	398
11–20 mm	4140	3.6	0.4	20	132	2160	66	7.5	381

[a] A 40 cm^2 mat section was taken from a 52–55 °C site in the effluent channel of Octopus Spring. The mat section was frozen and returned to Ames Research Center where it was dissected using apparent changes in the laminae as a guide. Mat sections were lyophilized and samples removed for elemental analysis using a Carlo Erba EA1108, and for lipid extraction using a one-phase, modified Bligh & Dyer method. Non-lipid contaminants were removed by the Folch method, the total lipid dried and weighed. Total lipid was resuspended in cold acetone; this procedure results in the precipitation of BHP and is used for the preparative separation of these compounds. This fractional crystallization method also precipitates phospholipids, and to some extent digalactosyl (DG) and sulfoquinovosyl (SQ) lipids; however, monogalactosyl lipids, as well as the remaining DG and SQ lipids, are recovered in the acetone-soluble material. Methylated fatty acids (FAME) were prepared using BF_3–methanol. BHP hopanols were prepared using a portion of the acetone-insoluble fraction by oxidization with periodic acid, then reduction with $NaBH_4$. Hopanols and FAME were further purified by TLC. After derivatization, hopanol-acetates and FAME were analysed by GC and/or GC-MS. Approximately 75% of the recovered hopanol product was a C_{32} compound in all layers. The remaining material was made up of two different C_{31} hopanols (one possibly a methylated C_{30} compound). Verrucosanol was present in 1–8 mm layers but insufficient for CSIA.

TABLE 5 Isotopic compositions of total organic carbon and individual lipids from isolated layers of a *Synechococcus–Chloroflexus* mat from Octopus Spring, Yellowstone National Park

				Hydrocarbons						*Wax esthers*								
Mat layer	*TOC*	*Hopanol*[a] C_{32}	*Phytanyl ether*	n-C_{17}	7-Me C_{17}	n-C_{18}	2-Me C_{18}	n-C_{19}	n-C_{20}	C_{30}	C_{31a}	C_{31}	C_{32a}	C_{32}	C_{33a}	C_{33}	C_{34a}	C_{34}
0–1 mm	−16.6	−27.3	ND	−31.5	−29.6	−21.2	−21.1	−23.6	−20.5			−19.1	−20.9	−19.2	−20.1	−19.8	−19.6	−19.3
1–2 mm	−16.7	−26.7	ND	−29.2		−21.6	−23.0	−23.8	−20.5	−19.6	−19.9	−20.1	−20.5	−19.8	−20.1			
3–4 mm	−17.1	−27.4	ND			−21.3	−22.8	−24.4	−19.9	−19.9	−20.3	−21.2	−20.5	−20.1	−20.8	−20.2	−20.6	−19.9
5–7 mm	−17.1	−27.0	ND							−19.3	−20.1	−19.7	−19.1	−19.4	−19.4	−18.8	−19.3	
8–11 mm	−17.5	−26.0	ND							−18.0	−19.4	−18.4	−18.7	−18.4	−18.8	−17.5	−18.2	
11–20 mm	−18.5	−25.7	ND							−17.9		−18.4	−18.8	−18.4	−19.0	−17.7	−18.5	

ND denotes analysis conducted but compound not detected. Carbon isotopic compositions are reported as delta values (parts per mil) where $\delta^{13}C$ (‰) = $[R_{sample} - R_{standard}/R_{standard}] \times 10^3$.
[a]Hopanol fractions were analysed as acetate derivatives and data corrected for the isotope effect of derivatization.

would be near −15‰ and slightly heavier than TOC. Given the high DIC content (5.2 mM) of current surface waters at this site and its $\delta^{13}C$ value of near 0‰, it seems very unlikely that we can discriminate between growth habits on the basis of isotopes alone. Some hydrocarbons (*n*-C_{17} and [mainly] 7-methylheptadecane) are attributable, on the basis of their structures, to the cyanobacterial component of the mat and at near −30‰ are very close to values measured for the same compounds in air-grown *P. luridum*. Others, such as *n*-C_{18} and *n*-C_{20} are similar to the *Chloroflexus* wax esters and may come from this organism. Evidently, with an intermediate isotope signature, *n*-C_{19} has mixed sources. Using the isotopic data for *P. luridum* we can estimate that hopanoids from cyanobacteria would have $\delta^{13}C$ values near −27‰. This is very similar to the measured values near −27‰ (Table 5) for the Octopus Spring surface mat, suggesting that cyanobacteria are the main source of bacteriohopanepolyols in these samples. We were unable to detect phytanyl ether lipids that would have indicated methanogen biomass in this particular mat environment. Further, the bacteriohopanepolyols yielded exclusively C_{32} derivatives when subjected to the periodate/ borohydride work-up procedure (Rohmer et al 1984) rather than the C_{30} and C_{31} derivatives or 3β-methylhopanoids commonly found in methanotrophs. Thus, through both their chemical structures and their isotopic signatures, the biomarkers suggest that methanogenesis–methanotrophy was not significant in carbon cycling in this case.

Relevance of the Octopus Spring data to interpretation of the fossil record

Some hydrocarbons (*n*-C_{17} and 7-Me C_{17}) from an Octopus Spring mat have structures and isotopic compositions in accord with analyses made using cultured cyanobacteria. The *Phormidium* culture experiments confirm suspicions that MAs and (especially) DMAs are diagnostic for cyanobacteria and, accordingly, oxygenic photosynthesis and suggest they may be useful as palaeo-pCO_2 indicators. Hopanoids from the Octopus Spring mat system are isotopically heavier than might be expected for a pure cyanobacterial source and probably have mixed sources. Evidence for methanotrophy, in the form of isotopically light hopanoids or 3β-methylhopanoids was not found. Wax esters from the *Chloroflexus* component of the mat system are distinctly isotopically heavy, although an absence of relevant isotope data from cultured organisms prevents these signatures being used to assign the mode of carbon assimilation in this case.

Summary

Our experiments indicate that structurally simple lipids such as *n*-alkanes and methylalkanes are diagnostic isotopic markers for biological processes in contemporary sediments. They add to the conviction that similar compounds in Proterozoic sediments constitute valuable environmental indicators. Issues which constantly thwart analysis of Precambrian organic matter are its instability at high temperature, susceptibility to microbial degradation and proneness to contamination. Hydrocarbons are a relatively stable group of compounds if they are trapped and sealed within the source rock or a nearby host. Crystallizing minerals along a migration pathway, or formed from mineralizing brines are renowned for trapping minute quantities of hydrocarbons. Thus, hydrocarbon-bearing fluid inclusions are a potential hunting ground for diagnostic information about the lipids of primitive organisms. Recent developments giving increased sensitivity to mass spectrometers and improved inlet systems will facilitate identification of these trace components. Isotopic analysis is also feasible with existing CSIA technology. Possibilities for completely new insights about early evolution are within reach.

Acknowledgements

We are grateful to Tsege Embaye, Anne Tharpe, Janet Hope, Lesley Dowling, Ian Atkinson and Natalie Crawford for technical support. Malcolm Walter and Graham Logan provided valuable comments on draft manuscripts. Financial support from the US National Aeronautics and Space Administration to L. L. J. and B. R. T. S. (Grant NAGW-4172 to B.R.T.S.) is gratefully acknowledged.

References

Bauld J, Brock TD 1974 Algal excretion and bacterial assimilation in hot spring algal mats. J Phycol 10:101–106

Brault M, Simoneit BRT, Saliot A 1989 Trace petroliferous organic matter associated with massive hydrothermal sulfides from the East Pacific Rise at 13 and 21° N. Oceanol Acta 12:405–415

Childress JJ (ed) 1988 Hydrothermal vents: a case study of the biology and chemistry of a deep-sea hydrothermal vent of the Galapagos Rift. Elsevier, New York (Deep Sea Research Series 35) p 1677–1849

Comita PB, Gagosian RB, Williams PM 1984 Suspended particulate organic material from hydrothermal vent waters at 21° N. Nature 307:450–453

Dobson G, Ward DM, Robinson N, Eglinton G 1988 Biogeochemistry of hot spring environments: extractable lipids of a cyanobacterial mat. Chem Geol 68:155–179

Estep M 1984 Carbon and hydrogen isotopic compositions of algae and bacteria from hydrothermal environments, Yellowstone National Park. Geochim Cosmochim Acta 48:591–599

Freeman KH, Hayes JM, Trendel J-M, Albrecht P 1990 Evidence from carbon isotope measurements for diverse origins of sedimentary hydrocarbons. Nature 343:254–256

Hayes JM 1983 Geochemical evidence bearing on the origin of aerobiosis: a speculative hypothesis. In: Schopf JW (ed) Earth's earliest biosphere: its origin and evolution. Princeton University Press, Princeton, NJ, p 291–301

Hayes JM 1993 Factors controlling ^{13}C contents of sedimentary organic compounds: principles and evidence. Marine Geol 113:111–125

Hayes JM, Kaplan IR, Wedeking KW 1983 Precambrian organic geochemistry, preservation of the record. In: Schopf JW (ed) Earth's earliest biosphere: its origin and evolution. Princeton University Press, Princeton, NJ, p 93–134

Hayes JM, Freeman KH, Popp BN, Hoham CH 1990 Compound-specific isotope analysis: a novel tool for reconstruction of ancient biogeochemical processes. In: Durand B, Behar F (eds) Advances in organic geochemistry 1989. Pergamon, Oxford, p 1115–1128

Hayes JM, Summons RE, Strauss H, Des Marais DJ, Lambert IB 1992 Proterozoic biogeochemistry. In: Schopf JW, Klein C (eds) The proterozoic biosphere: an interdisciplinary study. Cambridge University Press, Cambridge, p 81–133

Hefter J, Richnow JJ, Fischer U, Trendel JM, Michaelis W 1994 (-)-Verrucosan-2β-ol from the phototrophic bacterium *Chloroflexus auantiacus*: first report of a verrucosane-type diterpenoid from a prokaryote. J Gen Microbiol 139:2757–2761

Holo H, Sirevag R 1986 Autotrophic growth and CO_2 fixation of *Chloroflexus aurantiacus*. Arch Microbiol 145:173–180

Jones ML (ed) 1985 Hydrothermal vents of the eastern Pacific: an overview. Bull Biol Soc Wash 6:1–566

Kenig F, Damsté JSS, Dalen ACK, Rijpstra WIC, Huc AY, de Leeuw JW 1995 Occurrence and origin of monomethylalkanes, dimethylalkanes, and trimethylalkanes in modern and Holocene cyanobacterial mats from Abu Dhabi, United Arab Emirates. Geochim Cosmochim Acta 59:2999–3015

Laubier L (ed) 1988 Biology and ecology of the hydrothermal vents. Oceanologica Acta (Spec Vol 8)

Leif RN 1993 Laboratory simulated hydrothermal alteration of sedimentary organic matter from Guaymas Basin, Gulf of California. PhD thesis, Oregon State University, Corvallis, USA

Leif RN, Simoneit BRT, Kvenvolden KA 1992 Hydrous pyrolysis of n-$C_{32}H_{66}$ in the presence and absence of inorganic components. Am Chem Soc Div Fuel Chem 37:1748–1753

Logan GA, Hayes JM, Hieshima GB, Summons RE 1995 Terminal Proterozoic reorganisation of biogeochemical cycles. Nature 376:53–56

Peter JM, Simoneit BRT, Kawka OE, Scott SD 1990 Liquid hydrocarbon-bearing inclusions in modern hydrothermal chimneys and mounds from the southern trough of Guaymas Basin, Gulf of California. Appl Geochem 5:51–63

Rieley G, van Dover CL, Hedrick DB, White DC, Eglinton G 1995 Lipid characteristics of hydrothermal vent organisms from 9° N, East Pacific Rise. In: Parson LM, Walker CL, Dixon DR (eds) Hydrothermal vents and processes. Geological Society, Bath (Geological Society Special Series 87) p 329–342

Robinson N, Eglinton G 1990 Lipid chemistry of Icelandic hot spring microbial mats. Org Geochem 15:291–298

Rohmer M, Bouvier-Nave P, Ourisson G 1984 Distribution of hopanoid triterpenes in prokaryotes. J Gen Microbiol 130:1137–1150

Schidlowski M 1988 A 3,800-million-year old isotopic record of life from carbon in sedimentary rocks. Nature 333:313–318

Shiea J, Brassell S, Ward DM 1990 Mid-chain branched mono- and dimethylalkanes in hot spring cyanobacterial mats: a direct biogenic source for branched alkanes in ancient sediments? Org Geochem 15:223–231

Simoneit BRT (ed) 1990 Organic matter in hydrothermal systems: petroleum generation, migration and biogeochemistry. Appl Geochem 5:1–248

Simoneit BRT 1995 Evidence for organic synthesis in high temperature aqueous media: facts and prognosis. Orig Life Evol Biosphere 25:119–140

Summons RE, Jahnke LL, Roksandic Z 1994 Carbon isotopic fractionation in lipids from methanotrophic bacteria: relevance for interpretation of the geochemical record of biomarkers. Geochim Cosmochim Acta 58:2853–2863

Trewin NH 1994 Depositional environment and preservation of biota in the Lower Devonian hot-springs of Rhynie, Aberdeenshire, Scotland. Trans R Soc Edinb 84:433–442

Ward DM, Shiea J, Zeng YB, Dobson G, Brassell S, Eglinton G 1989 Lipid biochemical markers and the composition of microbial mats. In: Cohen Y, Rosenberg E (eds) Microbial mats, physiological ecology of benthic microbial communities. American Society for Microbiology, Washington, DC, p 439–454

Zeng YB, Ward DM, Brassell SC, Eglinton G 1992a Biogeochemistry of hot spring environments 2. Lipid compositions of Yellowstone (Wyoming, USA) cyanobacterial and Chloroflexus mats. Chem Geol 95:327–345

Zeng YB, Ward DM, Brassell SC, Eglinton G 1992b Biogeochemistry of hot spring environments 3. Apolar and polar lipids in the biologically active layers of a cyanobacterial mat. Chem Geol 95:347–360

DISCUSSION

Pentecost: I was very interested to hear about this depletion of ^{13}C in cyanobacteria when the pCO_2 is high. This has interesting implications for hot springs. At least in travertine-depositing springs, pCO_2 can reach atmospheric pressure. Does that imply that the associated cyanobacteria would be heavily depleted in ^{13}C?

Summons: You would expect them to show their maximum carbon isotope fractionation (i.e. be isotopically lightest) only when the CO_2 supply is unlimited. Organisms that are living in a low CO_2 environment would be expected to show zero fractionation if they were to consume all available inorganic carbon. Intermediate fractionations will be seen between these two extremes.

Pentecost: So in deposits of that nature one might pick that up and use it as an indicator of pCO_2?

Des Marais: We've made that measurement (Des Marais et al 1989). In a culture bubbled with 10% CO_2 in air, and with very slow growth, we observed up to 26‰ discrimination: that seems to be the maximum.

Stetter: We came across a specific lipid in an archaeum that might be used as a biomarker. In the tree of life, *Methanopyrus* is among the deepest and shortest phylogenetic branches. It's a methanogen and when we looked at its lipids, we were very surprised. All archaea so far discovered have phytanyl ether lipids. This organism, which is still rather primitive, uses 2,3-di-*O*-geranylgeranyl-*sn*-glycerol, the precursor lipid to this, as the dominating lipid in its membrane (Hafenbradl et al 1993). A biosynthetic pathway is commonly accepted to reflect its own phylogeny; if an organism uses a precursor, that is an additional primitive feature. I wonder if this could be a marker lipid: do you think it would be rapidly reduced?

Summons: This isoprenoid lipid is certainly unusual in having double bonds. However, these are not stable over geological time and would be eventually reduced or destroyed. Any remaining lipid could be recognized as archaeal, but not specifically as being from *Methanopyrus*.

References

Des Marais DJ, Cohen Y, Nguyen H, Cheatham M, Cheatham T, Munoz E 1989 Carbon isotopic trends in the hypersaline ponds and microbial mats at Guerrero Negro, Baja California Sur, Mexico: implications for Precambrian stromatolites. In: Cohen Y, Rosenberg E (eds)

Microbial mats: physiological ecology of benthic microbial communities. American Society for Microbiology, Washington DC, p 191–205

Hafenbradl D, Keller M, Thiericke R, Stetter KO 1993 A novel unsaturated archaeal ether core lipid from the hyperthermophile *Methanopyrus kandleri*. Syst Appl Microbiol 16:165–169

General discussion II

Marine hydrothermal systems

Simoneit: I want to make a few general comments and give an overview of the key literature about marine hydrothermal systems, which so far have not been covered in this meeting. They are found along the active tectonic areas of the earth and are defined as fluid flow regimes with thermal gradients at elevated temperatures (Holm & Hennet 1992, Humphris et al 1995). The different types of hydrothermal systems are spreading ridges, off-axis systems, back-arc activity, hot spots, volcanism and subduction. There are currently about 100 locations with known hydrothermal activity at various seafloor spreading centres (divergent plate boundaries) (cf. review by Rona 1988) and those with associated organic matter alteration and preservation have been extensively reviewed (e.g. Kennish et al 1992, Simoneit 1990, 1993a,b). Many fossil hydrothermal systems are preserved in the geological record (e.g. Scott 1985, Shepeleva et al 1990). Hydrothermal activity is a continuous process which can occur in one region over long geological periods (Holm & Hennet 1992, Humphris et al 1995). However, individual vent systems are generally active over briefer periods (decades to millennia, Peter et al 1991, Simoneit & Kvenvolden 1994).

Hydrothermal alteration of organic matter is generally strongly reducing, proceeding from immature organic detritus in these high temperature and rapid fluid flow regimes. Pressure in the marine systems (1 to >3 km water depth) maintains the fluid state. The minor components such as CO_2 and CH_4 are supercritical and the water is near critical to supercritical as the ambient temperature approaches 400 °C, making these fluids excellent organic solvents and reaction media. The agent of thermal alteration and mass transfer, hot circulating water (temperature range warm to >400 °C), is responsible for molecular alterations (primarily reductive), with concurrent product expulsion and migration from the sediment and rock sequences (Didyk & Simoneit 1989, Simoneit 1983). Organic compound alteration to other compounds predominates over destruction by complete oxidation in these systems. Hydrothermal alteration generally produces disequilibrium reaction products comprised mostly of reduced but also some oxidized species (e.g. methylcyclopentane vs. benzene, cholestane vs. Diels' hydrocarbon; Kawka & Simoneit 1987, Simoneit et al 1988, 1992).

The hydrothermal mounds of sediment-covered systems such as the Guaymas Basin are composed of mixed minerals (mainly sulfates, carbonates, silicates, sulfides and silica) and on the East Pacific Rise at 21°N the bare rock systems are comprised of massive sulfides. Tube worm casts and other fossils are found in both of these mound systems analogous to those described in the Cretaceous sulfide ores of the Samail ophiolite of Oman (Haymon et al 1984). Organic matter derived from the vent biota is preserved as fluid inclusions (Peter et al 1990) and disseminated in interstices. Product extraction and subsequent migration is extremely efficient and thus provides a mechanism for organic matter transport along a decreasing temperature gradient. The migration of hydrothermally altered organic compounds has been observed to occur as bulk phase, emulsion, and solution, with deposition of the less water soluble and less volatile products in the cooler zones of the system and venting of the more volatile and more soluble products in the fluids (e.g. Kawka & Simoneit 1987, Simoneit 1993b). The progressive alteration of biomarker natural products and their derivatives has been illustrated by Summons et al (1996, this volume).

References

Didyk BM, Simoneit BRT 1989 Hydrothermal oil in Guaymas Basin and implications for petroleum formation mechanisms. Nature 342:65–69

Haymon RM, Koski RA, Sinclair C 1984 Fossils of hydrothermal vent worms discovered in Cretaceous sulfide ores of the Samail ophiolite, Oman. Science 223:1407–1409

Holm NG, Hennet RJ-C 1992 Hydrothermal systems: their varieties, dynamics, and suitability for prebiotic chemistry. Orig Life Evol Biosphere 22:15–31

Humphris SE, Zierenberg RA, Mullineaux LS, Thomson RE (eds) 1995 Seafloor hydrothermal systems: physical, chemical, biological, and geological interactions. American Geophysical Union, Washington DC

Kawka OE, Simoneit BRT 1987 Survey of hydrothermally-generated petroleums from the Guaymas Basin spreading center. Org Geochem 11:311–328

Kennish MJ, Lutz RA, Simoneit BRT 1992 Hydrothermal activity and petroleum generation in the Guaymas Basin. Rev Aquatic Sci 6:467–477

Peter JM, Simoneit BRT, Kawka OE, Scott SD 1990 Liquid hydrocarbon-bearing inclusions in modern hydrothermal chimneys and mounds from the southern trough of Guaymas Basin, Gulf of California. Appl Geochem 5:51–63

Peter JM, Peltonen P, Scott SD, Simoneit BRT, Kawka OE 1991 Carbon-14 ages of hydrothermal petroleum and carbonate in Guaymas Basin, Gulf of California: implications for oil generation, expulsion and migration. Geology 19:253–256

Rona PA 1988 Hydrothermal mineralization at oceanic ridges. Can Mineral 26:431–465

Scott SD 1985 Seafloor polymetallic sulfide deposits: modern and ancient. Mar Min 5:191–212

Shepeleva NN, Ogloblina AI, Pikovskiy YI 1990 Polycyclic aromatic hydrocarbons in carbonaceous material from the Daldyn-Alakit region, Siberian platform. Geochem Int 27:98–107

Simoneit BRT 1983 Organic matter maturation and petroleum genesis: geothermal versus hydrothermal. In: The role of heat in the development of energy and mineral resources in the northern basin and range province. Geothermal Research Council, Davis, CA (Special Report 13) p 215–241

Simoneit BRT (ed) 1990 Organic matter in hydrothermal systems: petroleum generation, migration and biogeochemistry. Appl Geochem 5:1–248

Simoneit BRT 1993a Hydrothermal alteration of organic matter in marine and terrestrial systems. In: Engel MH, Macko SA (eds) Organic geochemistry. Plenum, New York, p 397–418

Simoneit BRT 1993b Aqueous high-temperature and high-pressure organic geochemistry of hydrothermal vent systems. Geochim Cosmochim Acta 57:3231–3243

Simoneit BRT, Kvenvolden KA 1994 Comparison of ^{14}C ages of hydrothermal petroleums. Org Geochem 21:525–529

Simoneit BRT, Kawka OE, Brault M 1988 Origin of gases and condensates in the Guaymas Basin hydrothermal system. Chem Geol 71:169–182

Simoneit BRT, Kawka OE, Wang G-M 1992 Biomarker maturation in contemporary hydrothermal systems, alteration of immature organic matter in zero geological time. In: Moldowan JM, Albrecht P, Philp RP (eds) Biological markers in sediments and petroleum. Prentice-Hall, Hemel Hempstead, p 124–141

Summons RE, Jahnke LL, Simoneit BRT 1996 Lipid biomarkers for bacterial ecosystems: studies of cultured organisms, hydrothermal environments and ancient sediments. In: Evolution of hydrothermal ecosystems on Earth (and Mars?). Wiley, Chichester (Ciba Found Symp 202) p 174–194

The limits of palaeontological knowledge: finding the gold among the dross

Andrew H. Knoll* and Malcom R. Walter†

*Botanical Museum, Harvard University, 26 Oxford Street, Cambridge, MA 02138, USA and †School of Earth Sciences, Macquarie University, NSW 2109 and Rix Walter Pty Ltd, 265 Murramarang Road, Bawley Point, NSW 2539, Australia

Abstract. Palaeontological interpretation rests on two interwoven sets of comparisons with the modern world. Palaeo*biological* interpretation relies on the placement of fossils within a phylogenetic and functional framework based primarily on the comparative biology of living organisms. Analogy to currently observable chemical, physical and taphonomic processes enables palaeo*environmental* inferences to be drawn from geological data. In older rocks, comparisons with the modern Earth can become tenuous, limiting palaeontological interpretation. The problem reaches its apogee in Archaean successions, yet pursuit of multiple lines of evidence establishes that complex microbial communities, fuelled by autotrophy and, likely, photoautotrophy, existed 3500 million years ago. Although Archaean palaeontology has to date focused on silicified coastal sediments, improved understanding of Earth's earliest biosphere may depend on the development of alternative environmental and taphonomic analogies. Spring precipitates and hydrothermal metal deposits are promising candidates. Terrestrial organisms may be of limited value in interpreting such fossils as may be found on Mars, although some points of comparison could prove general. Given limited opportunities for exploration, proper choice of environmental analogy is critical. Spring precipitates constitute excellent deposits for addressing questions of biology on another planet.

1996 Evolution of hydrothermal ecosystems on Earth (and Mars?). Wiley, Chichester (Ciba Foundation Symposium 202) p 198–213

A principal lesson of the fossil record is that ancient ecosystems differed significantly from the current biosphere. Yet, all palaeontological interpretation rests on two interwoven sets of comparisons with the modern world. Whether it be the morphological remains of the classical fossil record, tracks and burrows that preserve aspects of behaviour, or biomarker molecules extracted from organic matter in sediments, meaningful palaeobiological interpretation begins with the placement of fossil remains in a phylogenetic and functional framework based principally on the comparative biology of living organisms. The assignment of a fossil to an extant phylum, class or order constitutes a strong statement about its development, physiology and anatomical or cytological organization. Problematica — fossils that cannot be related to extant taxa — may be amenable to functional or ontogenetic interpretation, but many insights that accompany reliable phylogenetic placement will remain beyond our grasp.

Environment provides a second basis for ancient–modern comparisons. Sedimentological and geochemical data illuminate the range of ancient environments in which a fossil population lived, thereby constraining physiological and ecological interpretation. If depositional environment is obscure or has no close analogue in the modern world — temperate poles, for example — palaeontological interpretation may be sharply limited.

As we examine ever older terrestrial sediments, both biological and environmental analogies become more difficult to draw, constraining palaeontological understanding. Older fossils commonly document animals and plants only distantly related to their modern counterparts; in rocks older than about 600 million years (i.e. for most of the known geological record), fossils are commonly simple in morphology and not diagnostic for a specific taxonomic group. Sedimentological data, particularly those that record chemical rather than physical aspects of ancient environments, also become further removed from Recent analogues. Indeed, in terms of climate, environmental chemistry, mineral precipitation, and oceanic circulation, the present might better be viewed as an end-member state of the biosphere than as the key to all doors on the past (Wright 1992, Bottjer et al 1995, Grotzinger & Knoll 1995). To make matters worse, a metamorphic veil shrouds much of the Earth's oldest rock record.

Despite these problems, one fact shines through: life leaves a distinctive fingerprint that can be identified in even the oldest, most environmentally foreign and most highly altered sedimentary records of the Earth's surface. Dross may abound, but there is gold to be found by the patient, the careful and the creative.

Spring deposits of the past 15 million years

The physical, chemical, and biological conditions of silica and carbonate precipitation observable in active springs provide a framework for interpreting sedimentological and geochemical data for older deposits (e.g. White et al 1988, Chafetz & Folk 1984). The taphonomy of spring preservation — that is, the processes by which organisms become fossilized in precipitates and the relationships of preserved elements to ambient communities — must also be established by actualistic studies (e.g. Walter et al 1976, Zavarzin et al 1989).

Palaeobiological interpretation of younger spring deposits is relatively straightforward because most fossils bear a close biological relationship to extant species. Thus, in studies of a caldera-bound hydrothermal system in Kamchatka, Zavarzin et al (1989) were able to relate sub-Recent microfossils in siliceous sinters and sulfides to microbial communities observable locally in the still-active system. In the Ries impact crater, Germany, Late Miocene carbonate spring mounds similarly preserve both microorganisms (Arp 1995a) and arthropods (Arp 1995b) that are easily interpretable in terms of modern spring dwellers. Plant and animal remains in these younger spring deposits have distinctive morphologies that ally them closely to living species. In contrast, many of the microorganisms preserved in these deposits lack phylogenetically diagnostic morphologies, elevating environmental analogy to a principal role in interpretation. This reliance on environmental analogy even in relatively young deposits underscores the potential difficulties of palaeobiological interpretation in much older rocks that may have formed beneath a less oxic atmosphere or in oceans unlike those found at present (see below).

Rhynie chert (Devonian)

The remarkable silicified peats of Rhynie, Scotland, are discussed in more detail elsewhere in this volume. Here, we consider them briefly to develop our theme of biological and environmental analogy in palaeontological interpretation. The physical setting of Rhynie cherts is reasonably interpreted by analogy to modern gold-bearing hot springs (Rice & Trewin 1988, Trewin & Rice 1992), and this provides a framework for understanding the exceptional preservation of its contained early land community. Vascular plants and closely related embryophytes dominate the assemblage (Kidston & Lang 1917, 1920a,b, 1921a,b, Edwards 1986), but arthropods, algae, fungi, cyanobacteria and even lichens (Taylor et al 1995) are recognizably preserved (Fig. 1).

Most Rhynie fossils do not belong to extant species or genera, but morphological and anatomical characters permit their placement within a broader phylogenetic framework established on the basis of comparative biology. For example, Rhynie plants display character combinations distinct from those found in any living plant, but these characters include features that individually or in combination are diagnostic for embryophytes and, within them, tracheophytes (vascular plants). Preserved sporangia, cuticle, guard cells and conducting tubes can be interpreted by functional analogy to living plants. Thus, Rhynie fossils provide important insights into both early embryophyte diversification and the organographic basis of complex morphology in plants. As the Rhynie assemblage demonstrates, fossils need not be conspecific with living organisms to sustain palaeobiological interpretation. None the less, interpretation is greatly strengthened when they share enough characters with extant species to establish their approximate point of branching in a phylogenetic tree.

Kotuikan Formation (Mesoproterozoic)

As discussed by Walter (1996, this volume), spring deposits have been recognized in the Proterozoic geological record, but few are known to be fossiliferous. Modern springs, however, provide an appropriate taphonomic and sedimentological analogy for more widespread fossil assemblages preserved in Proterozoic carbonates and silica precipitated from seawater.

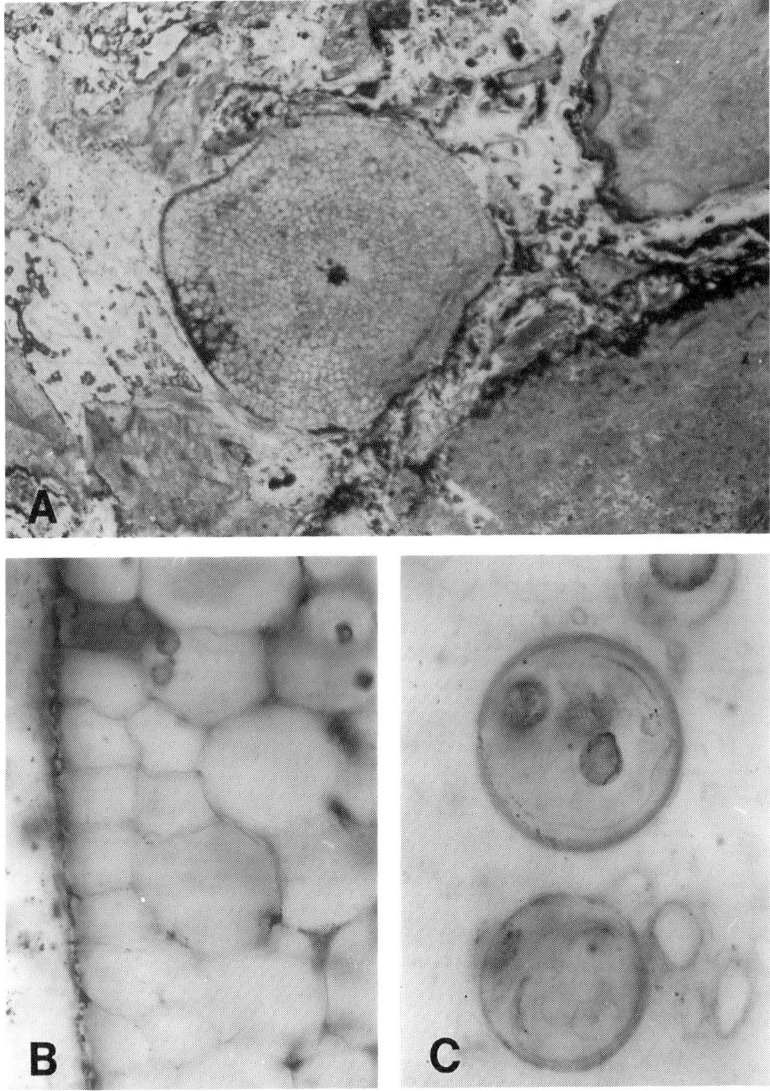

FIG. 1. Silicified fossils from the Devonian Rhynie chert, Scotland. (A) Cross-section of *Rhynia gwynne-vaughnii* (× 15). (B) Higher magnification of *Rhynia* stem, showing cellular detail of cortex, cuticle and fungi (× 250). (C) Fungi in decomposed *Rhynia* stems (× 600).

Carbonates in Archaean and older Proterozoic marine successions differ from those of younger eras in containing widespread seafloor precipitates whose closest textural analogues are modern alkaline lake and spring deposits (Grotzinger 1989, 1994). Before ca. 1800 million years ago, macroscopic crystal fans commonly nucleated on or within the seafloor in environments ranging from supratidal to basinal (Grotzinger 1994, Sami & James 1994). Seafloor encrustations persist through the Mesoproterozoic Eon (1800–1000 million years ago), but are largely restricted to peritidal environments (Kah & Knoll 1996). Such precipitates are rare in younger marine carbonate successions, although they are reprised in the distinctive 'cap carbonates' that overlie Neoproterozoic glaciogenic rocks and in Late Permian 'reefs' (Grotzinger & Knoll 1995).

The Mesoproterozoic Kotuikan Formation, northern Siberia, illustrates the potential of precipitated rocks to preserve fragile microorganisms over long time intervals (Sergeev et al 1994; Fig. 2). The preservation of microbial morphologies is all the more remarkable because many of the fossils contain almost no organic matter; they are preserved as casts and moulds in carbonates that lithified before decay reached an advanced state (Fig. 2D,E). The assemblage includes colonies of ensheathed coccoidal cells, trichomes and akinetes diagnostic for particular groups of cyanobacteria (Fig. 2C), as well as simple spheroids that defy systematic interpretation (Fig. 2B).

Carbonate spring precipitates provide a taphonomic analogy for understanding Kotuikan preservation, if not a precise environmental analogue for these tidal flat carbonates. As in recent springs, delicate microorganisms are preserved by encrustation, highlighting the importance of rapid lithification in microbial fossilization. Although rapid carbonate lithification is necessary for preservation in the short term, long-term preservation in the Kotuikan Formation and many other Proterozoic carbonates is ensured by chert. Recognizable microfossils occur only in precipitates replaced during early diagenesis by silica; in unreplaced carbonates, recrystallization and dolomitization have obliterated microbial casts and moulds.

The Kotuikan Formation can be used to illustrate a more general point about Precambrian palaeontology. Although permineralized microfossils are restricted to silicified carbonates that comprise less than 5% of the formation's total thickness, fossils also occur in carbonaceous shales (Veis & Vorobyeva 1992, Sergeev et al 1994), and stromatolites — the trace fossils of microbial mat communities — are widespread in the lower half of the formation (Komar 1966; Fig. 2F). Further, carbon and sulfur isotopic signatures for carbonates, organic matter and pyrite throughout the formation preserve a geochemical record of biological activity (Knoll et al 1995). Biomarker molecules have not been recognized in Kotuikan bitumens, but they do occur in broadly coeval deposits from Australia, where they have documented the former presence of archaebacteria and other microorganisms not preserved morphologically (Summons & Walter 1990). The conclusion that life was abundant and diverse when the Kotuikan succession was deposited owes as much to these sedimentary and biogeochemical proxies as it does to microfossils *per se*. Proterozoic sedimentary rocks preserve *multiple* lines of evidence about ancient life, and all contribute to palaeobiological interpretation.

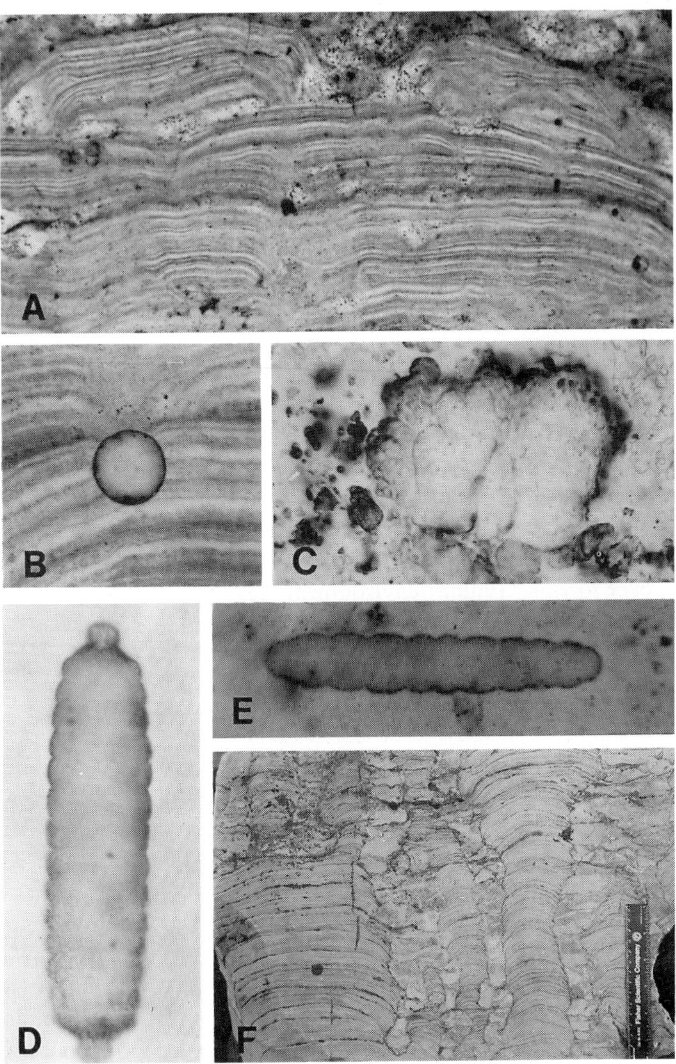

FIG. 2. Silicified carbonate precipitates and microfossils from the Mesoproterozoic Kotuikan Formation, Siberia. (A) Laminated precipitate structures, with micron-scale laminae preserved by early diagenetic precipitation (× 100). (B) *Myxococcoides grandis*, a coccoidal fossil of uncertain systematic position, preserved in the precipitated laminite shown in A (× 500). (C) *Eoentophysalis belcherensis*, a cyanobacterium closely related to living *Entophysalis* (× 300). (D, E) Cyanobacterial trichomes (*Filiconstrictus cephalon* and *C.* ex gr. *majusculus*) preserved as casts and moulds in early lithified peritidal sediments (× 1000). (F) Biologically accreted stromatolites in sub-tidal facies of the Kotuikan Formation (note 15 cm scale in lower right).

Gunflint Iron Formation (Palaeoproterozoic)

The Gunflint Iron Formation provides a second and somewhat older example of fossilization by mineral precipitation, in this case silica. Facies within the Gunflint Formation have been interpreted as subaerial splash deposits (Walter 1972) and spring precipitates (M. R. Walter & S. M. Awramik, personal observation). Indeed, in their seminal monograph, Barghoorn & Tyler (1965) pictured Yellowstone sinters as modern counterparts of Gunflint stromatolites. Yet, environmental interpretation remains uncertain because fossiliferous Gunflint lithologies contain mineralogical and sedimentological features that make choice of analogy difficult — abundant hematite, siderite and other iron-bearing minerals; primary silica; and facies associations that suggest marine deposition.

Environmental uncertainty is magnified by microfossils that are abundant but difficult to interpret metabolically. Within stromatolitic facies, the overwhelmingly dominant fossils are simple cocci and filamentous sheaths that could have been formed by a phylogenetically wide assortment of bacteria (Fig. 3A). In consequence, only generalized palaeobiological conclusions can be drawn with confidence, but these include the important observation that life was ubiquitous in Gunflint time. As in younger deposits, the microfossil record is augmented by stromatolites and biogeochemical data. However foreign Gunflint environments may appear, they harboured diverse and metabolically disparate prokaryotes and, possibly, simple eukaryotes, as well.

Warrawoona Group (Early Archaean)

Early Archaean (3000–3500 million years old) sedimentary rocks amenable to palaeontological investigation are known from only a few localities in southern Africa and Australia (Schopf 1994). Of these, the Warrawoona Group, Western Australia, has provided the clearest window on early life. Warrawoona successions contain abundant chert and organic carbon, but microfossils are exceedingly rare. More than a decade of investigation has turned up only a handful of assemblages containing a limited diversity of simple cocci and filaments (Schopf 1994; Fig. 3B). The morphologies of Warrawoona fossils are consistent with their interpretation as cyanobacteria (Schopf 1994), but they are also open to other interpretations. Indeed, in rocks of this antiquity, it is difficult to rule out the possibility that preserved fossils belong to extinct groups not closely related to present day bacteria or archaea. The absence of a compelling modern analogue for Warrawoona environments further complicates biological interpretation.

Stromatolites have been reported from Warrawoona sections (Walter et al 1980, Walter 1983), and one might think that these would spark more confident palaeobiological interpretation; however, they, too, have been subject to conflicting opinions. Tufa-like precipitation structures that mimic the domes and columns accreted by microbial mats occur in many later Archaean and Palaeoproterozoic carbonates (Grotzinger 1989, 1994). Although microbial processes (e.g. sulfate reduction) may have played a role in the nucleation of these structures, growth appears to have been

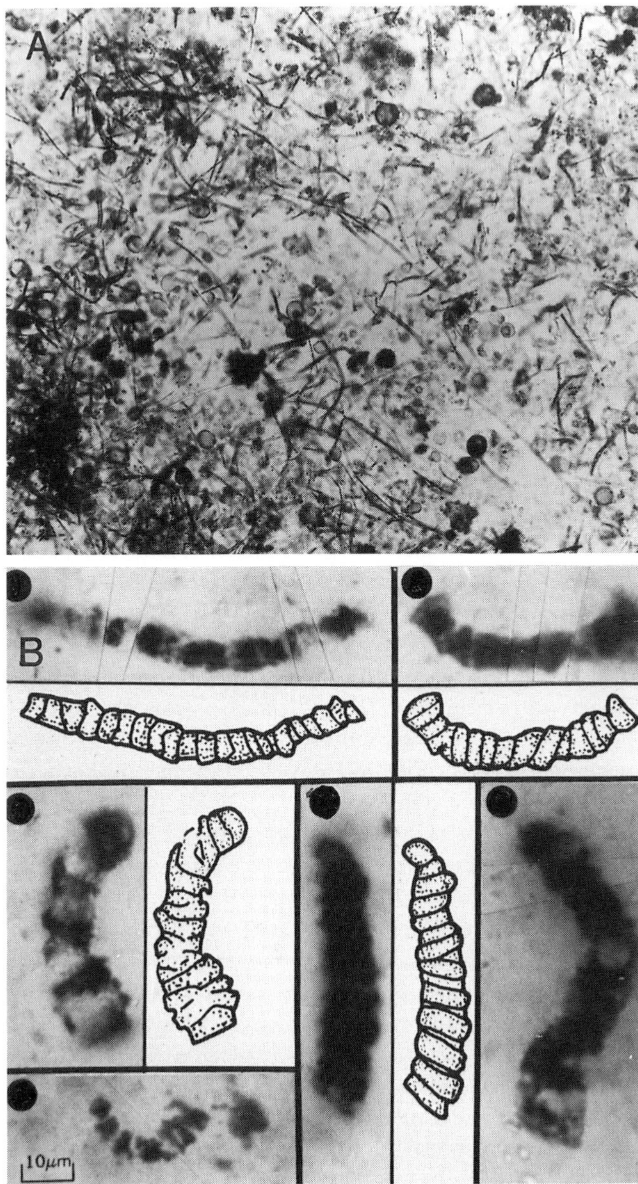

FIG. 3. (A) Fossiliferous chert of the Palaeoproterozoic Gunflint Iron Formation, Canada, showing abundant cocci and filaments of uncertain systematic affinities (×400). (B) Filamentous microfossils from the Early Archaean Warrawoona Group, Australia (×800; from Schopf 1993).

principally a physicochemical phenomenon. Under favorable circumstances, microscopically observed fabrics serve to differentiate tufa-like precipitates from mat-built stromatolites, but in the absence of petrographic data, or in the event of marked recrystallization, uncertainty lingers.

Lowe (1994) specifically suggested that Warrawoona stromatolites are abiotically precipitated laminites whose domal and pseudocolumnar morphologies are a product of small scale sedimentary deformation. Lowe did not supply petrographic data in support of his interpretation and did not cite earlier reports of microfabrics that may support a biological origin (Walter et al 1980, Walter 1983). Lowe's 'structural' observations might also be debated; however, his challenge to the widely accepted interpretation of Warrawoona stromatolites as biological does underscore the fact that, in the absence of preserved mat-building microorganisms, laminated precipitates are uncertain guides to ancient biology (Grotzinger 1994).

Fractionation of carbon isotopes between carbonates and organic carbon provides a third line of approach, suggesting that the Warrawoona sea supported autotrophically, and probably photoautotrophically, driven ecosystems (Schopf 1994). Biomarker molecules have not been recovered from rocks of this antiquity, but exceptionally low $\delta^{13}C$ values in younger Archaean organic matter (Hayes 1983) and unusual ^{13}C enrichment in carbonate nodules from coeval shales (Dix et al 1995) provide evidence for methanogenic archaea early in Earth's history.

As with the other lines of evidence, some uncertainty attaches to biogeochemical interpretations. Nonetheless, when all lines of evidence are considered simultaneously, invaluable nuggets of palaeontological gold emerge from the dross of the Early Archaean record: 3500 million years ago life had already taken root on Earth. More than that, organisms had diversified to form complex microbial ecosystems fuelled by autotrophy and, very likely, photosynthesis.

To date, most Archaean palaeontology has explicitly or implicitly proceeded from the view that silicified sediments provide the best facies for palaeontological investigation. This is a very narrow environmental paradigm that may severely and unnecessarily limit palaeontological study of Archaean rocks. Success in improving our understanding of early life may depend on the development of additional or alternative sedimentological analogies. Spring deposits provide one such possibility, given the widespread distribution of seafloor precipitates in Archaean successions. Sulfide chimneys and other metalliferous precipitates provide another (De Wit et al 1982). Vent-associated precipitates are particularly attractive targets for palaeontological investigation because molecular phylogenies (Kandler 1994) and some scenarios for the origin of life (e.g. Wächterhäuser 1988) identify hydrothermal environments as the cradles of biological evolution on Earth.

Mars

Palaeontological experience on Earth has important implications for the exopalaeobiological exploration of Mars.

First, the phylogenetic framework used to interpret terrestrial fossils of all ages is likely to be of limited use in interpreting such fossils as may exist on Mars. Barring planetary infection (difficult to demonstrate regardless of its potential feasibility), our best hope for interpretation may lie in the probability that some preservable aspects of biology will prove to be general. For example, isotopic fractionation may characterize autotrophic metabolism, whatever its origin. Given the limited number of routes likely for chemical evolution (DeDuve 1995), independently evolved biotas may also share common structural or metabolic molecules. Behavioural responses to environmental stimuli (taxis) may leave similar records in the orientation of fossils (e.g. Knoll 1989). Even cell shapes might bear comparison, although only limited metabolic interpretation could be attempted on this basis.

Choice of environmental analogy is critical, because the opportunities for exopalaeontological exploration are likely to be limited. Experience in Precambrian palaeontology on Earth suggests that precipitated minerals, epitomized by spring deposits, can both preserve delicate microfossils and record aspects of environmental conditions at the planetary surface. In some ways, prospects for the exopalaeontological exploration of Martian spring precipitates are better than they are for very old sediments on Earth, because many terrestrial successions have undergone severe diagenesis and metamorphism unlikely to have affected sedimentary deposits on Mars.

Perhaps most important, palaeontological experience underscores the need for multiple lines of investigation in any exopalaeontological exploration of Mars (Walter & Des Marais 1993). Collected samples may or may not yield recognizable microfossils; laminated precipitates may or may not be distinguishable as biogenic (and they almost certainly will not be interpretable in the absence of petrological data); analysis of kerogens may or may not yield biologically diagnostic molecules; isotopic compositions may or may not yield compelling evidence for biogeochemical cycles. However, a broad approach to the exploration of ancient Martian sedimentary rocks insures that if life ever evolved on the Red Planet, we will discern its inevitable fingerprint. Regardless of the success of palaeobiological probes, such an approach will fit best with other mission strategies and ensure that the precious opportunity to learn about the early development of Mars will be maximized.

Acknowledgements

Research leading to this paper for supported in part by the NASA Exobiology Program. We thank J. W. Schopf for supplying photographs of Archaean microfossils.

References

Arp G 1995a Lacustrine biotherms, spring mounds, and marginal carbonates of the Ries-impact crater (Miocene, southern Germany). Facies 33:35–90

Arp G 1995b Ein Diplopode (Tausendfüssler i.e.S.) aus den lakustrinen Karbonaten des Nördlinger Rieses (Miozän, Süddeutschland): Morphologie und Integumentstruktur. Paläontol Zeitschr 69:135–147

Barghoorn ES, Tyler SM 1965 Microorganisms from the Gunflint chert. Science 147:563–577

Bottjer DJ, Campbell KA, Schubert JK, Droser ML 1995 Palaeoecological models, non-uniformitarianism, and tracking the changing ecology of the past. Geol Soc Spec Publ 83:7–26

Chafetz HS, Folk RL 1984 Travertines: depositional morphology and bacterially constructed constituents. J Sediment Petrol 54:289–316

DeDuve C 1995 Vital dust: life as a cosmic imperative. Basic Books, New York

de Wit MJ, Hart R, Martin A, Abbot P 1982 Archean abiogenic and probable biogenic structures associated with mineralized hydrothermal vent systems and regional metasomatism, with implications for greenstone belt studies. Econ Geol 77:1783–1802

Dix GR, Thomson ML, Longstaffe FJ, McNutt RH 1995 Systematic decrease of high $\delta^{13}C$ values with burial in late Archaean (2.8 Ga) diagenetic dolomite: evidence for methanogenesis from the Crixás Greenstone Belt, Brazil. Precambr Res 70:253–268

Edwards DS 1986 *Aglaophyton major*, a non-vascular land plant from the Devonian Rhynie chert. J Linn Soc 93:173–204

Grotzinger JP 1989 Facies and evolution of Precambrian carbonate depositional systems: emergence of the modern platform archetype. SEPM Spec Publ 44:79–106

Grotzinger JP 1994 Trends in Precambrian carbonate sediments and their implication for understanding evolution. In: Bengtson S (ed) Early life on Earth. Columbia, New York (Nobel Symp 84) p 245–258

Grotzinger JP, Knoll AH 1995 Anomalous carbonate precipitates: is the Precambrian the key to the Permian? Palaios 10:578–596

Hayes JM 1983 Geochemical evidence bearing on the origin of aerobiosis: a speculative hypothesis. In: Schopf JW (ed) Earth's earliest biosphere: its origin and evolution. Princeton University Press, Princeton, NJ, p 291–301

Kah L, Knoll AH 1996 Microbenthic distribution of Proterozoic tidal flats: environmental and taphonomic considerations. Geology 24:79–82

Kandler O 1994 The early diversification of life. In: Bengtson S (ed) Early life on Earth. Columbia, New York (Nobel Symp 84) p 152–160

Kidston R, Lang WH 1917 On Old Red Sandstone plants showing structure from the Rhynie chert bed, Aberdeenshire. I. *Rhynia gwynne-vaughani* Kidston & Lang. Trans R Soc Edinb 51:761–784

Kidston R, Lang WH 1920a On Old Red Sandstone plants showing structure from the Rhynie chert bed, Aberdeenshire. II. Additional notes on *Rhynia gwynne-vaughani* Kidston & Lang, descriptions of *Rhynia major* sp. and *Hornea lignieri* n., g., n. sp. Trans R Soc Edinb 52:603–627

Kidston R, Lang WH 1920b On Old Red Sandstone plants showing structure from the Rhynie chert bed, Aberdeenshire. III. *Asteroxylon mackiei*, Kidston & Lang. Trans R Soc Edinb 52:643–680

Kidston R, Lang WH 1921a On Old Red Sandstone plants showing structures from the Rhynie chert bed, Aberdeenshire. IV. Restorations of vascular cryptograms, and discussion of their bearing on the general morphology of the Pteridophyta, and the origin of the organisation of land-plants. Trans R Soc Edinb 52:831–854

Kidston R, Lang WH 1921b On Old Red Sandstone plants showing structures from the Rhynie chert bed, Aberdeenshire. V. The Thallophyta occurring in the peat-bed; the succession of the plants throughout a vertical section of the bed, and the conditions of accumulation and preservation of the deposit. Trans R Soc Edinb 52:855–902

Knoll AH 1989 The palaeomicrobiological information in Proterozoic rocks. In: Cohen Y, Rosenberg E (eds) Microbial mats: physiological ecology of benthic microbial communities. American Society for Microbiology, Washington DC, p 469–484

Knoll AH, Kaufman AJ, Semikhatov MA 1995 The carbon-isotopic compisition of proterozoic carbonates: successions from northwestern Siberia (Anabar Massif, Turukhansk Uplift). Am J Sci 295:823–850

Komar VA 1966 Stromatolity verchnedokembriiskih otlozhenii severa Sibiskoi platformy i ih stratigraficheskoe znachenie. (Stromatolites in Upper Precambrian deposits of the Northern Siberian Platform and their stratigraphic significance.) Nauka, Moscow

Lowe DR 1994 Abiological origin of described stromatolites older than 3.2 Ga. Geology 22:387–390

Rice CM, Trewin NH 1988 A Lower Devonian gold-bearing hot-spring system, Rhynie, Scotland. Trans Inst Mining Metallurgy B 97:141–144

Sami TT, James NP 1994 Peritidal carbonate platform growth and cyclicity in an Early Proterozoic foreland basin, upper Pethei Group, northwest Canada. J Sediment Res B 64:111–131

Schopf JW 1993 Microfossils of the Early Archean Apex chert: new evidence for the antiquity of life. Science 260:640–646

Schopf JW 1994 The oldest known records of life: Early Archaean stromatolites, microfossils, and organic matter In: Bengtson S (ed) Early life on Earth. Columbia, New York (Nobel Symp 84) p 193–207

Sergeev VN, Knoll AH, Grotzinger JP 1994 Palaeobiology of the Mesoproterozoic Billyakh Group, northern Siberia. Palaeontol Soc Mem 39:1–37

Summons RE, Walter MR 1990 Molecular fossils and microfossils of prokaryotes and protists from Proterozoic sediments. Am J Sci 290:212A–244A

Taylor TN, Hass H, Remy W, Kerp H 1995 The oldest fossil lichen. Nature 378:244

Trewin NH, Rice CM 1992 Stratigraphy and sedimentology of the Devonian Rhynie chert locality. Scott J Geol 28:37–47

Veis AF, Vorobyeva NG 1992 Riphean and Vendian microfossils of the Anabar Uplift. Izvestya RAN Ser Geol 1:114–130

Wächterhäuser G 1988 Before enzymes and templates: theory of surface metabolism. Microbiol Rev 52:452–484

Walter MR 1972 A hot spring analog for the depositional environment of Precambrian iron formations of the Lake Superior region. Econ Geol 67:965–980

Walter MR 1983 Archaean stromatolites: evidence for Earth's earliest benthos. In: Schopf JW (ed) Earth's earliest biosphere: its origin and evolution. Princeton University Press, Princeton, NJ, p 187–213

Walter MR 1996 Ancient hydrothermal ecosystems on Earth: a new palaeobiological frontier. In: Hydrothermal ecosystems on Earth (and Mars?). Wiley, Chichester (Ciba Found Symp 202) p 122–130

Walter MR, Des Marais DJ 1993 Preservation of biological information in thermal spring deposits: developing a strategy for the search for fossil life on Mars. Icarus 10:129–143

Walter MR, Bauld J, Brock TD 1976 Microbiology and morphogenesis of columnar stromatolites (*Conophyton*, *Vacerrilla*) from hot springs in Yellowstone National Park. In: Walter MR (ed) Stromatolites. Elsevier, Amsterdam, p 273–310

Walter MR, Buick R, Dunlop JRS 1980 Stromatolites 3400–3500 Myr old from the North Pole area, Western Australia. Nature 284:443–445

White DE, Hutchinson RA, Keith TEC 1988 The Geology and remarkable thermal activity of Norris Geyser Basin, Yellowstone National park, Wyoming. US Geol Surv Prof Paper 1456

Wright VP 1992 Early Carboniferous carbonate systems: an alternative to the Cainozoic paradigm. Sed Geol 93:1–5

Zavarzin GA, Karpov GA, Gorlenko VM et al 1989 Kalderniye mikroorganismi. (Caldera-dwelling microorganisms.) Nauka, Moscow

DISCUSSION

Farmer: What processes were responsible for the very early silicification in your late Proterozoic peritidal marine sequences from Siberia? Obviously, it was a fast and ongoing process, but much different than thermal spring environments where silica precipitation is driven by a drop in temperature. What kind of driving mechanisms are likely to have prevailed in peritidal marine environments?

Knoll: In the Precambrian oceans, in the absence of diatoms, radiolaria and to a lesser extent sponges, sea water was probably much closer to saturation with respect to amorphous silica than it is in modern oceans. The meeting of this relatively high concentration of silica in sea water with (quite possibly) highly saturated ground waters, might create the conditions under which this very early silicification would happen. On a broad environmental scale, it is clear that the Precambrian silicification was most abundant around the edges of the ocean — in tidal flats where some mild evaporative conditioning of the sea water might be important. Within these environments, silica is commonly nucleated on organic material. Incipient fossils provide a template for precipitation.

Nisbet: As geologists, we have been thrown the job of dating the earliest fossil record. Our problem is that it is very difficult to date the deepest divisions in the 'tree of life'. What we can say is that the modern carbon cycle essentially gets going prior to 3.5 Ga ago, because the carbon isotopes in carbonate are similar to modern ratios by that stage (possibly even by 3.8 Ga ago). So the cyanobacteria on the left-hand side of the tree of life shown by Karl Stetter (1996, this volume) must have evolved by 3.5 Ga ago, or possibly by 3.8 Ga ago. This dates everything in the tree below cyanobacteria as prior to that. Then we ask ourselves, how do we get deeper? Andrew Knoll is absolutely right, we should look for hot spring material, but the trouble is that we run out of geological record. At 3.5 Ga we have Barberton (South Africa) and the Pilbara (Western Australia); at 3.8 Ga we have Isua (Greenland), but it has metamorphosed so the isotopic record has been damaged; and at 4 Ga we just have a few fragments. Perhaps we can attempt to map out the tree at 3.5 Ga by looking for very high light fractionation of carbon, for instance — in other words, by trying to map out the methanogens on the right-hand side of the tree, and thereby showing that the whole bottom end of the tree existed by 3.5 Ga ago. But we don't have the geological record to date successfully anything biological prior to that, apart, perhaps, from cyanobacterial photosynthesis. There is one very speculative idea that we put forward that some diamonds might actually be mid-ocean ridge fossils, and that might actually get you a little bit further back into the methanogens (Nisbet et al 1994, Nisbet 1995).

Walter: I would like to comment on the pessimism of that statement. As you well know, very few people actually work on early Archaean sedimentary rocks from a sedimentological perspective. These are early days, and I hope before too long you'll be forced to swallow your words. And already we have Roger Buick's new succession beneath the Warrawoona Group in the Pilbara, from 3.54 Ga ago. This could well provide significant new palaeobiological information. No doubt more such succesions remain to be discovered.

Knoll: Euan Nisbet suggested that by 3.5 Ga ago cyanobacteria were already around. While we certainly need an autotrophic mechanism of primary production in order to account for this 30‰ difference between the isotopic composition of carbonate and organic matter, and given the distribution of organic matter one would guess that photoautotrophy is likely, how can we distinguish in the record between cyanobacterial photosynthesis and types of photosynthesis that are present on earlier branches of the tree?

Des Marais: You could approach that using logic similar to that which Roger Buick used in the 2.7 Ga old fossils he studied in the Fortesque sequence in Western Australia. There was abundant productivity in an environment where sources of reducing power other than oxygenic photosynthesis were not available. If you could convince yourself that these shallow water deposits that have been mapped out in the 3.3–3.4 Ga old sediments really do represent photosynthetic communities, and that there was no locally available non-biological source of reducing power available to these organisms to make organic matter, then one could conclude that these communities depended on oxygenic photosynthesis. This would then be circumstantial evidence in favour of cyanobacteria, which are the only eubacteria to have oxygenic photosynthesis.

Shock: I think it would be useful to try and get away from photosynthesis as much as possible. There is a large 'darkened' part of the tree of life that doesn't care about photosynthesis. We would be more likely to find something preserved in and around a zone of mixing between fluids, and not necessarily near the surface manifestation of these deposits. In fact, UV radiation would probably have been more intense because of the lack of an ozone layer, scouring clean the surface rocks of any organic content. It seems to me that we should forget about photosynthesis if we want to learn about the deeper and thicker branches of the tree.

Des Marais: One of the advantages of cyanobacteria is that the cells are so large that even with a fair degree of recrystallization of the cherts, some morphology is still preserved. One of the main problems with looking for fossil thermophiles is their small size: silica grain sizes can approach microns in diameter, and when the crystal grain size begins to exeed the dimension of a cell, preservation becomes problematic.

Shock: That is further encouragement for us to stop relying on our eyes and go after molecular fossils.

Summons: Isn't there another potential problem with issues that rely on interpretation of the carbon isotopic signatures of carbonates of this age? If it is true that carbonates are near to 0‰, this requires there to be a corresponding fraction of the crustal carbon pool buried in the form of organic carbon so as to preserve isotopic mass balance. For this to have happened, there must have been appropriate tectonic processes in action to enable burial of the organic carbon that balances that carbonate carbon.

Walter: It only tells you about the flux, it doesn't tell you about the magnitudes.

Nisbet: But it is essentially telling you that the modern carbon cycle is going by then.

Jakosky: Andrew Knoll, in your last few remarks about the application to Mars, you suggested using a wide variety of approaches. That's important, but it has to be put into the broadest context of understanding Mars as a planet. The carbon is a great example: you can look for the carbon isotopes all you want and see whatever signature is there, but until you understand how the carbon isotopes vary due to non-biological processes it is not going to tell you anything. You can't just go and look for fossil evidence of Martian biology, you need to really understand Mars in the broader sense in order to know the context for what you're looking for.

Knoll: I agree completely. In fact, I have spent the last decade of my life trying to establish physical Earth history in the Late Proterozoic in order to interpret biology on Earth.

Pentecost: I wonder if you would like to reply to some observations I have made on cyanobacteria. I have noticed that thick, tough sheathes, which are hard to destroy even with strong acids, occur on cyanobacteria that are either intertidal or subaerial. Some of the pictures you have shown gave me the impression that these were the thick-sheathed varieties. Are cyanobacteria from particular environments being selected for good preservation?

Knoll: There's no question about that. The taphonomic window provided by silicified carbonates is heavily weighted towards the peritidal environments in which one will tend to find the kind of cyanobacteria you mention. There is however, another fossil record: the record of things that are preserved over a wide range of environments (from peritidal to deep basinal) in fine-grain siliciclastic deposits where they're preserved mostly as compressions. Rather delicate trichomes and cyanobacterial cells are preserved. Looking at the entire record, one sees both the easily preserved massive envelope producers that you mention and ones that don't have this characteristic.

Farmer: You have pointed out something important: that is, there are lots of rapidly mineralizing environments aside from thermal springs which would constitute much broader targets in terms of a search for evidence of an ancient biosphere on Mars. It's my observation that these lower temperature environments are often better at preserving organic matter. In terms of the broader questions about Mars, even if life never developed there, that organic record is still extremely important. The prebiotic chemistry that we are interested in finding out about on Mars may reside in these kinds of environments. Thermal springs are only one type of target for Mars, and some of these peritidal environments have equivalents in shoreline lake deposits. Although chemical precipitates are probably most favourable, even heavily cemented detrital sediments can constitute important targets in near-shore lake settings. And of course, as you have pointed out, fine-grained clay-rich shales (and I would add evaporites) are also potential targets in deeper basin settings.

References

Nisbet EG 1995 Archaean ecology: a review of the evidence for the early development of bacterial biomes, and speculations on the development of a global-scale biosphere. In: Coward MP, Ries AC (eds) Early Precambrian processes. Geol Soc Lond, Spec Publ 195:27–51

Nisbet EG, Mattey DP, Lowry D 1994 Are some diamonds dead bacteria? Nature 367:694
Stetter KO 1996 Hyperthermophiles in the history of life. In: Evolution of hydrothermal ecosystems on Earth (and Mars?). Wiley, Chichester (Ciba Found Symp 202) p 1–18

The role of remote sensing in finding hydrothermal mineral deposits on earth

Jonathan F. Huntington

CSIRO Exploration and Mining, Mineral Mapping Technologies Group, PO Box 136, North Ryde, NSW 2113, Australia

Abstract. The identification of surface mineralogical composition using hyperspectral sensors is now the major remote sensing opportunity for exploration geologists seeking refined vectors to potential ore-bearing hydrothermal systems. This involves no less than remote, visible and infrared spectroscopy of the molecular composition of geological materials from remote platforms using a large number of calibrated spectral bands. From field-portable systems to those flying in high-flying aircraft on the edges of space, it is now possible to define a long list of minerals and their weathering products detectable at these wavelengths. These include hydroxyl-bearing minerals such as hydrothermal clays, sulfates, ammonium-bearing minerals, phyllosilicates, iron oxides, carbonates, and a wide range of silicates. Indeed, even the chemical composition of micas and chlorites has been mapped remotely using subtle wavelength shifts in their diagnostic reflectance spectra, indicating varying degrees of Na, K, Al, Mg and Fe substitution. Spatial zones, relative abundances and assemblages of these minerals allow geologists to reconstruct the mineralogical, chemical and sometimes thermal disposition of ancient hydrothermal systems in their search for optimal drilling targets. Such minerals not only result directly from the hydrothermal processes involved but may also 'expose' older host rocks caught up in the process and brought to the surface. New microwave radar systems are also shedding new light on landscape processes, textures and structure and occasionally penetrating dry surface layers to reveal buried structures.

Evolution of hydrothermal ecosystems on Earth (and Mars?). Wiley, Chichester (Ciba Foundation Symposium 202) p 214–235

The major benefits of remote sensing in mineral exploration accrue from its relatively low cost (per sq km) and its capacity for non-invasive data collection in remote, poorly mapped and logistically challenging or politically inaccessible regions, compared with conventional ground-based techniques. Remote sensing provides the ability to focus on specific areas and downgrade the importance of others, thus optimizing subsequent exploration. To reap these benefits, remotely sensed data must be collected early in an exploration program, before costly ground methods are used.

Geological remote sensing is based on measuring the spatial and spectral characteristics of electromagnetic radiation (e.g. the sun or a radar beam) as it is differentially absorbed, reflected, scattered and/or emitted by surface geological materials. Between the radiation source (say the sun) and the target (a rock surface of interest) this radiation is subject to differential scattering and absorption by the atmosphere, which must be taken into account when interpreting what is actually sensed by some air- or space-borne instrument. Sensing can be passive, with the sun as a source, or active, when microwave or light energy is provided by an artificial source such as a radar or laser.

Rarely have sensors been built exclusively to sense geological parameters. Most of today's sensors are multi-functional in design. So far, none has been built to consider the specific problems of mapping the spatial distribution of the mineralogical composition of hydrothermally altered rocks. Developing systems that address specific end-user problems, rather than the boundaries of science or technological demonstration, will lead to a generation of sensors capable of yielding new geological knowledge. Improved geological understanding is the end; remote sensing merely the means. Fortunately, these sensing systems are getting smaller, lighter, more sensitive, cheaper and commercially deliverable.

A strategy for remote sensing of hydrothermal mineral deposits

Remote sensing is best achieved using a three-stage, hierarchical exploration strategy, with each stage representing an increase in complexity and capital cost: (a) field-based mineral mapping; (b) airborne mineral mapping; and (c) satellite mineral mapping. For greatest efficiency, one would start with space-based, regional-scale exploration of appropriate metallogenic provinces, defining areas for more detailed attention with aircraft systems which might have greater capabilities or sensitivity. From targets defined from airborne surveys, one may then seek confirmation and assess outcrop-scale mineralogical patterns in the field with hand-held field spectrometers, such as the PIMA-II shortwave infrared spectrometer. As discussed above, it is important to use these tools in such a manner as to optimize subsequent stages and minimize costs.

Characteristics of hydrothermal mineral deposits amenable to remote sensing

The major characteristics of hydrothermal mineral deposits amenable to remote sensing are their tectonic, structural and geomorphological settings and their lithological and mineralogical composition (Fig. 1). In remote-sensing parlance these translate to their spatial and spectral characteristics, respectively. Typically, spatial characteristics are best interpreted by skilled, manual, photogeological techniques applied to digitally enhanced Landsat, SPOT, AVHRR (advanced very high resolution radiometer) or SAR (synthetic aperture radar) satellite imagery and aerial photographs. In the very near future several commercial, high-spatial-resolution satellite systems will be providing spatial resolutions of between 1 and 3 m and stereoscopy almost equivalent to aerial photographs and capable of capturing mineralized veins and detailed geomorphological processes.

Today, spectral or compositional characteristics can be derived from semi-quantitative spectroscopic methods (Goetz 1989), where dependence on visual interpretation is

GEOLOGICAL CHARACTERISTICS	REMOTE SENSING CHARACTERISTICS
SETTING ● **TECTONIC SETTING:** Fracture zones above subduction zone. High angle normal fault zones related to continental margin rifting. Rhyolitic volcanic centres, caldera margins, geothermal systems ● **AGE:** Mainly Tertiary, but can be any age ● **SIZE:** Varies, but can be up to several km across ● **DEPTH:** Near surface ● **TEMPERATURE:** 150–300 °C? ● **STRUCTURES:** High angle structures, complex volcanic centres, hydrothermal brecciation, stockworks ● **ASSOCIATED ROCKS:** Sedimentary: Volcano-clastics: Intrusives: Felsic plugs and dykes	SPATIAL PROPERTIES ● Felsic volcanic rocks, caulderas, maars ● Small intrusions and felsic dykes ● Major lineaments, regional fracture traces, dilatant zones ● Normal faults, high angle structures ● Prominent (silicified) ridges
COMPOSITION ● **HOST ROCKS:** Rhyolite tuffs, hydrothermal breccia ● **ALTERATION:** Silica sinter, extensive silification, advanced argillic (alunitic), argillic (kaolinite), propylitic, ammonium clays and feldspar, Fe^{3+} ● **ZONING:** Top down—silica, alunite, kaolinite ● **ORES/METALS:** V. fine native gold, pyrite, realgar, stibnite, arsenopyrite, cinnabar, fluorite	SPECTRAL PROPERTIES ● Alunite bands near 1.76, 2.16, 2.32, 2.42 μm and shifts in these bands caused by NH_4 ● Clay bands near 2.2 μm caused by kaolinite, muscovite and smectites ● Buddingtonite bands near 2.02 and 2.12μm ● Fe^{3+} absorption bands near 0.9 μm. Ultra violet/visible shoulder of Fe oxides ● Jarosite bands at 0.43, 1.86 and 2.26μm ● Silica emission minimum near 9μm
GEOCHEMISTRY ● **MAJOR ELEMENTS**: Au ● **PATHFINDERS**: As, Sb, Hg, Ti, locally NH_3, W, F, Mo GEOPHYSICS ● **GRAVITY**: High ● **MAGNETICS**: High	REMOTE SENSING EXPLORATION STRATEGY ● Landsat TM for regional reconnaissance for 'clays' ● Thermal IR (TIMS) for silica emission minimum mapping ● Imaging spectrometers for identification of clay species, such as alunite, kaolinite, sercite, buddingtonite, and prophylitic, phyllic and argillic or advanced argillic alteration
REFERENCES Berger 1986, 1991, Erickson 1982, Johns et al 1989, 1990	REFERENCES Kruse et al 1988, 1990, Krohn 1986
(Examples: Paradise Peak, Nevada; Round Mtn. Nevada; many others throughout Nevada)	

FIG. 1. Geological and remote sensing characteristics of volcanic-hosted hot spring-type disseminated gold deposits.

reduced and increased use is made of digital spectral reference libraries of the expected minerals (Clark et al 1990).

Structural and geomorphological mapping

One of the most obvious characteristics of major hydrothermal ore-systems on Earth is their common association with very large structural dislocations, transverse fault systems, ground-preparing shear zones and volcanic edifices. Hardly a single major deposit is not close to or within major structures that can often be very easily mapped on satellite imagery such as Landsat, AVHRR, ERS-1, Radarsat or SPOT imagery. Along many parts of the much mineralized Pacific-rim province in Papua New Guinea, Indonesia, and the Philippines, Corbett & Leach (1995) have identified rifts, major transverse structures and dilatant zones between such structures that host many of the region's major ore deposits (Fig. 2). This sort of spatial and structural mapping is easily done using panchromatic imagery or, in cloud-covered areas, with various synthetic aperture radars (SARs) (Evans 1995). The latest, multi-attribute SARs, with variable frequencies, polarizations and incidence angles, are also proving valuable for studying landform processes in many types of terrain (I. Tapley, personal communication 1996), where mapping and deducing geomorphological processes can be a critical part of planning subsequent geochemical surveys.

Mineralogical or compositional mapping

Advantages of the spectroscopic approach are that one should have enough spectral resolution to sense molecular-level electronic and vibrational properties of materials and thus reduce the data to quantitative physical units that often permit actual identification, not just discrimination. Being calibrated and quantitative also means that results are more repeatable, can be compared with results from different times and places, and can also be unmixed when more than one material is involved (Boardman & Kruse 1994, Huntington & Boardman 1995).

The wavelength regions applicable to mineralogical mapping, and the geological materials that can be detected are listed in Table 1. In the visible and near infrared (VNIR) and shortwave infrared (SWIR) spectral regions we measure the spectral reflectance of materials as a function of wavelength. In the mid or thermal infrared (TIR) we can measure either spectral reflectance or the emission characteristics of materials. For sensing the products of hydrothermal alteration the SWIR spectral region is the most appropriate.

It is fortuitous that, of the minerals that result from hydrothermal alteration and weathering, most can be detected by current spectroscopic methods in the three wavelength regions mentioned above. Thus iron oxides that result from weathering or alteration can be detected in the VNIR with low-cost silicon sensing technology. Clay minerals and sulfates, many bleached and hard to identify by eye, can be detected in the

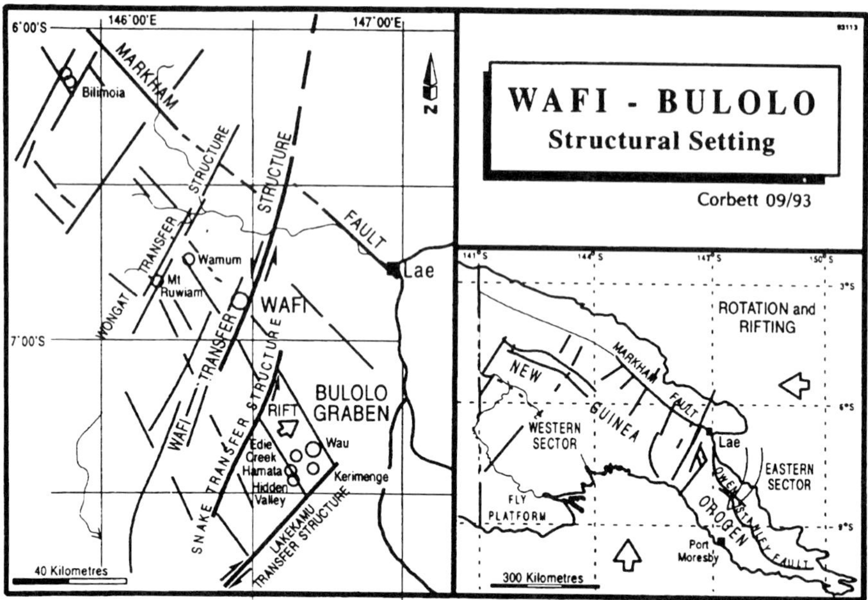

FIG. 2. Major structural features of Papua New Guinea and their association with large scale hydrothermal mineral deposits (from Corbett & Leach 1995).

SWIR (along with many carbonate minerals), while quartz and feldspars can be identified from their Si-O vibrations in the TIR region (Table 1).

Visible and near infrared spectral characteristics

In the VNIR, materials containing Fe^{2+}, Fe^{3+}, Mn, Cr and Ni produce different and identifiable spectra (Fig. 3 and Table 1). Thus maps can be made of the spatial distribution of iron-rich host rocks, lateritic duricrusts or iron species resulting from the oxidation of sulfides. Gossans, for example, in mineralized fault zones in the Pilbara region of Western Australia contain goethite, in preference to hematite, which occurs in the background iron-rich host rocks.

Variations in iron oxide species have also been mapped from different parts of weathering profiles, with hematite more prevalent in stable landscapes and goethite more common in active erosional or depositional environments (Fraser et al 1985). R. Clark (personal communication 1995) has recognized the unique spectral characteristics of ferrihydrite from AVIRIS (airborne visible infrared imaging spectrometer) data in polluted streams draining the Summitville Mining District in Colorado and proposed that this species has been formed by bacterial action.

Aluminium substitution in iron oxides also causes wavelength variation of the 850–950 nm iron oxide crystal field absorption, though no exploration case history of the

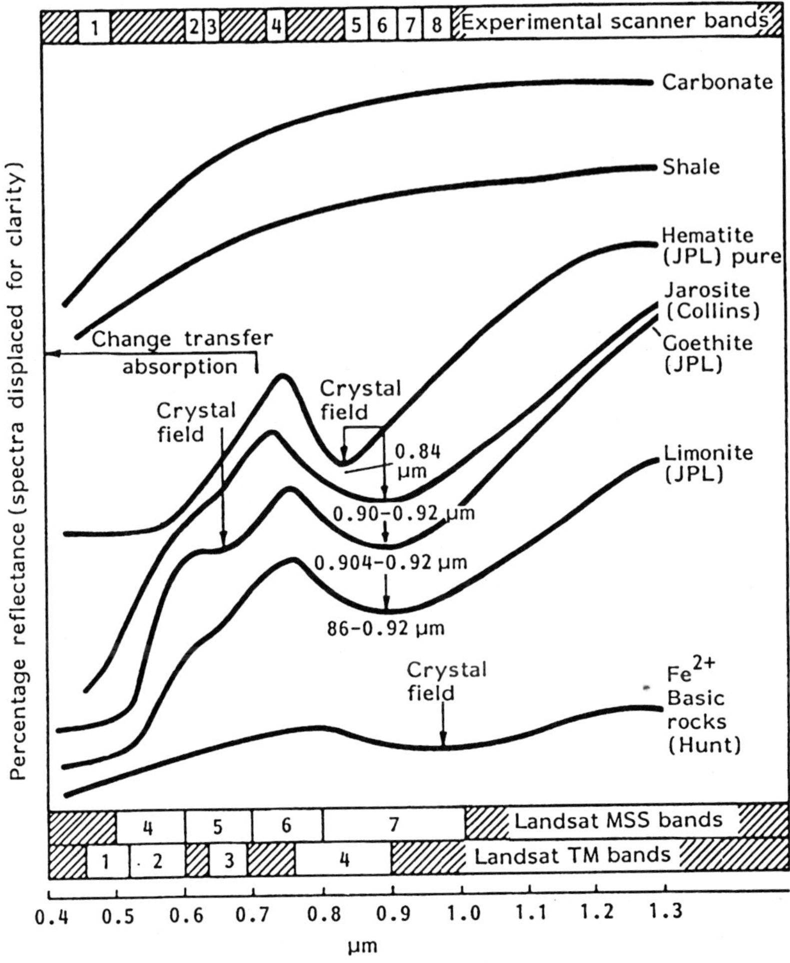

FIG. 3. Typical visible near infrared spectral characteristics of iron oxide-bearing phases.

TABLE 1 Wavelength regions applicable to mineralogical mapping

Name	*Wavelength range (µm)*	*Mineral groups / materials detectable*
Visible and near infrared (VNIR)	0.4–1.0	Iron oxides, REE-bearing minerals, green vegetation
Shortwave infrared (SWIR)	1–2.5	Sulfates, hydroxyl-bearing and carbonate minerals, dry vegetation
Mid or thermal infrared (TIR)	8–12	Silicates (quartz, feldspars, garnets, pyroxenes, olivine), carbonates

use of this characteristic has yet been documented. Such wavelength variation has been noted, however, in bauxite weathering regimes, interpreted by the author to be due to aluminium substitution. The VNIR is also the region of dominant chlorophyll absorption, and the latest unmixing algorithms allow the identification and removal of green vegetation from mixed pixels.

Shortwave infrared spectral characteristics

In the SWIR, the exploration objective is to identify and map the spatial distribution and zones of hydrothermal alteration minerals containing, primarily, hydroxyl groups (Figs 4, 5) and to use this knowledge, along with prevailing exploration models, to target potential mineralized sites and to guide further more detailed and expensive exploration. The technology for doing this in the field and from airborne platforms is relatively straightforward (Tables 2 and 3). Space-borne capabilities in this wavelength region will be tested in 1996 with the LEWIS technology demonstrator satellite (Table 4). Apart from phyllosilicate, and Al(OH)- and Mg(OH)-bearing minerals, many OH-bearing sulfates, ammonium-bearing minerals (Krohn 1986) and the carbonate family also can be mapped with SWIR technology (see Fig. 5).

A major benefit of the quantitative nature of spectroscopic remote sensing is that mixed spectral signatures from pixels containing more than one mineral phase, or a mineral and vegetation, can now often be separated. This means that semi-quantitative mineral abundance maps (Fig. 6) can be produced free of the diluting effects of vegetation. This is currently possible in areas of up to about 50% vegetation provided there is a strong enough mineral signature. Further, it is possible to detect quite small proportions of some materials in the presence of other phases. Thus targets such as veins that are smaller than the pixel size, may be mapped if they contain spectrally contrasting materials.

In addition to mineral identification, the most recent advances have demonstrated widespread and mappable chemical substitutions, involving Na, K, Al, Mg and Fe, revealed in SWIR spectra of micas, alunites, carbonates, chlorites and biotites. Changes in cation composition give rise to measurable shifts in wavelength of the 'normal' spectral absorption features of these minerals (Fig. 7). From these data, new maps of the distribution of, for example, paragonite versus muscovite versus phengite mica composition have been made with both field and airborne spectrometers. Thus new knowledge is coming to light about the spatial distribution and significance of these phases in background rocks, in metamorphic hosts and potassium-enriched alteration zones around mineralization.

When compared with conventional analytical methods such as X-ray diffraction, for some mineral phases field or airborne infrared spectroscopy can yield comparable semi-quantitative results, for a larger number of samples, much faster and more economically (Fig. 8).

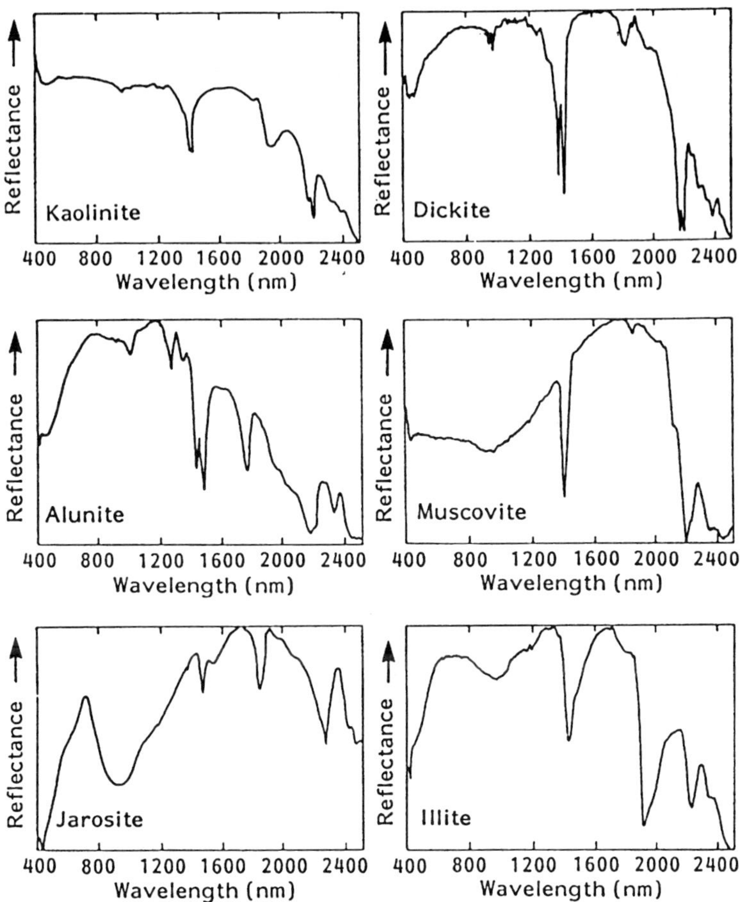

FIG. 4. Typical visible near infrared and shortwave infrared reflectance spectra of selected hydrothermal alteration minerals detectable by field and airborne remote sensing methods.

Thermal-infrared spectral characteristics

The TIR has been the traditional wavelength region used to study the emission from planetary surfaces. However, in the case of exploration for hydrothermal systems, it is not the first and most logical choice. What the thermal infrared is suited to, however, is the sensing of spectral characteristics based on Si-O vibrations in rocks and minerals and thus the discrimination of many hydrated silicate minerals, such as quartz, feldspars, garnets, pyroxenes, olivine, etc. (Fig. 9), which are not so readily sensed in the visible and shortwave infrared. Traditionally, TIR sensing is done passively, measuring the combined (and hard to separate) temperature and emissivity of the surface. New

Micas	**Resistate minerals**	**Carbonates**
Muscovite	Topaz	Calcite
Paragonite	Tourmaline	Dolomite
Phengite	Garnets	Ankerite
Illite	**Chlorites**	Siderite
Biotite	Mg-chlorite	Magnesite
Phlogopite	Fe-chlorite	**Clays (smectites)**
Al(OH) minerals	**Sulfates, arsenates**	Montmorillonite
Pyrophyllite	Alunite	Saponite
Mg(OH) minerals	Jarosite	**Clays (Kandites)**
Epidote	Gypsum	Kaolinite
Talc	Scorodite	Dickite
Antigorite	**Fe oxides**	Halloysite
Amphiboles	Haematite	**Silica**
Tremolite	Goethite	Hydrothermal quartz
Riebeckite	**REE-bearing minerals**	Opaline silica
Actinolite	Neodymium	**Feldspars**
Hornblende	**NH_4-bearing minerals**	**Zeolites**
	Buddingtonite and ammonium illite	

FIG. 5. Selected minerals that can be identified spectroscopically with field and airborne mineral mapping technologies.

methods are now making this easier. Furthermore, people are now experimenting with hyperspectral sensors at these wavelengths (Whitbourn et al 1994) that not only provide more spectral and mineralogical detail, but also make the separation of temperature and emissivity or reflectance even easier.

Operational tools of remote sensing

The real issue with modern remote sensing technology is not so much what can be achieved scientifically (a very great deal), but what can be commercially delivered to those users who need the resulting information. Thus it is worth reviewing the range of instruments that can or soon will be able to contribute to commercial sensing of alteration and weathering mineralogy (Tables 2–4) . Some of the instruments listed are still of limited accessibility, but will either soon be followed by commercial versions or can be accessed by negotiation for limited operational use.

TABLE 2 Selected field portable spectrometer systems potentially capable of contributing significantly to new geological remote sensing knowledge

Instrument	*Country*	*Company*	*Note*
PIMA II	Australia	Integrated Spectronics	Lab-quality SWIR
IRIS	USA	GER Inc	VNIR/SWIR
FieldSpec	USA	Analytical Spectral Devices Inc.	VNIR/SWIR
POSAM	Japan	Dowa Mining	SWIR

Field

Identification of hard-to-recognize and bleached alteration minerals is now commercially and routinely possible with a range of hand-held or field instruments (Table 2). From such instruments spectral libraries are being built to verify the identification of minerals from airborne sensors as well as contributing new knowledge about the mineral species and their chemical variations present in mineralized hydrothermal systems. Because these devices are both relatively cheap and lightweight, very many more samples (hundreds to thousands) can be measured than could previously be studied with X-ray diffraction. Thus our understanding of the spatial patterns of

TABLE 3 Selected airborne systems potentially capable of contributing significantly to new geological remote sensing knowledge

Instrument	*Country*	*Company*	*Note*
TIMS	USA	Daedalus/NASA	6 band thermal infrared
Daedalus 1268	USA	Daedalus Enterprises	12 band multispectral
Geoscan Mark II	Australia	Specterra Systems	24 band multispectral
MIVIS	USA	Daedalus Systems	102 band hyperspectral
AVIRIS	USA	NASA JPL	224 band hyperspectral
SFSI	Canada	Borstaad & Associates	SWIR hyperspectral
HYDICE	USA	US Navy	206 band hyperspectral
TRWIS III	USA	TRW Inc	384 band hyperspectral
GERIS 79, DAIS 7915	USA	GER Inc	79 VNIR/SWIR bands
FIMS	China	Academica Sinica	72 band hyperspectral
MIRACO$_2$LAS	Australia	CSIRO	100 band hyperspectral thermal infrared
OARS/GIMMS	Australia	CSIRO	256 band hyperspectral profiler

TABLE 4 Selected space-borne systems capable of contributing significantly to new geological remote sensing knowledge

Instrument	*Country*	*Date*	*Note*
Current			
Landsat 4–5 TM	USA	1982 (4) 1984(5)	30 m multispectral
SPOT 1–3	France	1990(2) 1995(3)	20 & 10 m multispectral
AVHRR	USA	onwards (NOAA-13 latest)	Regional structure
JERS-1	Japan		SAR and multispectral
ERS-1/2	Europe	1995(2)	SAR—regional structure
RADARSAT	Canada		SAR—regional structure
Future			
Lewis HSI	USA	1996	384 band hyperspectral technology demonstrator
Landsat 7	USA	1997	Multispectral
EOS/MODIS	USA	1998	Multispectral
EOS/ASTER	USA/Japan	1998	Multispectral
SPOT 4/5	France	1997/1999	Multispectral
ARIES-1	Australia	1999	Hyperspectral
Space Imaging Inc.	USA	?	High spatial resolution
Orbimage	USA	?	High spatial resolution
Earth watch	USA	1996 (Earlybird)?	High spatial resolution

hydrothermal alteration minerals at the surface, from underground and in drill cores has been greatly improved. Fig. 8 illustrates just one of many relationships that are now being documented with field-portable spectrometers which is leading to improved understanding of the geology of hydrothermal mineral deposits.

Airborne

As previously discussed, great benefits accrue when new geological information can be gathered from the air. Airborne scanners, imaging and profiling spectrometers and synthetic airborne radars (SARs) are now becoming more common and can be considered when planning exploration or geological surveys (Table 3). For example, NASA's multi-frequency, multi-polarization, multi-incidence-angle AIRSAR sensor is today the most sophisticated SAR system and is just now revealing new insights into regolith

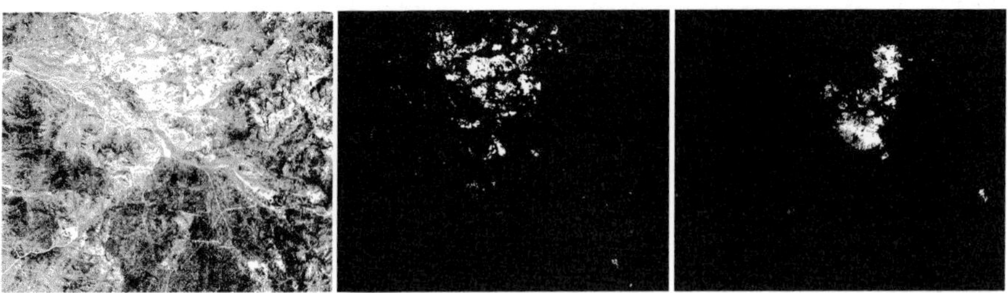

FIG. 6. Unmixed kaolinite (right) and muscovite (centre) abundance images relative to a panchromatic AVIRIS image (left). Abundance: white, 100%; black, 0%.

processes and, in some instances, buried structures resulting from its ground penetration (Evans 1990).

In the mid or thermal infrared region only a few sensors exist that can map hydrothermal silica concentrations and discriminate other lithologies (Kahle & Rowan 1980, Kahle & Goetz 1983, Watson et al 1990). These include NASA's thermal infrared multispectral scanner (TIMS), and CSIRO's mid-infrared CO_2

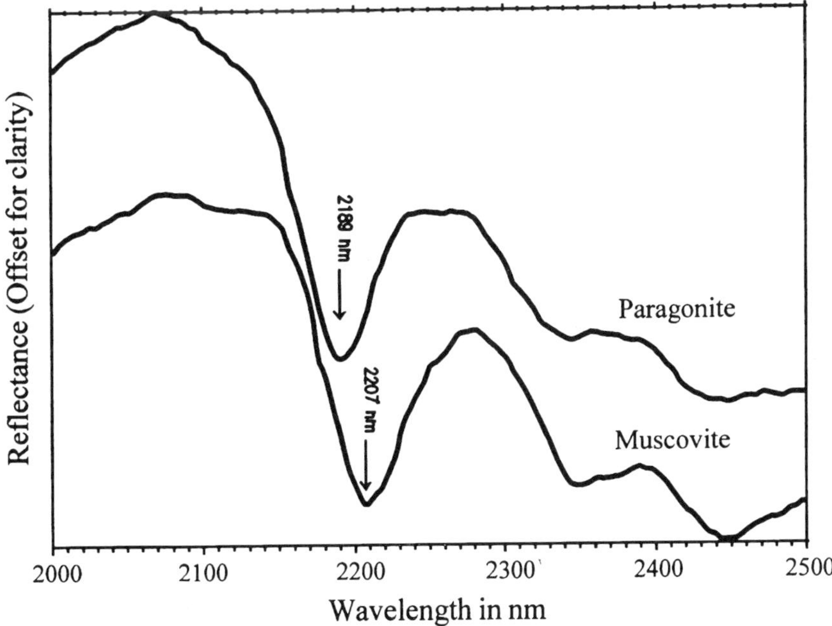

FIG. 7. Shortwave infrared spectra of Na-rich mica (paragonite) and K-rich mica (muscovite) which have been detected from different parts of alteration systems by both field and airborne systems.

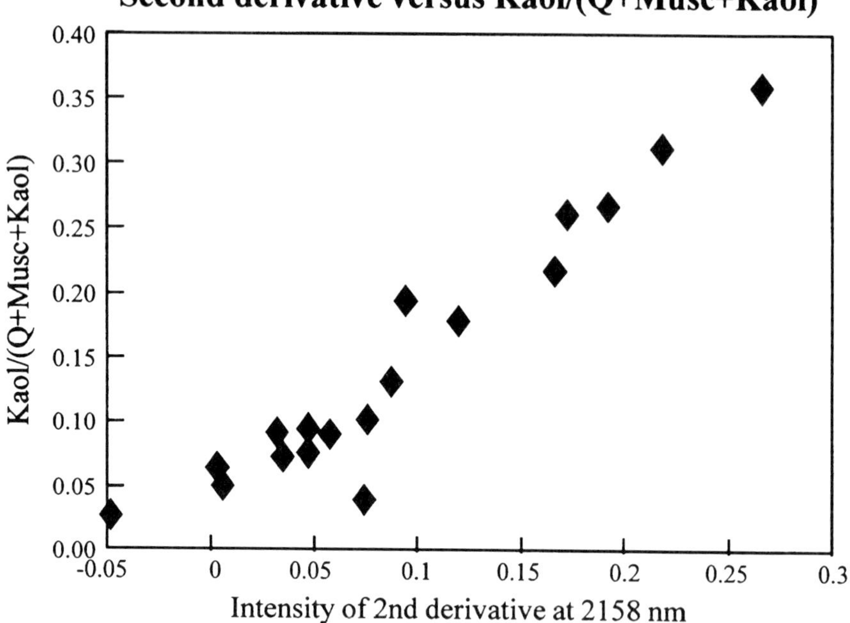

FIG. 8. An example of the quantitative relationship between spectral parameters (x axis) and semi-quantitative XRD (y axis), possible with field-portable shortwave infrared spectrometer systems.

laser spectrometer (MIRACO$_2$LAS), the first active hyperspectral thermal infrared reflectance spectrometer (Whitbourn et al 1994). The latter instrument can discriminate and identify a wider group of silicate minerals — such as quartz, feldspars, clays, garnets, talc, pyroxenes and amphiboles — than existing broadband sensors hampered by the difficulty of separating the temperature and emissivity information about the surface.

The most sophisticated airborne imaging spectrometer for remote and semi-automatic mineralogical identification is NASA's AVIRIS which flies at an altitude of 20 000 m and records 224 spectral bands for each 20 m pixel. With the latest atmospheric correction, end-member selection and unmixing software, it is now possible to make a variety of mineral maps (Fig. 6) depicting hydrothermal mineral species and, in the case of muscovite mica, chemical composition (Boardman & Kruse 1994, Huntington & Boardman 1995). The level of sophistication of these methods is amply illustrated by Fig. 10 which correlates a spectrum of kaolinite taken in the field in about 30 s and one from the same area taken with AVIRIS from 20 000 m in a few thousandths of a second!

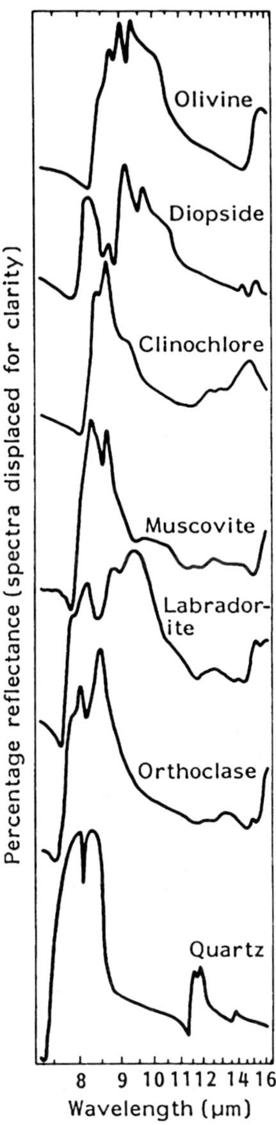

FIG. 9. Typical thermal infrared (TIR) reflectance spectra of selected silicate minerals targeted by TIR remote sensing instruments.

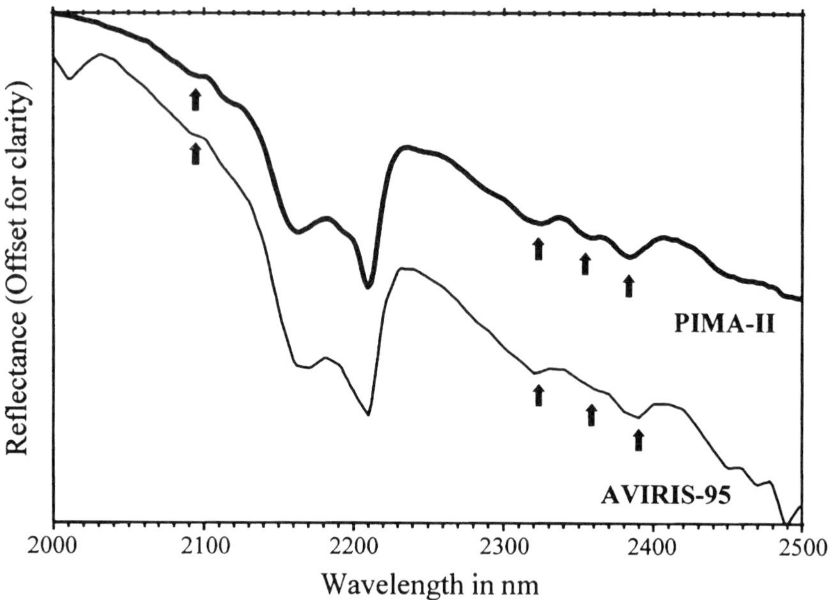

FIG. 10. Shortwave infrared spectra of kaolinite from the same area, taken in the field with a PIMA-II spectrometer and from 20 000 m with the AVIRIS imaging spectrometer (after correction for illumination and atmospheric effects).

Spaceborne

Present operational satellite capabilities to map hydrothermal alteration are limited to the broad bands of the Thematic Mapper sensor on the Landsat satellites. While Landsat is useful for regional reconnaissance it can really only pick up generic clay concentrations (and then normally only the Al[OH] clays) and offers nothing to help locate concentrations of Mg(OH) clays, chlorites, epidote and carbonates common in propylitic alteration systems. This situation may change soon with the launch of higher spectral resolution sensors with a wider range of narrow bands, such as the Lewis hyperspectral instrument (HSI) in 1996, the ASTER satellite in 1998, and the proposed exploration-specific ARIES satellite in 1999. The attraction of these new technologies is that they are all getting smaller, lighter, more sensitive and cheaper to construct and launch.

Martian sensing opportunities

From the foregoing discussion a number of significant opportunities can be envisaged for improved sensing of the Martian surface, during exploration for the signs of previous hydrothermal processes. On unmanned robotic Martian Rovers we can envisage

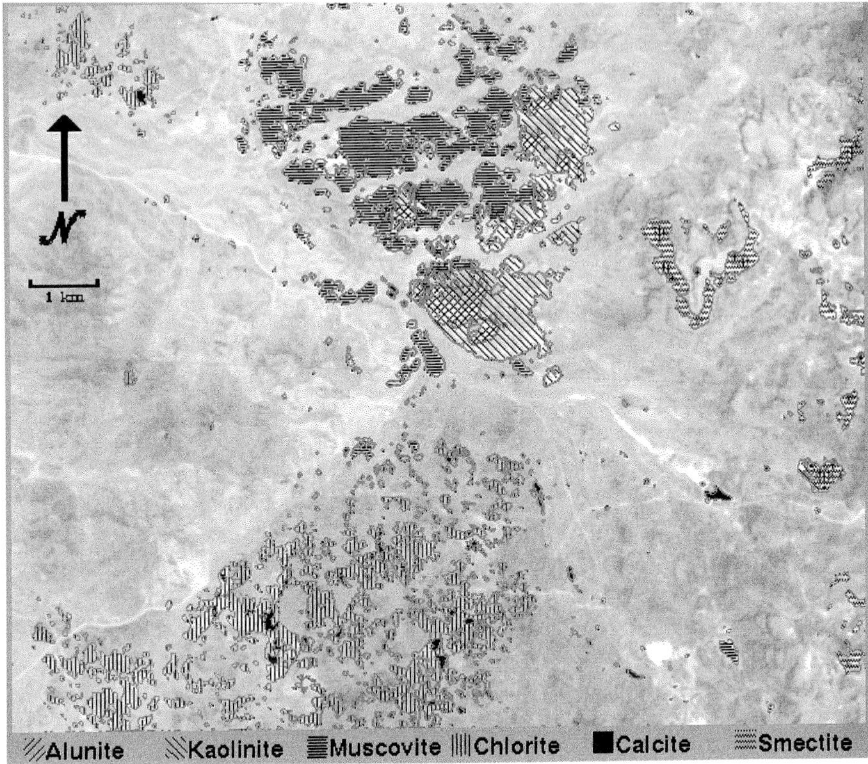

FIG. 11. Semi-automatically generated mineral alteration map of the Oatman, Au/Ag Mining District in Arizona. While advanced argillic, argillic and phyllic alteration zones are defined, mineralization is largely associated with the quartz/calcite veins.

incorporating small PIMA-like grating or Fourier-transform infrared spectrometers (these could be considerable smaller than today's already compact models), to seek out the spectral signatures of hydrous alteration minerals.

Optimal search for signs of hydrothermal products could be carried out with a visible, near infrared and shortwave infrared (400–2500 nm) imaging spectrometer to search places where the Martian surface has been disrupted by erosion or volcanic activity (thereby exposing older possibly altered materials). The ideal way to do this would be with an orbiting imaging system. However, if time is not a critical driver, a hyperspectral profiling system that can progressively build-up an interpolated map of the surface would allow one to construct a much smaller, lighter, cheaper and more sensitive spectrometer with an attendant lower, manageable data-rate. Such a system is currently being built for Earth-based mineral exploration to take advantage of existing geophysical survey aircraft flights and to reduce the costs of the data to the user.

Multi-attribute SARs of the AIRSAR type could also generate important new information about the Martian regolith and the geomorphological and tectonic processes that have acted on it.

Conclusions

Today's geological remote sensing is evolving from the broad-band sensing of spatial features, such as structure and geomorphology and limited lithological contrasts (used in the 1970s and early 1980s), into the use of non-invasive spectroscopy to identify and better discriminate the composition of geological materials, primarily alteration and rock-forming minerals and their weathering products. Thus the mining and exploration community now has the opportunity to increase the effectiveness of its exploration, add significant new knowledge and produce more information-rich geological maps (exemplified by Fig. 11) using operational mineralogical mapping based on the principles of reflectance spectroscopy.

For exploring for the hydroxyl-bearing clays, sulfates and carbonate minerals that occur in and typify hydrothermal alteration systems, the SWIR region of the electromagnetic spectrum (1000–2500 nm), not TIR, is the simplest, cheapest and most cost-effective part of the spectrum to concentrate on, from both information content and technological perspectives. Translating current scientific knowledge of how to do this into commercial practice on Earth will potentially lead to radical changes in geological mapping and exploration efficiency. There seem to be few scientific impediments to why this could not also be carried out on other planets, given a strong enough motive.

Acknowledgements

Much of the research drawn on in this paper has resulted from over twenty years of industry-sponsored research conducted through the Australian Minerals Industry Research Association Ltd (AMIRA) who, along with many companies, are gratefully acknowledged. Greg Corbett and Terry Leach are also gratefully acknowledged for permission to reproduce Fig. 2. Dr Fred Kruse (AIG) kindly provided Fig. 1 and Dr Ian Tapley, (CSIRO Division of Exploration and Mining) provided the most up-to-date information on AIRSAR data capabilities. Many of the recent AVIRIS results resulted from collaborative research with my colleague Dr Joe Boardman of AIG, in Boulder Colorado, who I also thank. Responsibility for the interpretations, opinions and compilation is, however, my own.

References

Berger BR 1986 Descriptive model of hot-spring Au-Ag. In: Cox DP, Singer DA (eds) Mineral deposit models. US Geol Surv Bull 1693:143–144

Berger BR 1991 A historical perspective on the nature and genesis of epithermal gold-silver deposits. Econ Geol Monogr 8:249–263

Boardman JW, Kruse FA 1994 Automated spectral analysis: a geological example using AVIRIS data, northern Grapevine Mountains, Nevada. Proceedings 10th Thematic Conference on geological remote sensing, San Antonio, 9–12 May 1994, p 407–418

Clark RN, King TV, Klejwa M, Swayze GA 1990 High spectral resolution reflectance spectroscopy of minerals. J Geophys Res 95:12653B–12680B

Corbett GJ, Leach TM 1995 Southwest Pacific Rim gold/copper systems: Structure, Alteration and Mineralization workshop proceedings, Townsville, 20–21 May. Corbett Geological Services, Sydney

Erickson RL 1982 Characteristics of mineral deposit occurrences. US Geol Survey Open-File Report 820795

Evans DL (ed) 1995 Spaceborne synthetic aperture radar: current status and future directions. A report to the Committee on Earth Science, Space Studies Board, National Research Council. NASA Technical Memorandum No. 4679

Fraser SJ, Huntington JF, Green AA, Stacey MR, Roberts GP 1985 Discrimination of iron oxides and vegetation anomalies with the MEIS-II narrow-band imaging system. Proceedings 4th ERIM thematic conference on Remote Sensing for Mineral Exploration, San Francisco, April 1–4 1985, p 233–253

Goetz AFH 1989 Spectral remote sensing in geology. In: Asar A (ed) Theory and applications of optical remote sensing. Wiley, Chichester

Huntington JF, Boardman JW 1995 Semi-quantitative mineralogical and geological mapping with 1995 AVIRIS data. Proceedings 3rd international symposium on Spectral Sensing Research 95 (ISSSR), Melbourne, 26 Nov–1 Dec 1995, in press

Johns DA, Thomason RE, McKee EH 1989 Geology and K-Ar geochronology of the Paradise Peak mine and the relationship of pre-Basin and Range extension to early Miocene precious metal mineralization in west central Nevada. Econ Geol 84:631–640

Johns DA, Nash JT, Clark CW, Wultange WH 1990 Geology hydrothermal alteration and mineralisation at the Paradise Peak Gold-silver-mercury deposit Nye County Nevada. In: Program with abstracts, geology and ore deposits of the Great Basin. Geological Society of Nevada, Reno, NV, p 81–82

Kahle AB, Goetz AFH 1983 Mineralogic information from a new airborne thermal infrared multispectral scanner. Science 222:24–27

Kahle AB, Rowan LC 1980 Evaluation of multispectral middle infrared aircraft images for lithological mapping in the east Tintic Mountains, Utah. Geology 8:234–239

Krohn MD 1986 Spectral properties (0.4–25 microns) of selected rocks associated with disseminated gold and silver deposits in Nevada and Idaho. J Geophys Res 91:767B–783B

Kruse FA 1988 Use of Airborne Imaging Spectrometer data to map minerals associated with hydrothermally altered rocks in the northern Grapevine Mountains, Nevada and California. Remote Sens Environ 24:31–51

Kruse FA, Kierein-Young KS, Boardman JW 1990 Mineral mapping at Cuprite Nevada with a 63 channel imaging spectrometer. Photogram Eng Remote Sens 56:83–92

Watson K, Kruse FA, Hummer-Miller S 1990 Thermal infrared exploration in the Carlin trend, northern Nevada. Geophysics 55:70–79

Whitbourn LB, Hausknecht P, Huntington JF, Connor PM, Cudahy TJ, Phillips RN 1994 Airborne CO_2 laser remote sensing system. Proceedings of the 1st international Airborne Remote Sensing conference and exhibition: Applications Technology and Science, Strasbourg, September 12–15, p II-94–II-103

DISCUSSION

Jakosky: Let me make a couple of comments on the application of remote sensing techniques to Mars. With Mars you get your wish of starting at the biggest scale and working your way down. People have been looking at telescopic spectra in the visible and near-infrared for 20 years. More recently, there have been a limited amount of reflectance spectra from orbit via the Russian Phobos spacecraft. When Mars Global Surveyor goes into orbit around Mars, we will get 3 km thermal emission spectra. Then we have a lander in two years that will have not full, high resolution spectra, but enough points to give some spectral information. The data analysis has been notoriously problematic. The reflectance spectra don't resemble anything. Part of this may be the spatial resolution where we are looking at spots hundreds of kilometres across.

Huntington: I suspect that the low spatial resolution (large pixel size) has a lot to do with these poor results, but I also suspect that the data analysis methods are far from optimum. For example, how many spectral channels were used in these measurements?

Jakosky: Comparable to NASA's AVIRIS, which has 224 spectral channels between 400 and 2500 nm.

Huntington: Not at thermal emission wavelengths, though?

Jakosky: No, just in the visible and near-infrared.

Huntington: One of the problems with a traditional thermal emission measurement is the difference between the emissivity of a mineral or a rock and its temperature. For several decades people have found it incredibly difficult to separate these two properties, and we are only now figuring out how to do it better. One answer is to use a large number of spectral channels (>20) to better deconvolve the emissivity and black-body curves. As a result, we are now able to make much better emissivity images and generate better emission spectra than in the past.

I acknowledge that a great deal has been done with thermal emission spectroscopy but, if we are to seek evidence of hydrothermal alteration, then we should be looking for the sulfate, carbonate and hydroxyl-bearing minerals that result from hydrothermal processes. These are most easily sensed in the near or shortwave infrared region between 1 and 2.5 μm. They are not easily sensed in the 8–14 μm thermal or mid-infrared region, for the reasons Bruce Jakosky has referred to. The thermal emission wavelengths are vital, however, for mapping the silicate minerals, the feldspars, pyroxenes and non-hydrothermal quartz, etc., since these are not distinguishable in the near or shortwave infrared when fresh.

Jakosky: There is some recent work in the thermal infrared that has 100–150 channels. As you get down to higher spatial resolution, people expect that the problem will go away and we'll begin to identify specific minerals. Whether that's at 1 km, 100 m or 10 m isn't clear. I think there are going to be problems down to even 10 m, because there's so much dust in the atmosphere and on the surface. Even when you get down to a 10 m or 5 m scale, 50–90% of your pixel might be filled with dust, and only a small part might be whatever is poking through the dust from the underlying surface. I'm not saying it's not going to work, but it hasn't been proven yet at the spatial scale of interest here, and at the spectral and compositional scale of identifying specific minerals and their abundance in a given location.

Huntington: Have shortwave infrared spectral reflectance measurements been made of the surface of Mars, between 2 and 2.5 μm, at say less than 100 m spatial resolution? If you are looking for all hydrothermal systems, all those hydroxyl-bearing minerals that

I listed have really interesting and busy reflectance spectra that allow one to identify what mineral species are present.

Jakosky: On the short side of 2 μm there's CO_2 in the atmosphere that absorbs. On the long side they have seen down to very low numbers and there is zero evidence of clay mineral absorption features. There is an absorption feature at 2.36 μm that is a combination of atmospheric CO_2 and the mineral scapolite.

Huntington: We have the technology; there is no question about that. The problem is whether these hydrothermal materials are present, and if they are, whether they are exposed or covered. We have problems on Earth with weathering rinds and wind-blown carbonate dust. Clay coatings and desert varnish can completely change the spectral signatures of rocks.

Jakosky: The other problem is that on Mars we don't have the luxury of being able to go down and pound on a rock with a hammer to see what we really have. At best, we'll get a couple of landings over the next decade that might yield minerological information. I'm not sure how fatal that's going to be.

Huntington: Exploration is similar in that sense, because in mineral exploration one does not want to or cannot go there before the event. If remote sensing has to be driven only by foreknowledge, then it doesn't have a future. The whole point about remote sensing is to use spectroscopy to identify what is present *without* having to go there, and, since we are dealing with molecular-level properties, to do it with a high degree of certainty. Otherwise it adds little commercial advantage.

Farmer: Perhaps these problems will go away with better spatial resolution infrared mapping, but where in the Mars Global Surveyor Program are we going to get that? I don't see higher resolution orbital instruments on the horizon and that worries me. We need the spatial resolution in the infrared remote sensing to find out what we have to deal with. We in the Mars community need to bring that pressure to bear on upcoming missions.

Carr: That's not quite true. The first steps are being taken. On the Russian mission Mars-96 we are going to have a visible and near-infrared spectrometer, and Mars Global Surveyor will carry a thermal emission spectrometer. The resolutions are not what we would like to detect hydrothermal deposits; nevertheless they're pretty good. We'll know after those missions, only two or three years from now, which technique is better suited.

Jakosky: It is not going to be possible to use imaging spectroscopy to map mineral abundances on Mars because of all these problems, but the issue of identifying them is a separate one — there, 100 m resolution with the types of instruments you have been talking about is probably feasible.

Huntington: A lot of it comes down to money, and we're also faced with that constraint in the exploration business. There's this terrible trade-off between making images *and* having detailed spectral information in many channels, and it is nigh on impossible to combine the two for a reasonable budget. But there is another alternative — and this is what the mining industry does all the time — and that is to take single-pixel measurements at low altitudes along single-pixel flight lines, without bothering to scan full images. These are made up later by interpolating or gridding the flight line data. We think that a major opportunity spectroscopy on Earth is to take advantage of the existing geophysical contracting industry and its infrastructure to gradually build up these single images that are only one pixel wide and then construct the interpolated images. The instruments used to do this can be much smaller, considerably cheaper and dramatically more sensitive. If you want to identify a weak mineralogical signal you can use this technique because you're not spending time building-up very detailed spatial images of the entire planet. It is an approach worth considering.

Shock: One reason remote sensing is successful on Earth is that we are seeing deeply into these systems because of uplift and erosion. It is not likely that there has been the same type of erosion on Mars. This might influence where people look on Mars. It might be better to look into the big canyon structures and larger impact craters to see if they have unearthed anything that would have been the subsurface hydrothermal system that wouldn't have the same kind of surface expression on the Earth.

Huntington: Mike Carr, in these turbulent-looking materials at the bottom of the Martian channels where everything has been mixed up, is it possible that the channels have gone deep enough to expose the underlying fresh rocks?

Carr: The flood channels are up to 2 km deep so they have excavated materials to that depth. To sample deeper you would have to look at impact ejecta or the walls of the canyons, which appear to be largely tectonic features.

Horn: Alteration minerals may be recognized in the sediments, too.

Shock: You should focus on those things and ignore the rest of the surface of the planet.

Walter: Isn't the problem with the Martian dust similar to the problem of working on Earth and stripping away the effects of vegetation and soil? Intuitively it would seem to me that the problems on Earth are even greater, but we deal with them successfully.

Huntington: Presumably the Martian Rover will give us a good definition of what this dust is, how thick it is and how much there is.

Jakosky: I think you are at the risk of solving the problem too easily because none of you are Martian spectroscopists. If this was a room filled with Mars spectroscopists they would be sitting here with their mouths agape trying to figure out where you have been for the last 20 years!

One other comment: what Everett Shock has suggested is exactly what Jack Farmer wants to do — high resolution, imaging spectroscopy of very specific locations that are candidate hydrothermal systems.

Farmer: I am sure Mike Carr has a list of thousands of these places.

Carr: As you know, we haven't done that exercise yet. But, in the next three years we are going to learn such a lot about the surface and its reflective properties — and the extent to which we can do the mineralogy. The dust problem is not necessarily universal. We know that the wind moves the dust around, and the albedo patterns change. There may be areas that are free of dust. We know that there are windswept areas in the lee of craters. But then there's the other problem of the weathering rind. Is there some alteration that's just a few microns thick that completely destroys our ability in the near infrared to discriminate between the different mineralogy? We don't know yet.

Farmer: The planet has gravity, things move downhill, they break, they fall open, they go through temperature cycling, and you get fresh rock exposed on a regular basis: it's just a question of looking in the right place.

Jakosky: We need to get there and find out with some of these instruments.

Exploration strategies for hydrothermal deposits

Robert A. Horn

INCO Limited, 2060 Flavelle Boulevard, Mississauga, Ontario, Canada L5K 1Z9

Abstract. With unlimited money the most certain strategy for finding most hydrothermal metal deposits would be by drilling to 5000 m at 50 m spacing. However, the cost would far outweigh the benefit of the discoveries. Geological knowledge and exploration techniques may be used to obtain the greatest benefit for minimum cost, and to concentrate human and material resources in the most economic way in areas with the highest probability of discovery. This paper reviews the economic theory of exploration based on expected value, and the application of geological concepts and exploration techniques to exploration for hydrothermal deposits. Exploration techniques for hydrothermal systems on Mars would include geochemistry and particularly passive geophysical methods.

1996 Evolution of hydrothermal ecosystems on Earth (and Mars?). Wiley, Chichester (Ciba Foundation Symposium 202) p 236–248

Value of exploration

The value of the total contained metal in the world's major sedex deposits is US$245 000 million, in volcanogenic massive sulfides $312 000 million and in major porphyry deposits $1 200 000 million; the gold deposits of the Canadian Superior Province alone have a contained value of $48 000 million (estimates are based on ore reserves quoted by Bertoni 1983, Sillitoe 1990, Goodfellow et al 1993 and Gilmour 1982). To place these astronomical numbers in the context of world economies, the total contained metal value of deposits formed by hydrothermal processes is around twice the annual GDP of the United Kingdom or four times that of Canada.

A normal business concept is that of expected value (EV) which is the product of the probability of a certain financial outcome and its value. A new business, if successful, may be worth $100 million, but only have a 50% chance of succeeding. Its EV is therefore $50 million. The same principle applies in exploration, although the mathematical probability of success — and therefore the EV — is uncertain. Nevertheless, EV may influence the exploration strategy.

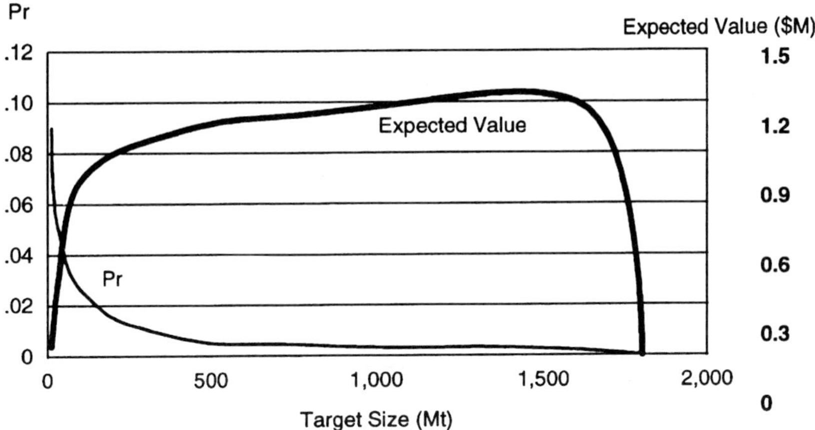

FIG. 1. Risk–reward curve, sedimentary copper exploration, plotting target size (million tonnes) of the ore deposit against the mathematical probability of success (Pr) and the expected value (after Horn & Telfer 1994).

If one or more ore bodies exist in an area under exploration, then it is a fair assumption, knowing nothing of any other influencing factors, that the chances of a discovery is a function of their frequency. The more deposits there are, the more likely it is that one will be found. Large ore bodies are very much rarer than small ones. The frequency distribution of ore deposit size is therefore skewed, usually in an approximately lognormal fashion. If the value of a deposit is a function of its size, which is again broadly true, then the frequency distribution of EV may be calculated. For all deposits studied by Horn & Telfer (1994), even without economies of scale, EV increases with the size of the deposit, which would indicate that theoretically it is a more profitable business to explore for large ore bodies than for small ones (Fig. 1). This conclusion is generally true also in the real world.

The miner has a different perspective on hydrothermal deposits from that of the scientist because exploration strategies assess economic and social as well as physical reality. But where they have a common interest is in increasing the fundamental understanding of hydrothermal processes—for the miner with the object of reducing the business risk and thereby to obtain a higher financial return on exploration investment, and for the academic to earn wealth of a higher order in academic recognition. The most significant areas of risk, and also the most difficult to quantify, are technical—the lack of a full understanding of the nature and mode of formation of ore accumulations, the interpretation of geology, and the ambiguities of the data from exploration techniques and procedures.

Ore deposit models and processes

Pirajno (1992) has classified hydrothermal deposits as follows:

Plutonic related, magmatic (Sn-W greisens).
Volcanic-plutonic to subvolcanic, magmatic to meteoric (porphyry Cu [Mo/Au] epithermal Au).
Sub-sea floor (volcanogenic massive sulfides).
Rifts and sedimentary basins (sedex Pb-Zn-Ag).
Basinal-diagenetic (Mississippi Valley type Pb-Zn).
Metamorphic (Archaean lode Au).

This classification contains almost all the world's economic metal deposits apart from the magmatic types such as the sulfide nickel ore bodies, platinum group metals in layered intrusions, mafic-hosted titanium and iron deposits and paleoplacers. Sedimentary copper deposits should logically be included with sedex Pb-Zn in rifts and sedimentary basin. The rare platinum group occurrences in hydrothermal veins (Watkinson & Melling 1992) may belong with the epithermal gold deposits. The economically important vein-unconformity uranium deposits are not readily classifiable in any of these categories.

A common practice among mining companies is to refer to ore deposit 'models' which attempt to incorporate the common geological features of a group of broadly similar ore deposits. The model is supposed to focus exploration on the most likely geological environment and to eliminate those with a lower probability to contain ore deposits from the search. Models have developed over the years in a haphazard manner, influenced to a large extent by studies and publications on particular deposits or mineral districts which often do not have global application. The concept of the deposit model has also created a tendency among economic geologists to regard ore types as specific domains with their own distinct characteristics rather than recognize the continuity of ore forming systems and the infinite possible variations of ensuing deposit types. In the planning of an exploration program, if the critical geological parameters are determined by the attributes of a particular district or mine they may be too restrictive and the risk of failure may be consequently increased because settings which are also suitable for the deposition of ore may be eliminated from the search.

If, as opposed to identifying the features of particular deposits, the entire hydrothermal system is well understood, then vectors may be derived, such as changing hydrothermal alteration style or distance from a hydrothermal vent, which may indicate the most probable direction for a particular ore type. Extrapolating the dynamic processes of ore deposition may also identify new mineralized geological settings not predictable from the limited possibilities of a static model.

That some apparently different ore deposits are deposited from the same type of hydrothermal system has been recognized for many years, while others appear to be similarly related to each other but have yet to be rigorously researched to determine that they are truly variations on the same theme. Examples of these cases are the well-established porphyry to epithermal system and the more speculative complex of the Kiruna type ironstone to sedimentary copper.

Lindgren (1913) recognized the vertical zonation of hydrothermal deposits generated by porphyry intrusions from deep high-temperature hypothermal mineralization through mesothermal veins up to the epithermal gold systems. The copper porphyries, the base and precious metal veins such as those at the Magma mine, Arizona, and the sediment-hosted gold deposits of the western USA Great Basin are all part of the same type of system, varying only in their depth and temperature of formation. Bingham Canyon, Utah, one of the world's largest porphyry copper deposits, which has been in production since the beginning of the century, appears related to the Barney's Canyon epithermal gold deposit which was discovered only in the 1980s. A similar relationship exists between the Ruth ore body and the gold deposits mined on its margins after the closure of the copper mine. Further to the north in Nevada, on the Carlin trend, high level epithermal gold deposits indicate the presence of a major deep-lying porphyry, detectable by magnetometry, situated to the south-east of the trend near the Gold Quarry deposit (Hildenbrand & Kucks 1988).

The Kiruna iron ore body in Northern Sweden has a magmatic component (Geijer 1931) which chokes the vent through which the hydrothermal metals were transported to be deposited on the sea floor as sedimentary ironstones (Parak 1975). It contains anomalous copper, zinc, lead and gold. The Viscaria copper deposit, hosted by sediments and basalts, is located so close to the iron mine that the two may be accessed from the same shaft. The conclusion that they are genetically related is difficult to avoid. Kiruna is, in many respects, similar to Olympic Dam in South Australia — an ironstone deposit but with economically recoverable values of copper, uranium and gold. Reeve et al (1990) have described the interpreted stages in the geological evolution of this ore body. Hydrothermal activity began with extensive brecciation of a Proterozoic granite, followed by alteration, metal mineralization and magnetite deposition and finally by shallow subsurface brecciation and associated hydrothermal eruptions which produced a composite hydrothermal crater with an ejecta apron. Ore minerals precipitated preferentially in hematite-altered breccia. Diatremes developed where phreatomagmatic activity was most pronounced. Fumarolic water with dissolved iron and barium was discharged into maar lakes and the metals precipitated with epiclastic sediments.

In the opinion of the writer, the Kalengwa mine in western Zambia is geologically very similar to Olympic Dam in its copper mineralization, its relationship to a maar lake, and the style of hydrothermal alteration. It seems likely that Kiruna, Olympic Dam and Kalengwa are related ore types formed from similar hydrothermal systems. If the mineralizing fluids had encountered shallow marine sediments rather than a maar lake then copper deposits of the Copperbelt or Kupferschiefer type could have been deposited. Lefebvre (1989) describes epigenetic copper mineralization at Kinsenda in Shaba southeast Zaire related to ferruginous conglomerates similar to Kalengwa. He proposes that the copper replaced pre-existing pyrite in a manner similar to that proposed for the Polish Kupferschiefer (Putmann et al 1987), the McArthur sedex zinc deposit in Queensland (Eldridge et al 1985) and for the copper ore at Mount Isa (Bell et al 1988). Annels (1989) has proposed a similar epigenetic origin for the Chambishi deposit in Zambia. If Kinsenda and Chambishi are epigenetic, then the synsedimentary origin of the entire Zambian–Zairian Copper belt is questionable. Olympic Dam and Kiruna may be parts of a hydrothermal system close to the intrusive and hydothermal vent with Viscaria, Chambishi and Kinsenda in distal marine sediments. Kalengwa and parts of Olympic Dam would be intermediate in their situation in maar lakes above the intrusion.

Exploration techniques

As a business principle, the cost of exploration must not exceed the EV of the project. That is to say the investment should not exceed the expected return. As a project progresses and the probability of success increases, then the EV also increases and greater exploration costs become economically justifiable. A porphyry copper deposit of 500 million tonnes may have a net present value of $300 million. If the probability of its discovery in the early phase of exploration is 0.001, then the EV, which is the maximum justifiable exploration cost at this stage, is $300 000. After evidence of mineralizing processes have been found the probability may improve to 0.01 with a consequent EV of $3 million, and at the drilling stage it may increase to 0.1, when the economically justifiable exploration costs would be $30 million. Techniques employed in exploration therefore range from relatively cheap regional geological, geophysical and geochemical surveys to more costly detailed investigations over restricted areas and finally diamond drilling.

Mineralogy and geochemistry

Modern exploration practice tends to be concerned less with the sources of water, metals and their ligands than with recognizable alteration which is the indirect evidence that hydrothermal fluids have passed through a rock. Hydrothermal alteration assemblages, or their metamorphic equivalents, have been used by exploration geologists as a field guide to mineralization for many years (Schwartz 1955, Roper 1976, Lydon 1988). Alteration patterns around ore deposits were probably recognized as far back as the Roman gold miners at the beginning of the first millennium. In more recent times, Lindgren (1926) documented the relationship of chlorite to ore at the United Verde mine in the Jerome district of Arizona, and Coats (1940) described for the first time propylitization related to silver gold mineralization in the Comstock district of Nevada.

Lowell & Guilbert's (1970) famous description of alteration around the San Manuel–Kalamazoo porphyry deposit introduced a systematic means of analysing the relative position of altered rocks based on the observed distribution of early potassic alteration at the core passing outward to younger phyllic, argillic and propylitic zones. Variations on this basic pattern have been recognised from many porphyry deposits. The mapping of alteration assemblages is now a standard technique in porphyry exploration.

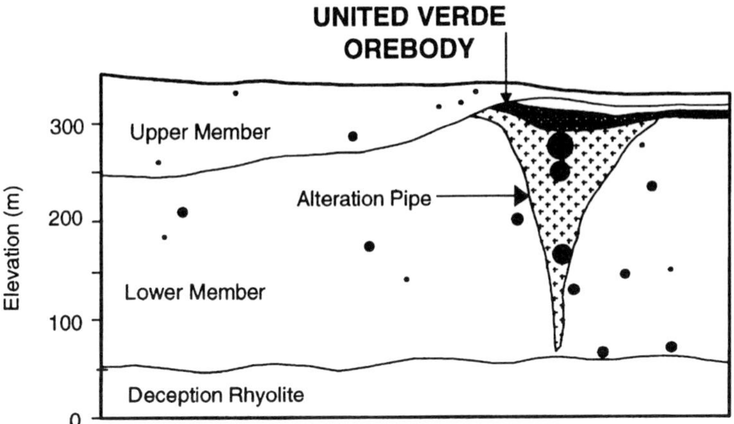

FIG. 2. United Verde Volcanogenic massive sulfide. Fe/Zr Pearce element ratios. (Stanley & Madeisky 1994).

Decalcification and silicification of siltstones in the epithermal gold deposits of the Great Basin of the USA (Coats 1940), dolomitization of limestones around the lead–zinc deposits of the Reocin district of Spain, tourmalinization at the Sullivan sedex lead–zinc mine, and chlorite footwall alteration in polymetallic deposits in Canadian shield are examples of characteristic alteration mineralogies easily recognizable in the field.

Chemical analysis is another means of measuring the type and degree of more subtle alteration due to the passage of mineralizing hydrothermal fluids. At Crandon, for example, interaction of ore fluid with wall rock has resulted in enrichment in SiO_2, Fe, K, F, S, Cu, Zn, As, Sb, Ba, Au, Hg, Pb, Bi, Se and Cd, and depletion in Al, Mg, Ca, Na, V and Sr (Lambe & Rowe 1987). The losses and gains in the chemistry of rocks through which hydrothermal solutions have passed may be measured by various alteration indices (Anderson 1968, Dube et al 1987), among which the Pearce element ratio (PER) in which the molar concentration of hydrothermally mobile elements are expressed as ratios with immobile elements, is one of the most effective in detecting small changes in the concentration of major elements (Fig. 2). The assumption is made that the rocks analysed are derived from a comagmatic suite. (Stanley & Madeisky 1994).

Geophysics and remote sensing

The physical properties of the magmatic intrusions which drive hydrothermal systems, changes in the properties of the hydrothermally altered rocks and the hydrothermal metallic minerals themselves may all be detected by satellite or airborne spectral scanning in the visible and infrared wavelengths, and ground or airborne geophysics. Table 1 shows the physical parameters measured by different geophysical techniques.

TABLE 1 Hydrothermal deposits: geophysical parameters

Method	*Measured parameter*	*Location of instrument*
IP	Electrical conductivity, chargeability	Ground
AMT	Electrical conductivity	Ground
EM, VLF	Electrical conductivity	Ground, airborne
Gravimetry	Density	Ground, ?airborne
Magnetometry	Magnetic susceptibility	Ground, airborne
α-Radiometry	α-Radiation	Ground
γ-Spectrometry	γ-Radiation	Ground, airborne
Seismic	Seismic velocity	Ground

Base metal sulfide minerals, which are economically the most significant component of hydrothermal deposits, may be detected by induced polarization (IP) and electromagnetic (EM) techniques because they are electrical conductors and are chargeable. In IP an electrical current is passed into the ground and the polarization induced in certain minerals, such as pyrite, is recorded. EM involves an electromagnetic transmitter and a receiver which detects eddies in the electromagnetic field caused by conductive bodies such as some massive sulfides. VLF (very low frequency) uses submarine communication signals from naval bases around the world as a source. The audio magnetotelluric (AMT) method also measures the affect of electromagnetic radiation, but from natural sources — cosmic or terrestrial rather than from a transmitter. Magnetometry and gravimetry record, respectively, magnetic response and density. Alpha radiometry detects α-particles usually emitted from decaying radon gas. Gamma spectrometry measures the spectrum of γ-ray emissions from which it may be determined whether the source is the unstable isotope of potassium, uranium or thorium.

Epithermal gold deposits in the Great Basin of Nevada are often concentrated near zones of anomalously high magnetism many of which are known to be caused by magnetite in granitic intrusions (Hildenbrand & Kucks 1988); the high density of which may also give a gravimetric anomaly. The presence of other conductive hydrothermal minerals such as clays, if they are present in sufficient quantities, may reduce the bulk electrical resistance of the rock, allowing the alteration zone to be mapped by IP, EM or AMT. In some deposits such as Lupin in the Northwest Territory of Canada, brittle fracturing of iron formations and the consequent increase in permeability may provide access for sulfidic gold-bearing hydrothermal fluids which sulfidize magnetite to produce pyrite and at the same time precipitate gold. The geophysical pattern over such a deposit may show a reduction in the magnetic response and a possible IP or EM conductive anomaly over the sulfides (Reed 1991).

Different geophysical techniques should be integrated with geology and geochemistry. Les Mines Selbaie in Quebec, a polymetallic, epigenetic, hydrothermal deposit

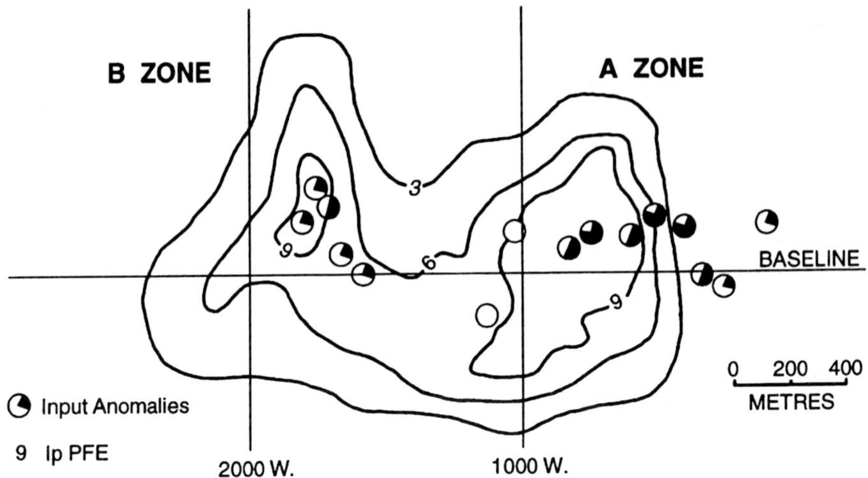

FIG. 3. Les Mines Selbaie, Quebec. Induced polarization (Ip) and input electromagnetic anomalies. (Reed 1991.)

containing some 50 million tonnes of mineralized rock, was mapped using a variety of geophysical techniques. The mineralogy consisted of the metallic sulfides, pyrite, chalcopyrite, sphalerite and galena, together with sericite, chlorite and other silicates. Airborne time-domain EM located the conductive massive pyrite towards the centre of the alteration zone. IP showed the disseminated pyrite in a halo around the focus of the system (Fig. 3). The increase in bulk rock density due to the addition of copper, lead and zinc is measurable by gravimetry. The Hope Brook gold deposit in Newfoundland is a 5 km-long structurally controlled zone of silicification and pyritization cutting a succession of volcanic and sedimentary deposits. Alumina alteration minerals include andalusite, pyrophyllite, kaolinite and sericite (McKenzie 1986). An IP survey identified the silicification zone by its high resistivity. The ore responds as weakly anomalous lower resistivity with an IP chargeability anomaly from the disseminated pyrite (Reed 1991).

In recent years, down-hole transient EM has been used effectively to seek conductive pyritic bodies in the vicinity of drill holes. AMT's is a technique which uses naturally occurring electromagnetic radiation. It has a depth penetration far deeper than the 500 m maximum of EM and is being increasingly used both as a direct method of finding conductive sulfides and for seeking alteration halos. Where elements with radioactive isotopes such as potassium and uranium are concentrated in hydrothermal environments, γ-ray spectrometry may be used to map their distribution. At the Cabacal gold deposits in Mato Grosso, Brazil, γ-emissions due to potassium in hydrothermal sericite and biotite cause an anomalous zone running parallel to the distribution of the buried mineralization (Barreira 1987).

Hydrothermally altered rocks may be identified from earth-orbiting satellites by analysing the spectral response of visible, infrared and thermal spectral channels. Combinations and ratios of channels and other manipulations of the digital data may enhance the character of geological structural features and identify iron or clay alteration, for example.

Many of the exploration methods described above which are effective on Earth may be equally applicable in the peculiar conditions of the Martian surface. However, it would appear unlikely that some geological environments, such as exhalites which require tranquil marine basins, would have been developed on Mars. Ore types such as the Errington base metal deposit, which is situated in sediments filling the interpreted Sudbury meteorite crater in Ontario, Canada, while rare on Earth may be more common on Mars. It is likely that this derived from hydrothermal convection cells driven by heat from a magmatic intrusion caused by the impact of the meteorite.

Alteration patterns around hydrothermal heat sources will be similar to those found in terrestrial systems if the controlling factors, such as heat gradient, chemical nature of the fluids and pressure, are essentially the same on both planets. This would apply particularly to deep hydrothermal systems and less to high level epithermal deposits which would tend to be less oxidized than those on Earth because of the planet's aridity and low atmospheric oxygen content.

Geochemical exploration techniques should be readily applicable to Mars. Methods such as PER described above should be adaptable to Martian conditions assuming that the rocks analysed are comagmatic.

The passive geophysical methods — γ-spectrometry and magnetometry — which record physical properties of the rocks without the need for an energy source, would provide the most effective and least costly method of exploration because they would not require the extra weight and energy supply of electrical or electromagnetic transmitters or seismic vibrators. Magnetometry, which is probably the most cost efficient technique used on Earth would be ineffective, however, because of the lack of a magnetic field on Mars.

Gamma spectrometry could provide measurements of the thorium, uranium and potassium content of surface rocks. It could only be used effectively on the surface from an aircraft such as a very large balloon or airship (rigid-framed machine constructed in the vacuum of space is a possibility) with instruments suspended at most 100 m or so from the ground.

Magnetotellurics is a ground system which maps electrical resistivity. It uses natural low frequency electromagnetic radiation and is able to penetrate to a depth of several kilometres.

VLF uses a signal that is high frequency by the standards of electromagnetic exploration techniques (its name refers to normal radio transmitters). While this technique does require transmitters, two stations would probably be adequete to cover Mars. It may be used either on the surface or airborne.

Other geophysical methods would need a power source. Electromagnetics would be effective in detailed mapping of the subsurface. Drill holes could also be probed to find possible nearby massive sulfide mineralization. Induced polarization may not be feasible on Mars. The Atacama desert of Chile presents unique problems for this method. The surface is so dry that it is difficult to pass any electrical current into the ground. Even when this is possible, the layer beneath the surface is saline and extremely conductive, short circuiting what little current reaches it. The Martian surface will probably be no better than the Atacama.

Alpha radiometry could be used to detect radon gas emissions which may be derived from hydrothermal uranium deposits. Techniques which measure emissions of other gases, such as mercury, helium, sulfur compounds and hydrocarbons, may also be valuable in locating hydrothermal minerals.

Seismic methods may use the natural energy of falling rocks to map shallow features such as the thickness of the regolith.

Acknowledgements

Advice and assistance in the preparation of this paper was provided by Professor M. Russell of Glasgow University and by colleagues R. J. Worsfold, R. A. Alcock, J. P. Golightly, E. F. Pattison and D. R. Burrows all of INCO Limited with whose kind permission it is published.

References

Anderson C 1968 Arizona and Adjacent New Mexico. In: Ridge J Ore deposits of the United States, 1933–1967. American Institute of Mining, Metallurgical and Petroleum Engineers (AIMM), New York

Annels AE 1989 Ore genesis in the Zambian Copperbelt, with particular reference to the northern sector of the Chambishi basin. In: Boyle RW, Jefferson AC, Jowett CW, Kirkham RV (eds) Sediment-hosted stratiform copper deposits. Geol Assoc Can Spec Pap 36:427–452

Barreira CF 1987 Geophysical response over the Cabacal 1 gold deposit. BP Mineracão unpublished report. In: Horn RA Uncertainty and risk in geological exploration. Proceedings of Exploration '87. Ontario Geological Survey, Queen's printer for Canada, p 52–59

Bell TH, Perkins WG, Swager CP 1988 Structural controls on development and localization of syntectonic copper mineralization at Mount Isa, Queensland. Econ Geol 83:69–85

Bertoni CH 1983 Gold production in the Superior Province of the Canadian Shield. Bull CIM 76:62–67

Coats R 1940 Propylitization and related types of alteration on the Comstock lode. Econ Geol 35:1–16

Dube B, Guha J, Rocheleau M 1987 Alteration patterns related to gold mineralization and their relation to CO_2/H_2O ratios. Mineral Petrol 37:267–291

Eldridge CS, Williams IS, Patterson DJ, Walsh JL, Both RA 1985 Ion microprobe determination of sulphur isotopic compositions of sulfides from the Mt Isa Pb/Zn ore body. Abstr Geol Soc Am 17:573

Geijer P 1931 The iron ores of the Kiruna type. Sveriges Geol Undersokning (Series C) 514:39–53

Gilmour P 1982 Grades and tonnages of porphyry copper deposits. In: Titley SR Advances in geology of porphyry copper deposits; southwestern North America. University of Arizona Press, Tucson, AZ, p 7–35

Goodfellow WD, Lydon JW, Turner RJW 1993 Geology and genesis of stratiform sediment-hosted (SEDEX) zinc-lead-silver sulphide deposits. In: Kirkham RV Sinclair R Thorpe WD, Duke JM (eds) Mineral deposit modeling. Geol Assoc Can Spec Pap 40:201–251

Hildenbrand TG, Kucks RP 1988 Total intensity magnetic anomaly map of Nevada. Map 93A Nevada Bureau of Mines and Geology

Horn RA, Telfer SM 1994 Exploration as a business. Rational minimum size of exploration targets. CIM Annual Meeting, Toronto. Candian Inst Mining, Metallurgy and Petroleum

Lambe RN, Rowe RG 1987 Volcanic history, mineralization and alteration of the Crandon massive sulphide deposit, Wisconsin. Econ Geol 82:1204–1238

Lefebvre J-J 1989 Depositional environment of copper-cobalt mineralisation in the Katangan sediments of southeast Shaba, Zaire. In: Kirkham RV, Sinclair R, Thorpe WD, Duke JM (eds) Mineral deposit modeling. Geol Assoc Can Spec Pap 40:401–426

Lindgren W 1913 Mineral deposits. McGraw-Hill, New York

Lindgren W 1926 Ore deposits of the Jerome and Bradshaw Mountains quadrangles, Arizona. US Geol Survey Bull 782

Lowell DJ, Guilbert JM 1970 Lateral and vertical alteration-mineralization zoning in porphyry ore deposits. Econ Geol 65:373–408

Lydon JW 1988 Volcanogenic massive sulphide deposits. 1. A descriptive model. In: Roberts RG, Sheehan PA (eds) Ore deposit models. Geoscience Canada Reprint Series 3, Geological Association of Canada, p 145–154

McKenzie CB 1986 Geology and mineralization of the Chetwynd deposit, southwestern Newfoundland, Canada. In: Macdonald AJ (ed) Gold '86, Toronto, p 137–148

Parak T 1975 Kiruna iron ores are not 'Intrusive magmatic ores of the Kiruna type'. Econ Geol 70:1242–1245

Pirajno F 1992 Hydrothermal mineral deposits. Springer-Verlag, Berlin

Putmann W, Hagemann HW, Merz C, Speczik S 1987 Influences of organic material on mineralisation in the Permian Kupferschiefer formation, Poland. In: Matavelli L, Novelli L (eds) Proceedings of 13th international conference on organic geochemistry. Advances in organic chemistry, vol 13. Pergamon Press, Oxford, p 357–363

Reed L 1991 Geophysics in mining exploration. Hailebury School of Mines. OE Walli Memorial Lecture. Course Notes

Reeve JS, Cross KC, Smith RN, Oreskes N 1990 Olympic Dam copper-uranium-gold-silver deposit. In: Hughes FE (ed) Geology of the mineral deposits of Australia and Papua New Guinea. Australasian Institute of Mining and Metallurgy, Brookfield, p 1009–1035

Roper MW 1976 Hot springs mercury deposit at McDermitt Mine, Humboldt County, Nevada. Soc Mining Eng AIME Trans 260:192–195

Schwartz G 1955 Hydrothermal alteration as a guide to ore. Econ Geol (Fiftieth anniversary volume) p 300–323

Sillitoe RH 1990 Geology of the Andes and its relation to hydrocarbon and mineral resources. In: Ericksen GE, Pinochet MTC, Reinemund JA Geology of the Andes and its relation to hydrocarbon and mineral resources. US Geol Survey, Reston, VA, p 285–311

Stanley CR, Madeisky HE 1994 Lithogeochemical exploration for hydrothermal ore deposits using Pearce element ratio analysis. In: Lintz DR Alteration and alteration processes associated with ore-forming systems. GAC Short course notes. University of British Columbia, Vancouver, p 193–212

Watkinson D, Melling D 1992 Hydrothermal origin of platinum-group mineralization in the low temperature copper sulphide-rich assemblages, Salt Chuck intrusion. Alaska Econ Geol 87:175–184

DISCUSSION

Carr: Most of the techniques that you were talking about were, in a sense, local. You are doing very detailed searches over fairly small areas. The problem we have with Mars is that we are starting with the whole globe, and somehow we have to narrow the search. What are the best techniques for narrowing down from a global perspective to where one can start to use some of these lower scale techniques?

Horn: The way we do it on Earth is on the basis of the geological characteristics of the ore bodies we are looking for. When we look at a chunk of North America, for example, it is easy to identify the Nevada Great Basin as a good location by using geological models. From our point of view, the blind intrusions — those that are not exposed at surface but could drive hydrothermal systems — could be a lot more interesting than those that come out onto the surface. Any one of your calderas would be a logical place to go.

Henley: I thought it was a pity that you took us relatively quickly through the risk analysis issue. If I was involved heavily in Mars exploration, I would first identify what I wanted to find, and in this case it would be looking for signs of life. Therefore you have already set the stakes, because if you persuade NASA and other governments to put millions of dollars into going there and you don't find life on Mars, your credibility has gone down. If you do find life on Mars, there is the expected value of getting more money to go again. Thus you can actually set some kind of value on the process and therefore some kind of value–cost relationship concerning the kinds of techniques and research strategies to use, in much the same way as we do for looking for porphyry coppers, for instance.

Horn: I think you are right. It came as a surprise to me that the porphyry copper deposits in the world have a value higher than the GDP of Canada. They are immensely valuable. With expected value, when you first start out the risk is higher, and because the risk is high the expected value is low. Logically, as a business investment you can only spend up to the expected value. As you gain knowledge, then the probability increases, and with it the amount you can justify spending — it is an incremental process. Unfortunately, Mars exploration has big bucks at the front end, and as a business decision it would therefore be a problem.

Walter: So what is the key knowledge concerning Mars that needs to be gained, apart from the ore deposit modelling?

Horn: The single biggest feature in controlling hydrothermal deposits is structure. This is something that we can record on Mars — we can actually see it. The lack of large features running across the landscape surprises me. On Earth, you see some gigantic structures. For instance, there's one structure that comes tearing down from the Indian ocean, goes down to Mozambique and skirts around the bottom of Africa.

Walter: Don't forget about Valles Marinaris, which is 5000 km long.

Horn: That is the sort of structure I would go for. Once I found such a large structure I would work along it and look for evidence of alteration by geophysical methods.

Walter: Where does mineral mapping fit in?

Horn: In looking for hydrothermal focus along the structure. This cuts down the area of the search.

Huntington: Is there a gravity map for Mars? This would tell you a lot about the internal structure.

Carr: There's a crude one, but we're going to get a much better one from Mars Global Surveyor.

Horn: You would need pretty good detail. The maps I was showing were flown at about 100 m.

Jakosky: Jack Farmer has been talking about the possibility of using a balloon-borne payload to do similar small-scale remote sensing on Mars.

Carr: The problem with balloons on Mars is that the atmosphere is so thin that they would need to be very large and would only be able to carry small payloads — we're talking total payload mass of 10 kg.

Horn: The instrumentation is getting smaller. GPS is going to be important, too. If you have very good GPS you could have airborne gravity surveys, which would be very useful.

Henley: It might be worth pointing out that there is a key difference between science-driven and discovery-driven exploration. Science-driven exploration attempts to accumulate the maximum amount of information essentially because it *may* be useful, but in the discovery-driven approach you are driven by the desire to keep the shareholders happy. Under that circumstance of course, rather than disperse the of money available, you have to rapidly focus the available money on a particular target. In a Martian exploration program, one might want to define the key societal value as seeing whether there was or is life on Mars, and then target all the funds on the most likely opportunity to find it.

Water on early Mars

Michael H. Carr

US Geological Survey, MS-975, 345 Middlefield Road, Menlo Park, CA 94025, USA

Abstract. Large flood channels, valley networks and a variety of features attributed to the action of ground ice indicate that Mars emerged from heavy bombardment 3.8 Ga ago, with an inventory of water at the surface equivalent to at least a few hundred metres spread over the whole planet, as compared with 3 km for the Earth. The mantle of Mars is much drier than that of the Earth, possibly as a result of global melting at the end of accretion and the lack of plate tectonics to subsequently reintroduce water into the interior. The surface water resided primarily in a porous, kilometres-thick megaregolith created by the high impact rates. Under today's climatic conditions groundwater is trapped below a thick permafrost zone. At the end of heavy bombardment any permafrost zone would have been much thinner because of the high heat flows, but climatic conditions may have been very different then, as suggested by erosion rates 1000 times higher than subsequent rates. Water trapped below the permafrost periodically erupted onto the surface to form large flood channels and lakes. Given abundant water at the surface and sustained volcanism, hydrothermal activity must have frequently occurred but we have yet to make the appropriate observations to detect the results of such activity.

1996 Evolution of hydrothermal ecosystems on Earth (and Mars?). Wiley, Chichester (Ciba Foundation Symposium 202) p 249–267

Mars is the only planet or satellite, other than the Earth, for which we have evidence that liquid water has been abundant at the surface. Temperatures at the surface now are such that the ground is permanently frozen to kilometre depths yet the evidence for liquid water is both abundant and persuasive. Huge flood channels have been eroded into the surface, the oldest terrains are extensively dissected by valley networks that resemble terrestrial river valleys, and water appears to have episodically ponded to form large lakes. In addition, there are numerous features that are best explained as the result of the presence of ground ice. Yet only sparse amounts of water can be detected today. Liquid water has not been observed anywhere, water ice has been detected only at the north pole, and the Martian atmosphere contains only minute amounts of water vapour. If significant amounts of water are present, it must be hidden below the surface as groundwater or ground ice. Furthermore, climate modelling suggests that it is very difficult to obtain conditions at the surface such that liquid water would be stable, particularly on early Mars when the energy output of the Sun is thought to have been lower than today. The preferential location of seemingly water-worn features in the most ancient terrains is therefore particularly puzzling.

Water is of interest not only from the perspective of planetary evolution but also for biology. Liquid water is universally recognized as a essential for life. The apparently large water inventory and the possibility that past climatic conditions allowed liquid water at the surface, raise the possibility that life could have started there. Sites of hydrothermal activity are commonly mentioned among locations where life may have started here on Earth. If water was abundant on early Mars, as appears likely, then hydrothermal activity would have been common irrespective of the climate, since the high heat flows expected at that time would have resulted in high rates of volcanism and, inevitably, interaction of volcanic products with the water and ice near the surface. Long-lived lakes and possible fluvial features are of additional biological interest.

This paper addresses two major aspects of the water story: the total water inventory and the climate history. For both topics, the geological evidence is reviewed and then attempts are made to reconcile the geological evidence with geochemical arguments and modelling studies. A plausible case, consistent with all the evidence, can be made that Mars possesses — or did possess in the past — a substantial inventory of water at the surface. However, the story on climate history is much less clear, and considerably uncertainty remains as to when and how major climate changes took place.

The paper is based largely on geomorphic evidence and on evidence gleaned from the SNC meteorites (for Shegotty, Nakhla and Chassingny; their places of discovery) which are believed to have come from Mars. The case for a Martian origin for the SNC meteorites is compelling (McSween 1994). All are basaltic rocks. Most have radiometric ages from 0.15 to 1.3 Ga, although a recently discovered one has a 4.5 Ga whole rock age. Several were observed to fall and so are known to be meteorites. Being basaltic and having young ages they had to originate from a body that was volcanically active between 0.15 and 1.3 Ga ago, and there are very few potential candidates within the Solar System. All the SNC meteorites have similar oxygen isotope patterns which are distinctively different from other meteorites, and different from the Earth and the Moon. They have elemental abundance patterns that indicate that they came from a body that is more enriched in moderately volatile elements than is the Earth. Gases enclosed in some of the meteorites have almost identical isotopic and chemical compositions to gases in the Martian atmosphere as measured by Viking. The ratio of the abundances of noble gases to each other, and ratio of $^{40}Ar : ^{36}Ar$ are especially diagnostic (Wiens & Pepin 1988). In addition, the D/H ratio of water in the Martian atmosphere is 5.2 times the terrestrial value, and the same enrichment is found in some of the water extracted from the meteorites (Watson et al 1994). The case for a Martian origin is thus very compelling. The meteorites are believed to have been ejected from Mars during one or more large impact events. The ejection velocities were such as to allow them to escape from Mars into heliocentric orbits from which they were captured by the Earth, probably after several million years in orbit around the Sun.

Geological evidence for a large surface inventory

Outflow channels present the least ambiguous evidence for large amounts of water at the surface in the past (Fig. 1). These enormous channels start full size, have few if any

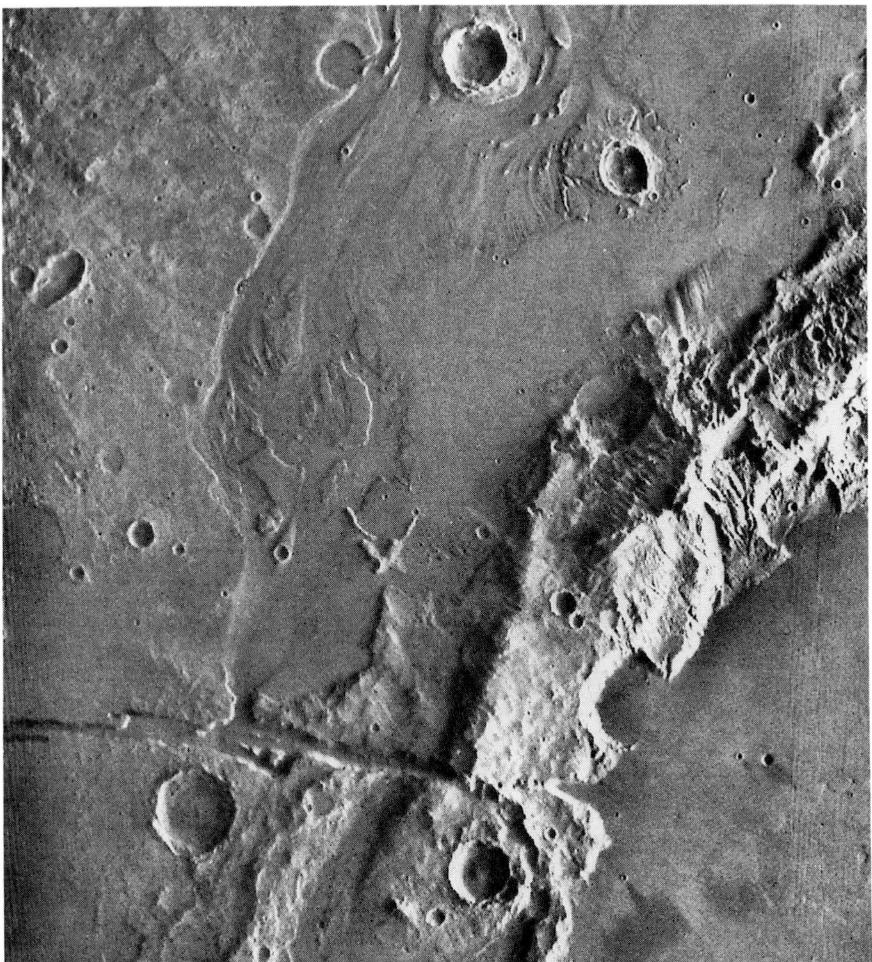

FIG. 1. The source of the large outflow channel Mangala Vallis. The channel originates at a graben and continues northwards for several hundred kilometres. At the top of the picture, longitudinal scour and teardrop-shaped islands are clearly visible. Groundwater may have been able to reach the surface because of disruption of the permafrost seal by the graben. In addition, the graben may have served as a subsurface conduit for groundwater. The scene is 250 km across.

tributaries, and show numerous indications that they formed by large floods (Carr 1996). They occur in four large regions: around the Chryse–Acidalia basin, centred at 20°N, 45°W; in Elysium Planitia, centred at 30°N, 230°W; in the eastern part of the Hellas Basin, around 40°S, 270°W; and along the western and southern margins of Amazonis Planitia, which is centred at 20°N, 160°W. Where incised into the cratered highlands the outflow channels are generally a few tens-of-kilometres wide, but where they cross plains they may expand to erode swaths hundreds of kilometres across. The streamlined channel shapes, the teardrop-shaped islands and the striated floors are the most distinctive features of the channels, and cause them to closely resemble large terrestrial flood features. Other attributes of terrestrial flood channels found in the Martian channels include inner channels, cataracts, converging and diverging scour patterns, and etched or plucked zones. The bedforms and islands indicate that these sinuous depressions are indeed true channels that were once filled with fluid; not valleys formed by slow erosion. The resemblance between the Martian outflow channels and large terrestrial flood channels is so strong and there is so much other supportive evidence of water and ice that a flood origin for most of the outflow channels can hardly be doubted.

The floods were much bigger than even the largest known terrestrial floods. Estimates of the discharges made from the channel dimensions and slopes range from $10^7\ m^3\ sec^{-1}$ for some of the smaller channels around Chryse Planitia to $3 \times 10^9\ m^3\ sec^{-1}$ for Kasei Vallis, the largest outflow channel. For comparison, discharges of $10^7\ m^3\ sec^{-1}$ have been estimated for the two largest known floods on Earth, the Channeled Scablands in Eastern Washington and the Chuja Basin flood of Siberia (Baker et al 1993), and the discharge of the Mississippi River peaks at $10^5\ m^3\ sec^{-1}$. The amounts of water involved are much harder to estimate, but estimates are that individual flood events released as much as 10^5–$10^6\ km^3$ of water (Carr 1996).

Although the floods may have resulted from a variety of causes, the two most likely are massive release of groundwater from below a thick permafrost seal and catastrophic release of water from lakes. The survival of ancient, heavily cratered terrain over two-thirds of the Martian surface suggests that a deep, impact-brecciated zone is preserved over most of the planet, although in places it may be covered by a thin veneer of younger deposits. The brecciated zone formed during the period of heavy bombardment between 4.5 and 3.8 Ga ago, when impact rates were much higher than during Mars' subsequent history. The zone is likely to be porous and could have a substantial holding capacity for groundwater and ground ice. Clifford (1993) estimated the storage capacity to be the equivalent of 1–1.5 km spread evenly over the whole planet. At present, liquid water is stable only below the zone of permanently frozen ground near the surface. The base of this permafrost zone is controlled by surface temperatures and the heat flow and estimates are that at present the zone is roughly 6 km thick at the poles and 2 km thick at the equator (Clifford 1993). Early in Mars' history, at the end of heavy bombardment, heat flows would have been four to five times higher than at present (Schubert & Spohn 1990) and the permafrost seal would have been four to five times thinner for the same climatic conditions. Migration of groundwater would cause hydrostatic pressures to build below the permafrost seal in low lying areas. If the seal were broken, massive groundwater eruptions could occur. Breakout would be favoured where the permafrost seal was thin, as at low latitudes or near volcanoes. This mechanism requires that the groundwater be contained under high pressures by the permafrost seal and so implies mean annual temperatures well below 0°C. The ages of the outflow channels span most of Mars' history, but they formed predominantly in the first half of Mars' history, after the end of heavy bombardment.

If the Martian megaregolith acted as a groundwater aquifer below the permafrost, then depressions that extended to depths well below the permafrost zone would be susceptible to filling by water. In the Valles Marineris, which extend well below the depth of even today's permafrost, are thick stacks of layered sediments that have been taken as evidence of the former presence of lakes. Under present climatic conditions such lakes would form an ice cover and the lake level would be controlled by the level of groundwater in the adjacent terrain. The lakes could be long lived as ablation from the surface of the ice was compensated for by leakage of groundwater into the lake. Emergence of large flood channels from the lower ends of the canyon suggests that some of the lakes postulated to have been within the canyons drained catastrophically to the east (McCauley 1978).

If abundant water is present we would expect to see evidence of ice, at least in those places where ice is stable. At present, mean diurnal temperatures range from around 215 K at the equator to 150 K at the winter pole. On average, as measured by the Mars Atmospheric Water Detector experiment on Viking, the atmosphere holds 12 pp μm (precipitable micrometres) of water. For this amount of water distributed evenly throughout the atmosphere, the frost point temperature is close to 200 K. As a result, ground ice is unstable at all depths for latitudes lower than about 30°, because ground temperatures exceed the frost point for all or some part of the year (Farmer & Doms 1979, Fanale et al 1986). Any ground ice present will tend to sublime. The water would be lost to the atmosphere then be frozen out at the poles. If present climatic conditions are typical of much of the planet's history, then near-surface materials at low latitudes should be devoid of ice unless it has been replenished in some way. In contrast, at high latitudes, the mean annual temperature is lower than the frost point temperature so that below the depth affected by the annual thermal wave, down to the base of the permafrost, ice is stable. If the surface is water rich, then we should see evidence for ice at high latitudes but not at low latitudes.

Several features observed only at high latitudes have plausibly been attributed to movement of the near-surface materials as a result of the presence of ground ice (Fig. 2). The two most prominent such features are debris aprons and terrain softening (Squyres & Carr 1986). At the base of most steep slopes in the 30–50° latitude band are aprons of debris with convex upward surfaces and steep outer margins. They extend roughly 20 km away from the slope. Within valleys, longitudinal ridges form where aprons from opposing walls meet. The aprons are not found at low latitudes. The simplest explanation is that at high latitudes, where ground ice is expected, talus shed from slopes is mixed with ice. This lubricates the talus, allowing it to move slowly away from the slopes to form the aprons. At low latitudes, where no ice is expected, the talus is stable and no aprons form. At very high latitudes temperatures are so cold that strain rates are negligible for the basal stresses expected at the bases of the talus piles so that they do not move. Terrain softening is a term applied to the characteristic rounding or muting of terrain at mid to high latitudes. At latitudes less than about 30°, most

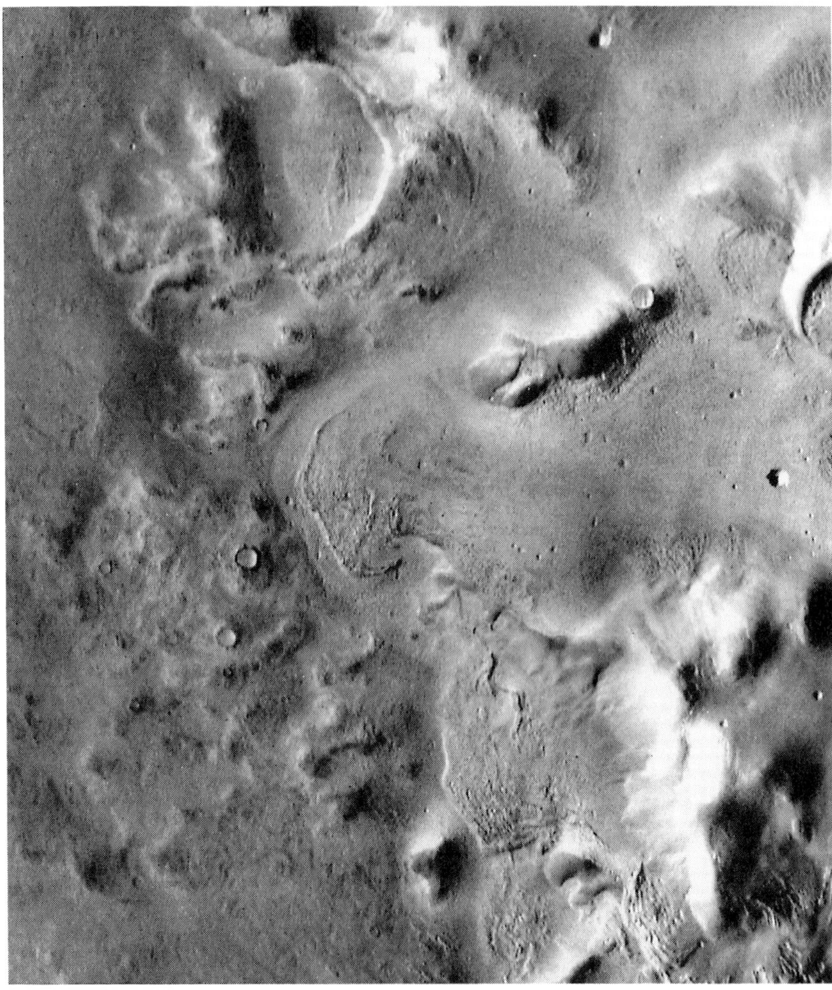

FIG. 2. Rounded massifs and debris aprons at 47°S, 247°W. Flow of near-surface materials at high latitudes where ice is stable has been attributed to the presence of ground ice. Here a glacier-like tongue of debris, 7 km across, has flowed away from high standing ground between isolated massifs. The rounding of the massifs is also attributed to ice-abetted flow.

primary features that formed after heavy bombardment, such as crater rims and fault scarps, are sharply defined, having suffered negligible amounts of erosion. At latitudes above about 30°, however, most crater rims and scarps are rounded, and the small crater population is largely missing. Massifs, such as those around impact basins, are rounded and, in places, are the source of glacier-like tongues of material. Although not proven, the simplest explanation of these differences is that downslope movement of the near surface materials at high latitudes is enhanced as a result of the presence of ground ice. Thus the presence of large amounts of water near the surface, as inferred from the outflow channels, is supported by the seemingly pervasive presence of ground ice at high latitudes where it is stable, and by the seeming absence of ground ice at low latitudes where it is unstable.

The outflow channels provide a means of estimating the amount of water near the surface. The large outflow channels around the Chryse basin have removed approximately 4×10^6 km^3 of material to form the outflow channels and chaotic terrain. If we assume that all the water that flowed through the channels carried the maximum reasonable sedimentary load, then at least 6×10^6 km^3 of water has passed through these channels — the equivalent of 40 m spread over the whole planet. If we further assume that water was evenly distributed over the whole planet, not just restricted to the Chryse basin, then we get a total inventory of roughly 400 m spread over the whole planet. The equivalent figure for Earth is 3 km. The 400 m estimate is little more than a reasonable guess. The number is too low if much of the water that flowed through the channels did not carry its maximum load, and if much of the groundwater remained in the ground, as was surely the case. The number is too high if groundwater was re-circulated in some way so that the same water passed through the channels more than once. Despite the uncertainties, it is difficult to see how the water inventory at the end of heavy bombardment could have been less than a few hundred metres spread over the whole planet.

Fate of the water erupted onto the surface

The fate of the water that passed through the channels is uncertain. A conservative view (Carr 1996) is that lakes formed at the ends of the channels and had volumes comparable to the volumes given above for individual floods (10^5–10^6 km^3). The lakes then froze in place. Most of the outflow channels terminate at high latitudes where ice is permanently stable at depths larger than a few metres. If buried below a thin layer of dust, therefore, the lakes would form stable ice deposits. Many of the peculiarities of the low-lying northern plains could be attributed to such deposits. Lakes that form at low latitudes would sublime over geological time and the water would ultimately be frozen out at the poles. Water sublimed from low-latitude lakes and that fraction sublimed from the high-latitude lakes may now be locked in the layered deposits observed at the poles. This model assumes that throughout most of the post-heavy-bombardment history the Martian climate was similar to that of the present.

A much more radical suggestion has been made with respect to the amounts of water involved in floods and their climatic consequences. Baker et al (1991) proposed that ocean sized bodies of water ($> 10^7$ km^3) formed episodically in the northern plains as a result of the large floods. The proposal was made mainly to explain three observations: (1) the presence of valley networks on surfaces that formed late in Martian history; (2) the presence of seemingly glacial features of similar age at high southern latitudes (Kargel & Strom 1992); and (3) the presence of linear features, interpreted as shorelines, at high northern latitudes (Parker et al 1993). Baker et al (1991) suggested that the oceans formed as a result of several large floods occurring simultaneously, possibly with one flood triggering another. According to the model, eruption of the large volumes of water caused massive release of CO_2 as a result of degassing of CO_2 dissolved in the water and release of CO_2 stored in the regolith. The enhanced CO_2/H_2O atmosphere caused global warming, thereby temporarily creating conditions under which fluvial and glacial features could form. The oceans slowly disappeared by processes not clearly explained and a long frozen period ensued, until the next flooding episode. They envisage, therefore, a history characterized mainly by climates similar to today's except for brief but dramatically different oceanic episodes. The plausibility of this proposal depends critically on the uniqueness of the interpretation that various geomorphological features are caused by fluvial, glacial and shoreline processes. Unfortunately, geomorphology is an inexact science and alternative interpretations are always possible.

The geological record only extends back to the end of heavy bombardment. The outflow channels, most of which formed shortly after this time, indicate that the planet emerged from heavy bombardment with a total near-surface inventory of at least a few hundred metres of water. How this inventory changed between that time and the present is unclear. Volcanic outgassing after heavy bombardment probably added only slightly to the inventory. Greeley & Schneid (1991), by measuring the volume of volcanic rocks erupted during this period and assuming terrestrial values for the water content of the rocks, estimated that 60 m of water had been added to the surface inventory. This, however, is almost certainly an overestimate since the SNC meteorites are much drier than terrestrial basalts. Water is currently being lost to space from the upper atmosphere. The present loss rates are very small mainly because the present atmosphere contains very little water. At present loss rates, the equivalent of only 3.6 m spread over the whole planet would be lost over the life of the planet (Yung et al 1988). If the likely increases in water content of the atmosphere during periods of high obliquity are taken into account, losses could be as high as 70 m over the life of the planet (Jakosky 1990) in the same units. Losses would have been significantly higher if the planet experienced major climatic excursions: if the planet were warmer, the water content of the atmosphere would increase substantially and losses would increase correspondingly.

Oxygen isotope and D/H ratio measurements on SNC meteorites provide additional constraints on how the inventory has evolved. Karlsson et al (1992) showed that oxygen in water driven off from SNC meteorites is isotopically distinct from oxygen in the silicates, and both are isotopically distinct from terrestrial oxygen. They suggested that the SNC meteorites were altered by crustal waters after their eruption. The oxygen in the silicates is representative of the oxygen in the mantle as acquired by the planet during accretion, unaltered by subsequent processes. The oxygen from water is, however, contaminated with crustal oxygen whose isotopic composition has been affected by processes such as cometary infall and losses of oxygen from the upper atmosphere. Oxygen in the mantle and the crust have seemingly evolved independently since accretion, most likely because the lack of plate tectonics prevents crustal materials being mixed back into the interior.

Separate evolution of crustal and mantle water is also suggested by D/H ratios in water extracted from SNC meteorites. The D/H ratio in the present atmosphere is 5.2 times that of terrestrial water (Bjoraker et al 1989). The enrichment is caused by preferential loss of hydrogen with respect to deuterium from the top of the atmosphere, a consequence largely of diffusive separation of hydrogen and deuterium above the homopause (Yung et al 1988). Water extracted from SNC meteorites has values of D/H that range from unenriched to 5.2 enriched, which again is consistent with water from two sources that have evolved independently, an unenriched mantle source and an enriched crustal source (Watson et al 1994).

The same D/H enrichment of 5.2 observed in the SNC meteorites and the present atmosphere implies that the surface reservoir undergoing fractionation has not been enriched, within experimental errors, in the last 1.3 Ga, the age of the meteorites. From the present loss rates and the experimental errors, Donahue (1995) estimated that the reservoir of water undergoing enrichment must be the equivalent of at least 25 m spread over the whole planet otherwise we would see differences between the 1.3 Ga water from the meteorites and present day water. Moreover, since this whole 25 m is enriched a lower limit of 280 m can be placed on the amount of surface water originally at the end of accretion, which is consistent with the geology.

Low estimates of the global inventory

Geochemical arguments have been made in support of much lower estimates for the total water inventory than those derived from geological arguments. Anders & Owen (1977), for example, argued that most of the noble gases present on the planet at the end of accretion or subsequently outgassed from the interior must still be present in the atmosphere today. Taking terrestrial values for the Ar : H_2O ratio, they estimated that the equivalent of only 10 m of water spread over the whole planet was at the surface. McElroy et al (1977) used a similar argument to estimate the water inventory from nitrogen. They calculated how much nitrogen had to be present on early Mars, given the enrichment of atmospheric nitrogen in ^{15}N and present loss rates. Taking terrestrial values for the N/H_2O ratio their estimates of the water inventory ranged from 3–133 m. In retrospect both these estimates are clearly in error since they did not take into account losses of nitrogen and noble gases by impact erosion of the atmosphere which could have been substantial on early Mars. During the process the noble gases and nitrogen would be lost preferentially with respect to water because much of the water would have been in or on the surface, not in the atmosphere.

On the basis of the low water content of the SNC meteorites and a homogeneous model for Martian accretion, Dreibus & Wänke (1987) concluded that Mars retained a total of only 130 m of water, it being all in the interior at the end of accretion, and that only a small fraction of this would have outgassed to the surface. The SNC meteorites are indeed dry rocks. They mostly contain 200–500 ppm water as compared with

1500–10 000 ppm for terrestrial oceanic basalts. In addition, oxygen isotope data and the D/H ratio of water extracted from the meteorites indicates that a significant fraction of what little water is in the meteorites is surface water and not mantle water (Karlsson et al 1992, Watson et al 1994), so that the mantle of Mars appears to be significantly drier than that of the Earth. Dreibus & Wänke (1987) suggested that the two planets accreted from materials that were in chemical disequilibrium, some materials being reduced and containing metallic iron, other materials being oxidized and containing water. During global differentiation to form the core the metallic iron reacted with water to form hydrogen which was driven off the planet, leaving both the surface and the interior very dry. They attributed the higher water content of the Earth's mantle to acquisition of some of the oxidized water-containing components after the core formed. Evidence for this late acquisition is the excess of siderophile elements in the mantle over that expected if it equilibrated with the core. Because the Martian mantle appears not to have a comparable siderophile anomaly, they suggested that Mars did not acquire a late veneer, and so was left with both a drier surface and a drier interior than the Earth.

Carr & Wänke (1992), however, suggested two main possibilities whereby Mars could have been left with abundant water on the surface despite the dry interior and lack of a siderophile anomaly. Both planets may have acquired a late volatile-rich veneer after global fractionation but only the Earth was left with evidence in form of a siderophile anomaly in the mantle because vigorous mantle convention and plate tectonics resulted in the stirring of near surface materials into the interior. Impact rates and noble gas patterns provide additional support for this late addition (Chyba 1990, Owen et al 1992). The second possibility is that towards the end of accretion the Earth acquired a steam atmosphere as a result of release of volatiles from the impacting materials. This caused the surface to melt, at which time water in the atmosphere could dissolve into the molten interior (Abe & Matsui 1985). The water released on impact during the accretion of Mars did not form a steam atmosphere but condensed on the surface because Mars is smaller and further from the Sun.

Climate change

Two types of climate change have been discussed with respect to Mars: quasi-periodic changes induced by variations in the orbital and rotational motions, and long term changes. Unless the poles are major repositories of CO_2 which could be released during periods of high obliquity, which is unlikely, climatic changes induced by planetary motions are probably small, and they will not be discussed further. The evidence for major changes in climate in the past is mainly the presence of valley networks, and indications of past high rates of erosion.

The valley networks have been taken as evidence of warmer climates in the past because they are widely believed to have formed, like terrestrial river valleys, by slow erosion of running water. They differ from the outflow channels in that they are only 1–3 km wide, have tributaries, and form coherent drainage networks (compare Figs 1 and

FIG. 3. A typical scene from the cratered highlands that have survived since the end of heavy bombardment 3.8 Ga ago. Craters are in various stages of preservation and valleys wind between the large craters. The valleys are mostly flat floored, 1–3 km across, and form open drainage networks. The scene is 280 km across.

3). Under present climatic conditions, small streams, even if they could be initiated, would rapidly freeze so that flow of stream-fed rivers could not be sustained to cut the kilometre wide, tens-of-kilometre-long valleys that have been described. Clearly, the climatic conditions required to form the valleys depend crucially on whether the Martian valleys are indeed close analogues of river valleys on Earth. Like terrestrial river systems, the valley networks have tributaries that converge downstream. They are, however, smaller and simpler than their terrestrial counterparts. The Martian networks rarely have more than 50 branches, large areas between the branches are undissected, and the longest path through an individual network is generally under 100 km. In the uplands, where most of the valleys occur, drainage densities are 10 to 100 times less than the most sparsely dissected terrestrial material. Not only are the areas between the branches of a network undissected but individual networks are separated by undissected areas so that there has been little competition between nearby networks.

The origin of the valley networks is controversial. Most observers have concluded that they are close analogues of terrestrial river valley networks — that is, they formed as a result of slow erosion by streams much smaller than the valleys themselves (Mars Channel Working Group 1983). Arguments about origin have been largely concerned with the source of the water and the relative roles of run-off as a result of precipitation and groundwater sapping. An alternative, minority view is that the networks formed mostly by mass wasting aided by movement of groundwater, and that surface streams were rare (Carr 1995). This view is supported by the lack of tributaries smaller than the typical 1–3 km-wide valleys, the lack of stream channels on the valley floors, the uniform rectangular or U-shaped cross-section from source to mouth, the presence of longitudinal ridges down the length of some valleys, and the low drainage densities. Carr (1995) suggested that the valley networks are indeed drainage features, but what have drained off are the surface materials themselves, aided by the presence of interstitial water. According to this model, the role of surface streams is minor, although occasional floods — much smaller than those that formed the outflow channels — could have occurred. The valleys extend themselves headward as a consequence of convergence of groundwater flow there. The valleys formed mostly early in Mars' history because high heat flows and possibly different climatic conditions allowed liquid water close to the surface. Younger valleys are preferentially found on steep slopes and/or where high heat flows are expected, because these conditions favour mass-wasting. Whatever the origin of the valley networks, however — whether by surface run-off, groundwater sapping or mass wasting — water appears essential for their formation.

A fluvial origin of the valley networks almost certainly requires climatic conditions very different from those of today. Streams fed by precipitation require surface temperatures above freezing to provide water as precipitation, or, if the precipitation is ice, to provide meltwater. Spring-fed streams could survive for short distances before freezing, but discharges would have to be large, with streams several metres deep, to sustain flow for the tens of kilometres (and in some cases 100–200 km) required to explain the valleys. Moreover, enormous volumes of water are required to cut valleys by slow erosion. Gulick & Baker (1993) estimate that 1000 : 1 ratios of water to rock materials removed are typically required. Such large volumes of water imply some groundwater replenishment mechanism. This almost certainly requires warm climates so that water brought to the surface can re-enter the groundwater system and is not prevented from doing so by a thick permafrost.

If the valley networks are indeed fluvial and require warmer climatic conditions, then the climate history of the planet has been very complex. While most of the networks are in the cratered uplands that date from the end of heavy bombardment, they appear to have a wide spread of ages. These ages are difficult to determine but, on the basis of intersection relations, as many as 30% formed after the end of heavy bombardment. Although most of these formed in era just after the end of heavy bombardment, they continued to form slowly, or episodically, throughout Martian history as shown, for example, by the dissection of relatively young volcanoes such as Alba Patera. Thus, if fluvial, the valley networks imply warm climatic episodes late in Martian history, as emphasized by Baker et al (1991), when they proposed their episodic ocean hypothesis. If the valley networks are not fluvial but formed as a result of mass-wasting aided by the presence of groundwater, as described above, then they may have been able to form under climatic conditions not very different from those of today. They would have formed more readily early in Mars' history because the higher heat flow allowed liquid water to be at shallow depths, despite cold surface temperatures. According to this model, valley formation persisted to later dates on volcanoes because high local heat flows and easily erodible ash deposits are expected there. Mass wasting is also expected to operate more efficiently on steep slopes such as crater or canyon walls, where younger valley networks are also found.

While the climatic implications of valley networks may be controversial the changes in erosion rates at the end of heavy bombardment are unambiguous, and it is difficult to see how they could not be the result of climate change. Survival of craters at the Viking Lander 1 landing site and on crater floors that formed just after the end of heavy bombardment (Carr 1996) show that the average erosion rate for most of Mars' history has been very low, around $10^{-2}\,\mu m\,yr^{-1}$, as compared with typical terrestrial rates of 10–1000 $\mu m\,yr^{-1}$. Because of the low erosion rates, crater populations on post-bombardment surfaces at low latitudes are almost perfectly preserved down to crater diameters of around 100 m, as are most of the details of their ejecta blankets. In contrast, heavily cratered surfaces show a wide range of degradation, and craters as large as tens of kilometres across are commonly shallow and rimless and have no traces of ejecta. Erosion rates during late heavy bombardment are difficult to determine because of lack of knowledge of the cratering rates, but Carr (1996) and Craddock & Maxwell (1993) estimated the rates to be in the range of 0.1–10 $\mu m\,yr^{-1}$, 10–1000 times the subsequent rates and at the bottom end of terrestrial rates.

The extremely low erosion rates for most of Mars' geological history and preservation of fine surface details that are over 3.5 Ga old is difficult to reconcile with young fluvial features requiring recycling of abundant water through the atmosphere. Baker et al (1991), in proposing their episodic ocean hypothesis, recognized this issue and suggested that the warm, marine episodes that they were invoking were very short and separated by long periods of geomorphic quiescence. According to the non-fluvial hypothesis for the origin of the valley networks, erosion rates are low throughout most of Mars' history simply because the surface has been permanently frozen. The dramatic change in erosion rates at the end of heavy bombardment, around 3.8 Ga ago, is strongly indicative of climate change. Early modelling of greenhouse warming on Mars suggested that a 2–3 bar CO_2/H_2O atmosphere would raise surface temperatures to over 273 K, thereby allowing liquid water at the surface. It was suggested, therefore, that on early Mars a thick CO_2/H_2O atmosphere created warm conditions that resulted in high erosion rates and formation of the valley networks, and that this atmosphere subsequently collapsed as a result of formation of carbonates. We should therefore expect to see young carbonate deposits close to the surface. This simple model now appears untenable because (1) the valley networks have a spread of ages, (2) thick carbonate deposits have not been detected, and (3) subsequent modelling has cast doubt upon the magnitude of greenhouse warming possible on early Mars, at a time when the energy output of the Sun is thought to have been lower. Kasting (1991) showed, for example, that for a Sun luminosity of 0.7 times that of the present Sun — approximately that predicted for 4 Ga ago — greenhouse warming by a CO_2/H_2O atmosphere cannot raise surface temperatures above about 215 K, no matter how thick the atmosphere, because of cloud formation. Other greenhouse gases such as methane, ammonia and sulfur dioxide have such short lifetimes in the atmosphere that they are unlikely to have been able to maintain high surface temperatures unless there was an abundant, continuous supply.

The climate story is thus very uncertain. The climate since the end of heavy bombardment is likely to have been like today's for much of the time. Short, significantly warmer episodes have been suggested to explain the young valley networks and possible glacial and shoreline features, but other interpretations of these features consistent with present climates are possible. Much higher erosion rates prevailed during the era of heavy bombardment than during later epochs, and warmer climates are the most likely explanation, but climate modelling has been unable to explain how such warmer conditions might have been achieved.

Hydrothermal activity

Hydrothermal activity is a natural consequence of volcanic activity in a water-rich environment. Given that Mars has a water- or ice-rich surface and has been volcanically active, hydrothermal activity must have occurred. The history of Martian volcanism is one of declining rates and increased localization to specific provinces. Immediately after heavy bombardment, volcanism was widespread as evidenced by extensive volcanic plains, particularly in the northern hemisphere. With time volcanism largely became restricted to three main regions: Tharsis, Elysium and Hellas. Whereas the volcanic constructs in these three regions, particularly those in Tharsis, dwarf their terrestrial counterparts, the magma production rates since the end of heavy bombardment have been much lower than on the Earth (Greeley & Schneid 1991). The rate of decline of volcanism probably paralleled the decline in heat flow, which is estimated to have declined in level roughly four to five times since the end of heavy bombardment.

The SNC meteorites provide unequivocal evidence that Mars underwent global differentiation 4.5 Ga ago (Shih et al 1982). It follows that the near-surface heat flow was very high 4.5 Ga ago and declined rapidly during heavy bombardment so that high and declining rates of volcanism are expected during this period (Schubert & Spohn 1990). Much of the evidence of such volcanism has, however, been destroyed as a result of the high rates of impact.

We have no direct, unambiguous evidence of hydrothermal activity on Mars either today or in the past. Although Mars is probably still volcanically active, as indicated by the young ages of both the SNC meteorites and some of the surfaces on the large volcanoes, the average rate of volcanism since heavy bombardment, normalized to the surface area, is only about 2% of the terrestrial rate. Hydrothermal activity is difficult to detect on Earth from orbit, so the failure to detect such activity on the much less active Mars is not unexpected. Indeed, the high-resolution infrared sensors and spectrometers needed to detect thermal and compositional anomalies have yet to be flown.

Although hydrothermal activity has not been detected, a variety of morphological features have been attributed to volcano–ice interactions. Table mountains in Iceland and Antarctica have been attributed to eruption of lavas into or under ice, and somewhat similar, flat-topped hills on Mars have been attributed to the same process (Allen 1979). Several broad, irregular ridges in volcanic regions, particularly Elysium, have been interpreted as Moberg ridges, which form by fissure eruptions under ice. Squyres et al (1987), for example, attributed many linear ridges and flow-like features in southern Elysium to a combination of Moberg ridge formation and intrusion of sills into ice-rich ground. The start of numerous large outflow channels, adjacent to volcanoes, is strong evidence that volcanic activity has affected the movement of groundwater and the stability of ground ice, as is the emergence of channels from under individual lava flows. Several volcanoes have flanks that are densely dissected by valley networks, and Gulick & Baker (1990) suggested that hydrothermal activity played a major role in locally recycling the water. Thus, while hydrothermal deposits have not been unequivocally identified, there is abundant indirect evidence of water–volcano interactions.

At present we do not know how common hydrothermal deposits are at the Martian surface. The high rates of volcanism expected at the end of heavy bombardment suggest that terrains that date from this era may be among those where hydrothermal deposits are most likely to be found. However, the record from this era is confused by the masking effects of impact. Ironically, the volcanic regions of Tharsis and Elysium may not be the best places to look, because volcanic activity sustained for thousands of millions of years may have largely depleted the water in these regions. This supposition is supported by the dominant style of volcanism, which appears to have been large scale effusion of fluid lava rather than pyroclastic activity, which would have been expected if water had played a stronger role. Location of the optimum places to search must await better geochemical and imaging data.

Summary

On Mars, liquid water has not been detected, ice has been detected only at the north pole, and the atmosphere contains only minute amounts of water. Despite this, the geological evidence is convincing that the planet has (or had in the past) a substantial near-surface inventory of water, the equivalent of a few hundred metres spread evenly over the whole planet. The bulk of this water must be present as ground ice or groundwater. Large flood features are evidence that groundwater periodically erupted onto the surface. The history of water action is controversial. One radical suggestion forwarded to explain young valley networks and other features interpreted as glacial or littoral, is that ocean-sized bodies of water formed episodically throughout Martian history. An alternative, more conservative view is that the large floods are a manifestation of the one-way transfer of water from high to low areas and represent an adjustment of the distribution of water to the present topography. According to this model, the lakes that formed at the end of the floods simply froze in place to form ice deposits which are present today. The climate history is also controversial. According to the episodic ocean model, the long periods of cold climate, similar to today's, were separated by several short, warm marine episodes. An alternative is that the climate since the end of heavy bombardment has been similar to today's climate, and that the valley networks and supposed glacial deposits are not true analogues to terrestrial fluvial and glacial features. Whatever the subsequent history, a dramatic climate change may have occurred at the end of heavy bombardment as indicated by the thousand-fold change in erosion rates. The water-rich surface and abundant evidence of volcanism indicate that hydrothermal activity must have occurred frequently, although we have not yet unambiguously detected hydrothermal deposits, nor do yet know where best to look for them.

References

Abe Y, Matsui T 1985 The formation of an impact generated H_2O atmosphere and its implications for the early thermal history of the Earth. J Geophys Res 90:545C–559C

Allen CC 1979 Volcano–ice interactions on Mars. J Geophys Res 84:8048–8059

Anders E, Owen T 1977 Mars and Earth: origin and abundance of volatiles. Science 198:453–465

Baker VR, Strom RG, Gulick VC, Kargel JS, Komatsu G, Kale VS 1991 Ancient oceans, ice sheets and the hydrological cycle on Mars. Nature 352:589–594

Baker VR, Benito G, Rudoy AN 1993 Paleohydrology of Late Pleistocene superflooding, Altay Mountains, Siberia. Science 259:348–350

Bjoraker GL, Mumma MJ, Larson HP 1989 Isotopic abundance ratios for hydrogen and oxygen in the Martian atmosphere. Bull Am Astronom Soc 21:990

Carr MH 1995 The martian drainage system and the origin of valley networks and fretted channels. J Geophys Res 100:7479–7507

Carr MH 1996 Water on Mars. Oxford University Press, New York

Carr MH, Wänke H 1992 Earth and Mars: water inventories as clues to accretional histories. Icarus 98:61–71

Chyba CF 1990 Impact delivery and erosion of planetary oceans in the early inner solar system. Nature 343:129–133

Clifford SM 1993 A model for the hydrologic and climatic behavior of water on Mars. J Geophys Res 98:10973–11016

Craddock RA, Maxwell TA 1993 Geomorphic evolution of the martian highlands through ancient fluvial processes. J Geophys Res 98:3453–3468

Donahue TM 1995 Evolution of water reservoirs on Mars from D/H ratios in the atmosphere and crust. Nature 374: 432–434

Dreibus G, Wänke H 1987 Volatiles on Earth and Mars: a comparison. Icarus 71:225–240

Fanale FP, Salvail JR, Zent AP, Postawko RS 1986 Global distribution and migration of subsurface ice on Mars. Icarus 67:1–18

Farmer CB, Doms PE 1979 Global and seasonal water vapour on Mars and implications for permafrost. J Geophys Res 84:2881–2888

Greeley R, Schneid BD 1991 Magma generation on Mars: amounts, rates and comparisons with Earth, Moon and Venus. Science 254:996–998

Gulick VC, Baker VR 1990 Origin and evolution of valleys on martian volcanoes. J Geophys Res 95:14325–14344

Gulick VC, Baker VR 1993 Fluvial valleys in the heavily cratered terrains of Mars: evidence for paleoclimatic change? Lunar Planet Inst Tech Rep 93-03:12–13

Jakosky BM 1990 Mars atmosphere D/H: consistent with polar volatile theory. J Geophys Res 95:1475–1480

Kargel JS, Strom RG 1992 Ancient glaciation on Mars. Geology 20:3–7

Karlsson HR, Clayton RN, Gibson EK, Mayeda TK 1992 Water in SNC meteorites: evidence for a martian hydrosphere. Science 255:1409–1411

Kasting JF 1991 CO_2 condensation and the climate of early Mars. Icarus 94:1–13

Mars Channel Working Group 1983 Channels and valleys on Mars. Geol Soc Am Bull 94:1035–1054

McCauley JF 1978 Geologic map of the Coprates quadrangle of Mars. US Geol Surv Misc Inv Map I-897

McElroy JF, Kong TY, Yung YL 1977 Photochemistry and evolution of Mars atmosphere: a Viking perspective. J Geophys Res 82:4379–4388

McSween HJ Jr 1994 What we have learned about Mars from SNC meteorites. Meteoritics 29:757–779

Owen T, Bar-Nun A, Kleinfeld I 1992 Possible cometary origin of heavy noble gases in the atmospheres of Venus, Earth and Mars. Nature 358:43–46

Parker TJ, Gorsline DS, Saunders RS, Pieri DC, Schneeberger DM 1993 Coastal geomorphology of the Martian northern plains. J Geophys Res 98:11061–11078

Schubert G, Spohn T 1990 Thermal history of Mars and sulfur content of its core. J Geophys Res 95:14095–14104

Shih CY, Nyquist LE, Bogard DD et al 1982 Chronology and petrogenesis of young achondrites, Shergotty, Zagami and ALHA 77005: late magmatism on the geologically active planet. Geochim Cosmochim Acta 46:2323–2344

Squyres SW, Carr MH 1986 Geomorphic evidence for the distribution of ground ice on Mars. Science 231:249–252

Squyres SW, Wilhelms DE, Moosman AC 1987 Large-scale volcano–ice interactions on Mars. Science 231:249–252

Watson LL, Hutcheon ID, Epstein S, Stolper EM 1994 Water on Mars: clues from deuterium/hydrogen and water contents of hydrous phases in SNC meteorites. Science 265:86–90

Wiens RC, Pepin RO 1988 Laboratory shock emplacement of noble gases, nitrogen, and carbon dioxide into basalt, and implications for trapped gases in shergottie EETA79001. Geochim Cosmochim Acta 52:295–307

Yung YL, Wen J, Pinto JP, Pierce KK, Paulsen S 1988 HDO in the martian atmosphere: implications for the abundance of crustal water. Icarus 76:146–159

DISCUSSION

Davies: Surely impacts would liberate water and heat, and don't we actually see examples where rivers channels start at the rim crests of impact craters?

Carr: One often sees craters where it looks as though water has actually come out of the crater itself, but it actually came out long after the formation of the crater.

Davies: So it's not that the heating from the impact liberates the water?

Carr: I doubt it, because of the time problem: in most cases there is a time interval between the formation of the crater and the flow, because the rim has to erode away.

Davies: You mentioned tributaries: how can tributaries exist in the absence of rain?

Carr: One can have ground water sapping: water comes out of a spring and the channel erodes headward fed by the spring water. This happens on Earth all the time.

Davies: I can see how tributaries might flow out of the main channel, but in the absence of rain how do small streams flow into the main channel from somewhere else?

Carr: The heads of the upper part of the channel are fed by a spring.

Davies: Are you envisaging a number of springs which discover each other?

Carr: With respect to the valley networks, a major controversy over the last 20 years has been sapping versus rain. The dominant opinion is overwhelmingly in favour of sapping.

Davies: How long did these flows carry on for?

Carr: The valley networks formed mainly at the end of heavy bombardment and continued for a short time after except on volcanoes and some very steep slopes such as the canyon walls where valley formation continued until much later. The 2 Ga old channels on volcanoes look quite different from the other ones in the old crater terrain.

Davies: Could outflow still happen today?

Carr: What we call outflow channels are these very big features carved by large floods. It is unlikely that they could form now because of this progressive draw-down of the global aquifer system. Consequently, the enormous hydrostatic pressures can no longer build, so there is no longer the instability. At the same time, the cryosphere is thickening, so it becomes much more difficult for the water to get out.

Nisbet: My main question concerns what happens on Earth. One of the characteristic features of deglaciation in the high northern hemisphere is the involvement of methane from methane hydrates. Methane is stored in what are known as clathrate hydrates, which form widely in the Arctic, both in Siberia and in Northern Canada. When that heats and the hydrate collapses, the methane is released and you get all manner of interesting structures. Large areas of the Arctic ocean floor are pock-marked. These pocks are up to 100 m or more across, and form when methane, hydrodynamically involved with water, comes out. Obviously you won't get that on Mars in any recent history, but had there been life on Mars a long time ago, there would have been a chance of extensive methane involvement. Is there any sign of anything like that?

Carr: I'm not familiar with the features you are referring to, so I couldn't recognize them on Mars.

Giggenbach: CO_2 forms clathrates, too; they could provide another means to store carbon and water.

Jakosky: There has been a long discussion of CO_2 clathrate hydrate on Mars. The problem in this context is that the temperature is too low and you're not going to cross that boundary and cause it to destabilize. In the polar regions there could be a substantial amount of clathrate hydrate. The poles could hold a significant amount of CO_2 and water together, and it's possible that that could be released. But this would have to be purely a polar phenomenon.

Farmer: Could you comment on these 'fretted' channel systems up on the northern planes and how those relate to hydrological history? I'm a little confused by them.

Carr: These channels are essentially wet debris flows. They are wide channels that go from the plains–upland boundary, deep into the uplands. They have flat floors and steep walls. On the floors you can see unequivocal evidence of mass wasting on a grand scale: the material has flowed off the walls and down the channels, moving tens of kilometres apparently without the help of surface streams. Probably what happened is that you had a mass wasting by flows that were liquid-water-lubricated at their base. Climatic conditions may have been similar to those of today but the higher heat flow allowed liquid water at shallow depths. The mechanism subsequently turned off because of the decline in heat flow. The 273 K isotherm no longer intercepted the base of the flow so the base froze. Flow then stablized.

General discussion III

Catastrophic fluvial systems and potential hydrothermal systems in the ancient history of Mars

Kuzmin: I would like to turn your attention to one more potential example of a hydrothermal system on Mars. Most active hydrothermal systems on Earth are related to regions of modern (or recent) volcanic activity where local geothermal gradients are high. On Mars, similar conditions could have existed in the distant past. Magmatic and tectonic processes within the ancient Martian crust were more widespread and more active during earlier periods of the planet's history than they have been during the last billion years. Thick permafrost layers which existed at that time may have provided water sources for areas of hydrothermal activity.

During my search for morphological evidence of hydrothermal activity within the old highland terrains of Mars, I found an example in the area known as Terra Cimmeria. This region is characterized by surface elements which are also typical of terrestrial hydrothermal systems, namely, volcanic landforms and signs of fluvial erosion. Fig. 1 (*Kuzmin*) shows a large (diameter about 30 km) volcano with a morphology very similar to the strato-volcanoes of Earth. This volcano has a large summit caldera (8 km diameter) with a flat, shallow floor. Its slopes are drained by a system of radially oriented grooves and valleys similar to features commonly found on terrestrial strato-volcanoes with hot springs on their slopes. Furthermore, it is remarkable that a surrounding palaeofluvial landscape is associated with this Martian volcano. A wide, flat, linear depression starts immediately at the base of the volcano and extends to the north. At a distance of about 30 km the depression becomes a narrow and sinuous fluvial drainage channel which, as it becomes wider and complicated by tributaries, continues further northwards for a distance of more than 300 km. It seems probable that hydrothermal systems generated by geothermal gradients within the volcanic construct served as the source area for this regional fluvial resurfaçing process. This potential example of a hydrothermal system on Mars is located on one of the planet's oldest surfaces which has an age of more than 3 Ga (Noachian period of Martian history).

Other examples of potential hydrothermal systems on Mars might be related to source areas within the large-scale dry valleys (outflow channels) previously formed by multi-episodic catastrophic floods. These valleys extend in lengths of up to a thousand kilometres, and their widths vary from some tens to hundreds of kilometres. Usually these outflow channels originate from collapsed depressions characterized by chaotic terrain. Others originate from grabens formed in artesian basins which

FIG. 1. *(Kuzmin)* A potential hydrothermal system on Mars. This is located in the highly cratered area of Terra Cimmeria, among the oldest of Martian terrains, with a surface age greater than 3 Ga. The combination of a volcanic construct with eroded slopes and a surrounding palaeofluvial lanscape observed on this slide might be considered as a morphological indication that hydrothermal processes existed in the planet's history.

discharged when provoked by intense magmatic and tectonic activity. Geothermal activity on Mars was likely to have been at its highest between 3 and 1.5 Ga ago, a time range that includes the Noachian and Hesperian epochs. So, in this ancient span of Martian history, numerous hydrothermal systems appeared and became active. As seen in Fig. 2 *(Kuzmin)*, one of the outflow channels, named Mangala Vallis, starts directly from one of the largest of the Memnonia Fossae grabens (18°S, 149°W). The graben area is interpreted to be the source area for artesian water emerging from an underground aquifer. This channel is located on the SW edge of the Tharsis volcono--tectonic mega-updome and extends to the north for a distance of about 1000 km. Interaction of an artesian basin with magmato-tectonic processes seems the probable reason for activation of hydrothermal processes (for example, hot spring activities) in the source area of Mangala Vallis during all periods of the outflow channel activity.

During the last one billion years of Martian history, the Amazonian period, hydrothermal systems may have been associated with regions of effusive volcanic activity along faults and around volcanic sheets. Thus far, morphological signatures of potential hydrothermal activity during this time have not been identified. A possible reason why no Amazonian period correlations between volcanic and fluvial activities have been identified is that fluvial activity on Mars, for the most part, had stopped. In addition, shallow surface features, including morphological manifestations of hydrothermal systems, were buried beneath thick lava plains deposited by widespread effusive events. It seems probable that most large-scale catastrophic water floods on the Martian surface coincided with major episodes of magmatic and tectonic activity during the Noachian and Hesperian periods. Hydrothermal processes on Mars are therefore likely to have been most active only in the ancient geological history of the planet.

Nisbet: Do the volcanoes on Mars of different ages have different topographies? Topography is viscosity-dependent, and from this you can back-guess the composition of the volcano. Is there a systemic change over time?

Carr: The thing that you notice most with respect to the volcanoes, is that the younger ones appear to be much larger than the older ones.

Nisbet: Presumably, that's because they're each from a single stationary plume source.

Carr: All the young volcanoes are very much like Hawaiian shield volcanoes. What's puzzling about the volcanoes that has implications for hydrothermal activity, is that the Hawaiian volcanoes go through a stage where there's a lot of pyroclastic activity, that is, explosive eruptions of ash, after building an initial shield. You don't see any evidence of that stage on the large volcanoes like Olympus Mons. The implication is that this more volatile-rich stage is not reached in those particular volcanoes. However, when you go to some of the older volcanoes you see quite a bit of evidence that could be interpreted as pyroclastic activity. When you go back to really ancient ones, particularly one called Tyrrhena Patera, it appears to just consist of enormous ash deposits, because they are very deeply eroded. I think there is a sort of rough progression that you can see in the volcanic style.

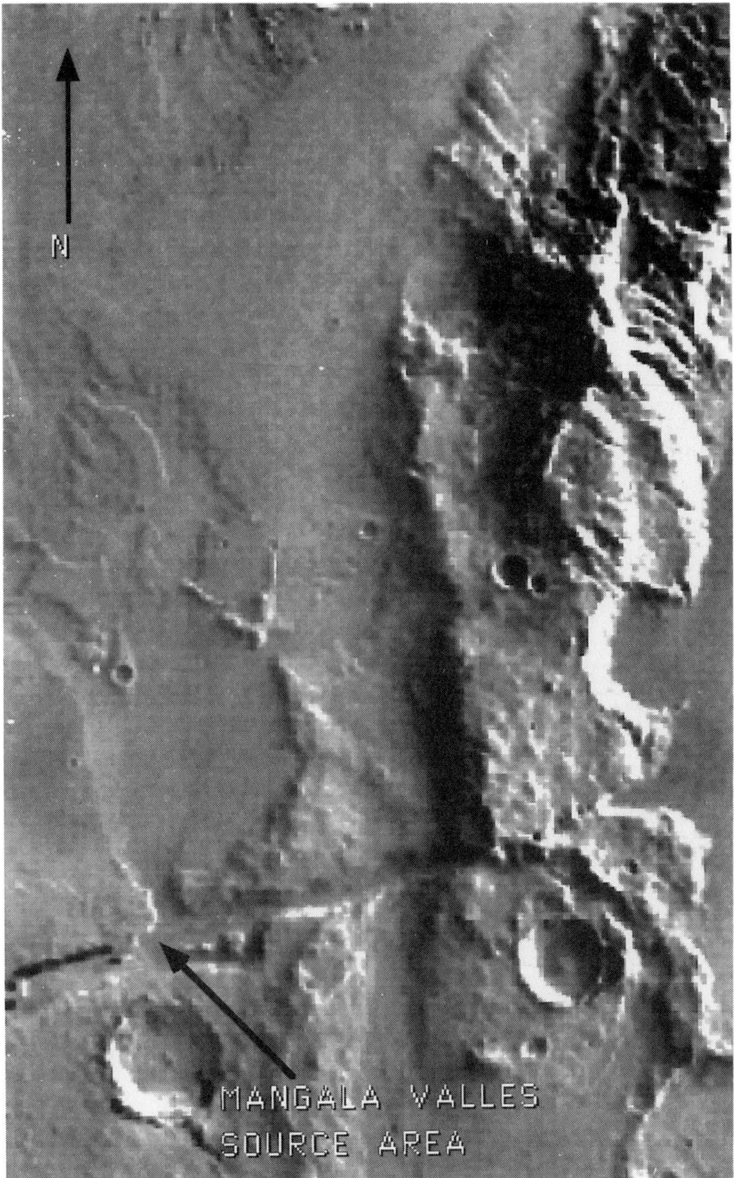

FIG. 2. *(Kuzmin)* Source area of the Martian outflow channel Mangala Vallis. The channel starts directly from a graben in the Memnonia Fossae area. Interaction between a regional artesian basin and underlying magmatic activity could have been the mechanism behind the development of hot spring processes here during the Hesperian period.

Nisbet: Presumably you're implying that the older ones are more volatile-rich volcanoes and the younger ones are basaltic shields?

Carr: My suspicion is that Tharsis, the area where these large young volcanoes are found, has basically become dehydrated. The water has flowed away to form some of the very large channels. The whole region has become dehydrated and you don't see the effects of the volatiles. In some of the other provinces this has not happened and the effects of water on the volcanism are much more evident.

Hydrothermal systems on Mars: an assessment of present evidence

Jack D. Farmer

NASA Ames Research Center, MS 239-4, Moffett Field, CA 94035-1000, USA

Abstract. Hydrothermal processes have been suggested to explain a number of observations for Mars, including D/H ratios of water extracted from Martian meteorites, as a means for removing CO_2 from the Martian atmosphere and sequestering it in the crust as carbonates, and as a possible origin for iron oxide-rich spectral units on the floors of some rifted basins (chasmata). There are numerous examples of Martian channels formed by discharges of subsurface water near potential magmatic heat sources, and hydrothermal processes have also been proposed as a mechanism for aquifer recharge needed to sustain long term erosion of sapping channels. The following geological settings have been identified as targets for ancient hydrothermal systems on Mars: channels located along the margins of impact crater melt sheets and on the slopes of ancient volcanoes; chaotic and fretted terranes where shallow subsurface heat sources are thought to have interacted with ground ice; and the floors of calderas and rifted basins (e.g. chasmata). On Earth, such geological environments are often a locus for hydrothermal mineralization. But we presently lack the mineralogical information needed for a definitive evaluation of hypotheses. A preferred tool for identifying minerals by remote sensing methods on Earth is high spatial resolution, hyperspectral, near-infrared spectroscopy, a technique that has been extensively developed by mineral explorationists. Future efforts to explore Mars for ancient hydrothermal systems would benefit from the application of methods developed by the mining industry to look for similar deposits on Earth. But Earth-based exploration models must be adapted to account for the large differences in the climatic and geological history of Mars. For example, it is likely that the early surface environment of Mars was cool, perhaps consistently below freezing, with the shallow portions of hydrothermal systems being dominated by magma–cryosphere interactions. Given the smaller gravitational field, declining atmospheric pressure, and widespread, permeable megaregolith on Mars, volatile outgassing and magmatic cooling would have been more effective than on Earth. Thus, hydrothermal systems are likely to have had much lower average surface temperatures than comparable geological settings on Earth. The likely predominance of basaltic crust on Mars suggests that hydrothermal fluids and associated deposits should be enriched in Fe, Mg, Si and Ca, with surficial deposits being dominated by lower temperature, mixed iron oxide and carbonate mineralogies.

1996 Evolution of hydrothermal ecosystems on Earth (and Mars?). Wiley, Chichester (Ciba Foundation Symposium 202) p 273–299

Hydrothermal systems develop wherever a fluid phase coexists with a heat source to drive convective energy loss. Hydrothermal systems consist of spatially confined, warm (~50 °C) to hot (> 500 °C) fluids that are in chemical disequilibrium with their host rocks (Pirajno 1992). These fluids alter, leach, transport and subsequently precipitate their primarily metallic mineral load in response to changes in physicochemical conditions. The solutes of hydrothermal systems are derived from primary (magmatic) and secondary (host rock) sources. An actively convecting hydrothermal system consists of a recharge system, a circulation cell, and a discharge system. Hydrothermal minerals are usually deposited at shallow depths in the crust along natural conduits, subsurface channels or fracture systems, and at sites of surface discharge. The history of hydrothermal activity resides in the rock record it leaves behind, although these processes also impact geochemical cycles and the composition of the atmosphere and hydrosphere (Des Marais 1996).

The most compelling evidence for hydrothermal activity on Mars derives from studies of SNC meteorites, objects believed to have come from Mars (see page 300). But in addition, geomorphic features visible in orbital images obtained during the highly successful Viking missions of the late 1970s, as well as the limited infrared spectral data obtained from the floors of rifted basins, or 'chasmata' on Mars are also suggestive of hydrothermal activity. What is presently needed for a definitive evaluation of the importance of past hydrothermal processes on Mars are broadly distributed, high spatial and spectral resolution compositional data that will allow us to identify discrete mineral deposits on the Martian surface.

The goals of this paper are to review evidence consistent with past hydrothermal activity on Mars, and to discuss the broad range of geological features that may have been formed by such processes. The environment and geological history of Mars differ substantially from those of Earth, thereby hindering direct comparative approaches to geological interpretation. However, depositional models derived from studies of ore deposits in similar geotectonic settings on the Earth provide an important starting point for discussing potential differences in hydrothermal mineralization on Mars. For comparative purposes, such models are outlined briefly under discussions of potential Martian hydrothermal features.

Data from SNC meteorites

The SNC (Shergottite, Nakhlite and Chassignite) meteorites comprise a geochemically and isotopically related group of objects that have bulk compositions similar to terrestrial basalts (for review see McSween 1994). With the possible exception of the Alan Hills 84001 meteorite, which has a crystallization age >4.5 Ga, the SNCs are geologically young, falling within the age range 1.3–0.18 Ga. The late crystallization ages and compositional range of the SNC meteorites suggests they were derived from a large thermally active, chemically differentiated, planetary-sized body (Pepin & Carr 1992). Model ages of SNCs indicate that differentiation of the Martian core and mantle were completed by the end of early accretion, and that similarly to the Earth, Mars was hot early in its history. Arguments for a Martian origin are based on the composition of atmospheric gases that were injected into glassy phases during impact. In both relative abundance and isotopic composition, extracted gases precisely match the Martian atmosphere within the measurement errors of Viking (McSween 1994).

Gooding (1992) showed that the SNC meteorites contain trace quantities (< 0.01 wt%) of primary hydrous minerals, including amphiboles and micas contained within glassy inclusions of primary igneous origin, as well as post-crystallization alumino-silicate clays, sulfates, carbonates, halides and ferric oxides formed through interactions with late-stage aqueous solutions. The post-crystallization assemblage indicates that oxidizing, hydrous solutions were present in Martian crustal environments sometime after 0.18 Ga.

Comparison of the oxygen isotope composition of secondary carbonates in Shergottites and Nakhlites indicates precipitation at low temperatures (see McSween 1994). However, the older Alan Hills 84001 meteorite contains carbonate spherules with $\delta^{13}C$ values of +41‰, suggesting they formed at elevated temperatures from groundwaters that readily exchanged with atmospheric reservoirs (Romanek et al 1994). This is consistent with D/H ratios for water extracted from SNC meteorites, which are enriched more than five times the terrestrial value, an observation attributable to escape to space of the lighter isotope (Watson et al 1994). But, the similarity of D/H values between SNC meteorites and the Martian atmosphere (determined spectroscopically) requires the operation of a highly effective mechanism of crustal–atmosphere exchange (Donahue 1995, Jakosky 1991). Plausible mechanisms include hydrothermal systems associated with intrusives, or large impacts (Jakosky & Jones 1994).

Heat sources for hydrothermal systems on Mars

As on Earth, impacts played a substantial role in shaping early Martian environments. Schubert et al (1992) suggested that impact heating during accretion raised mantle temperatures to near the solidus, and the core above the liquidus. Given the comparatively smaller size and heat capacity of Mars, the energy invested by impacts toward the end of heavy bombardment would have been a major heat source for hydrothermal systems.

On the Earth most internal heat originates by radioactive decay, being dissipated by the upwelling of mafic magmas along diverging plate margins, above deep mantle plumes, or by the production of intermediate and silicic magmas in subduction zones along convergent margins. Although Mars differs fundamentally in this respect (it never developed plate tectonics), crustal magmatism has still played an important role in the thermal evolution of the planet. The abundances of radiogenic isotope abundances (K, U, Th) in SNC meteorites are broadly similar to the Earth (McSween et al 1979) and it is quite possible that radioactive decay provided a longer-term heat source to sustain magmatism. Evidence for shallow magmatism and volcanism is widespread over the planet and, in the absence of crustal recycling, has produced the largest volcanoes in our solar system. Evidence from SNC meteorites indicates that magmatic activity occurred on Mars at least as late as 0.18 Ga (see McSween 1994).

Geological and atmospheric models supporting hydrothermal activity on Mars

In order to assess the importance of hydrothermal processes on Mars, we must also establish how much water was present at different times in the planet's history. What we know of the general hydrological history of Mars has been extensively reviewed elsewhere (Carr 1981, 1996a,b, this volume) and only selected aspects will be discussed here.

Early climate models for Mars (Pollack et al 1987) assumed an atmosphere of 1–3 bars of CO_2. However, more recent models by Kasting (1991) incorporate the effects of reduced solar luminosity which prevailed during early Martian history. Such models apparently do not yield enough greenhouse warming to produce liquid water at the surface. But, atmospheric warming can also be attained by introducing small amounts of alternative greenhouse gases such as CH_4, NH_3 and SO_2 (see Squyres & Kasting 1994), although the short residence times of these compounds in the atmosphere presents a problem for such scenarios (Carr 1996b, this volume).

Loss of a primitive CO_2-rich atmosphere on Mars is generally attributed to the formation of crustal weathering products and the precipitation of carbonate minerals. This idea receives support from geochemical models using a basaltic host rock, a meteoric water source, and low water/rock ratio which indicate that over a broad range of conditions, hydrothermal mineralization could have been a highly effective means for sequestering CO_2 in the Martian crust as disseminated carbonates (Griffith & Shock 1995).

Small, poorly integrated and highly localized channel networks having tributaries with amphitheatre-shaped headwalls, dominate the older, heavily cratered terranes on Mars. Lunar-calibrated cratering chronologies (Neukum & Hiller 1981) suggest that most of the small valley networks were formed toward the end of early bombardment (Carr 1996b, this volume). Most older Martian channels resemble terrestrial features created by spring sapping, and not those formed by surface run-off (Baker 1990). Some of the smallest valley networks visible in Viking images originate near the margins of impact crater melt sheets and have been attributed to hydrothermal outflows (Brakenridge et al 1985).

The localized nature of Martian valleys and their limited integration into drainage networks is difficult to explain by atmospheric precipitation models, which should result in a more uniform distribution over regional terranes (Gulick 1993). The great length of many Martian channels (hundreds of kilometres) requires sustained periods of erosion and a long-term hydrological cycle. But to sustain the headward erosion of valleys there must have been an effective mechanism for recharging local aquifers. Recent climate models for early Mars suggest sub-zero surface temperatures (-10 to -20 °C) and an extensive subsurface cryosphere. This appears to be inconsistent with extensive recharge through atmospheric precipitation. Alternatively, it has been suggested that recharge of Martian aquifers was maintained by hydrothermal convection (Squyres & Kasting 1994).

Geological evidence suggests that early outgassing of water on Mars was equivalent to several hundred metres depth over the surface (Carr 1996b, this volume). Much of this water may yet remain in the crust as ground ice and permafrost. A variety of geological features at higher latitudes (> 40°) have been attributed to the action of ground ice and support the concept of a widespread subsurface Martian cryosphere (Squyres & Carr 1986). Clifford (1993) presented evidence for an extensive global groundwater system capable of sustaining long-term interchanges with the atmosphere. Such interchanges would be significantly enhanced by hydrothermal circulation associated with localized igneous or impact heat sources. As mentioned before, isotopic data from SNC meteorites suggests an on-going mechanism for crustal–atmosphere exchange, with hydrothermal systems being a prime candidate (Jakosky & Jones 1994).

A number of geomorphic features on Mars have been attributed to releases of water by the magmatic heating and melting of ground ice. In contrast to the small valley networks which dominate the ancient cratered highlands on Mars, are much larger channels formed by catastrophic outflooding of subsurface water. These features are mostly found in younger Martian terranes, particularly around the margins of the large volcanic complexes at Tharsis (Carr 1981, 1996b, this volume). In the presence of a thick, confining layer of near-surface ground ice, outfloods of subsurface water are likely to have been focused at sites where the cryosphere was thinner because of locally higher heat flow as a result of magmatic intrusions and/or crustal thinning. In fact, many outflow channels originate within chaotic terranes interpreted to have formed where ground ice melted due to localized magmatic intrusions (Masursky et al 1986).

Hydrothermal prospecting on Mars

On Earth, the surface expression of hydrothermal activity varies greatly in relationship to such things as tectonic setting, geothermal gradient, geohydrology and secular variations in climate. We still do not yet understand the relative importance of these factors on Mars and how they have varied through time. Furthermore, with an average atmospheric density of only ~7.5 mb and an equatorial temperature range of about − 93 °C to + 13 °C, liquid water is unstable on the Martian surface and can only exist at depth, beneath a confining layer of ground ice. Thus, if active subsurface systems currently exist on Mars, they are only likely to be detectable at the surface as spatially confined anomalies in surface temperature or atmospheric composition.

The bulk composition of the Martian crust is similar to basalt, and the solutes of Martian hydrothermal systems can be expected to be comparatively enriched in Fe, Mg and Ca (similar to seafloor hydrothermal systems on Earth). Because of higher temperatures, and lower silica and water contents, mafic magmas usually rise to higher elevations in the crust than silicic magmas, a situation favouring near-surface hydrothermal systems. But at 53% of the Earth's diameter and with 38% the gravity, volatile loss and cooling on Mars should also occur with much greater efficiency, particularly in the presence of a porous and permeable megaregolith. Thus, surface hydrothermal temperatures on Mars are likely to have been lower, on average, than for comparable geological settings on Earth, with surface deposits being dominated by lower temperature mineral assemblages.

As noted previously, at this stage in Mars exploration what we can say about hydrothermal processes rests upon a comparatively small number of geochemical observations from SNC meteorites, and interpretations of a select number of geomorphic features observed in orbital images. This stands in marked contrast to the types and amounts of data that are usually available to explorationists searching for hydrothermal deposits on Earth (e.g. elemental abundances, surface mineralogy, magnetic intensities and gravity). Earth-based reconnaissance often begins with broadly-based geochemical sampling and high spatial resolution, hyperspectral, near-infrared mapping (1.0–3.0 μm range). The thermal emission spectrometer (TES) which will be flown to Mars in 1996 will provide useful information within the spectral range of 5–12 μm, although at a somewhat coarse spatial resolution of ~3 km/pixel (Christensen et al 1992).

Targeting specific sites on Mars for high resolution orbital imaging during upcoming missions is especially important because only a small fraction of the Martian surface is likely to be imaged at high resolution. The sites listed in Table 1 were selected to represent a broad range of geological features on Mars that may be shown related to past hydrothermal activity. Potential hydrothermal sites were chosen based on the concurrence of simple channel features (e.g. those formed by sapping, or outfloods of subsurface water), potential heat sources (e.g. impact craters, volcanic constructs, and shallow subsurface igneous intrusives), and/or anomalous albedo features that could reflect local mineralizaton. The discussion which follows is organized around

TABLE 1 Potential hydrothermal features and terranes for Mars

Name	*Location*	*Features*
Hadriaca Patera	32° S, 266° W	Sapping and outflow channels on volcanic slopes and caldera floor deposits
Apollinaris Patera	90° S, 186° W	Sapping channels on volcanic slopes and high albedo features near caldera rim
Cerberus volcanic plains	7° N, 200° W	Volcanic fissures with anomalous albedo features
NW Elysium Mons	25° N, 224° W	NW slope fissure system source for volatile-rich pyroclastics
Ismenius Lacus	35.5° N, 334° W	Small volcanic constructs associated with fretted channels
Aram Chaos	3° N, 20° W	Chaotic terranes and source areas for large outflow channels
Candor Chasma	5° S, 76° W	Rifted basin with albedo features attributed to mineralization
Margaritifer Sinus impact craters	24° S, 8° W	Head reaches of small, flat-floored ravines flanking impact craters

this general geotectonic framework, and provides a summary of pertinent observations for each site type. Where appropriate, brief summaries of depositional models for hydrothermal deposits found in similar geotectonic settings on Earth are included to help draw attention to the potential differences between meteoric/magmatic systems that dominate on Earth and the cryospheric/magmatic systems that seem to have dominated on Mars.

Potential magmatic hydrothermal systems on Mars

Channels on volcanic slopes

Hydrothermal circulation systems associated with large stratovolcanoes on Earth tend to be positioned deep within volcanic edifices. In contrast to Mars, where aquifers are likely to have been recharged by mostly subsurface processes (e.g. melting of ground ice by magmatic intrusions and hydrothermal convection), such systems on Earth are typically sustained by meteoric recharge on volcanic slopes, or by the lateral inflow of groundwater from adjacent areas (Henley 1996, this volume, see also Pirajno 1992). Water, drawn upward toward centrally located intrusives, emerges at high elevations on volcanic slopes, exiting near crater and caldera rims. Rising hydrothermal columns become progressively vapour dominated due to the decreasing influx of groundwater upslope. Because H_2S, CO_2 and other gases dominate the vapour phase, surface systems on high volcanic slopes are characterized by acid-sulfate springs, fumaroles and solfataras. Fumarolic vapours alter volcanic materials to clays and deposit sublimates of various minerals, including chlorides of alkali metals, ammonia and ferric iron, sulfates of various alkali metals and Ca, native sulfur (by oxidation of H_2S), and sulfides of Fe and Pb, with traces of Cu, Mn, Zn, As, Hg and Sn. In contrast, springs emerging at some distance from volcanic slopes tend to be neutral chloride or alkaline springs which deposit siliceous sinters, sulfates or carbonates (see Pirajno 1992). Obviously, volcanic slopes and caldera floors (see below) comprise important targets for high resolution infrared remote sensing on Mars. Identification of similarly-zoned mineral assemblages on the slopes of Martian volcanoes would provide compelling evidence for past hydrothermal systems on Mars and aid our understanding of the planet's hydrological history.

Hydrological processes have been invoked to explain incised valleys on the slopes of many Martian stratovolcanoes (Gulick & Baker 1990). For example, Hadriaca Patera (Lat. 32°S, Long. 268°W) is a low-relief highland volcano with a large (~75 km diameter) caldera (Fig. 1, large arrow). The flanks of the Hadriaca Patera are incised by numerous simple, non-branching channel forms, most of which have shallow, trough-shaped floors (Fig. 1, small arrow). The morphology of these channels indicates that volcanic slopes are comprised of layered pyroclastic deposits that are easily eroded. Channels are thought to have been initiated by density-driven pyroclastic flows, but subsequently enlarged by fluvial activity (Crown et al 1992).

Another example is Apollinaris Patera (8°S, 187°W; Fig. 2). This volcano possesses a large (80 km diameter) summit caldera, and a well-defined lobate deposit emanating from a breach in the caldera rim (Robinson et al 1993). Three types of channels are

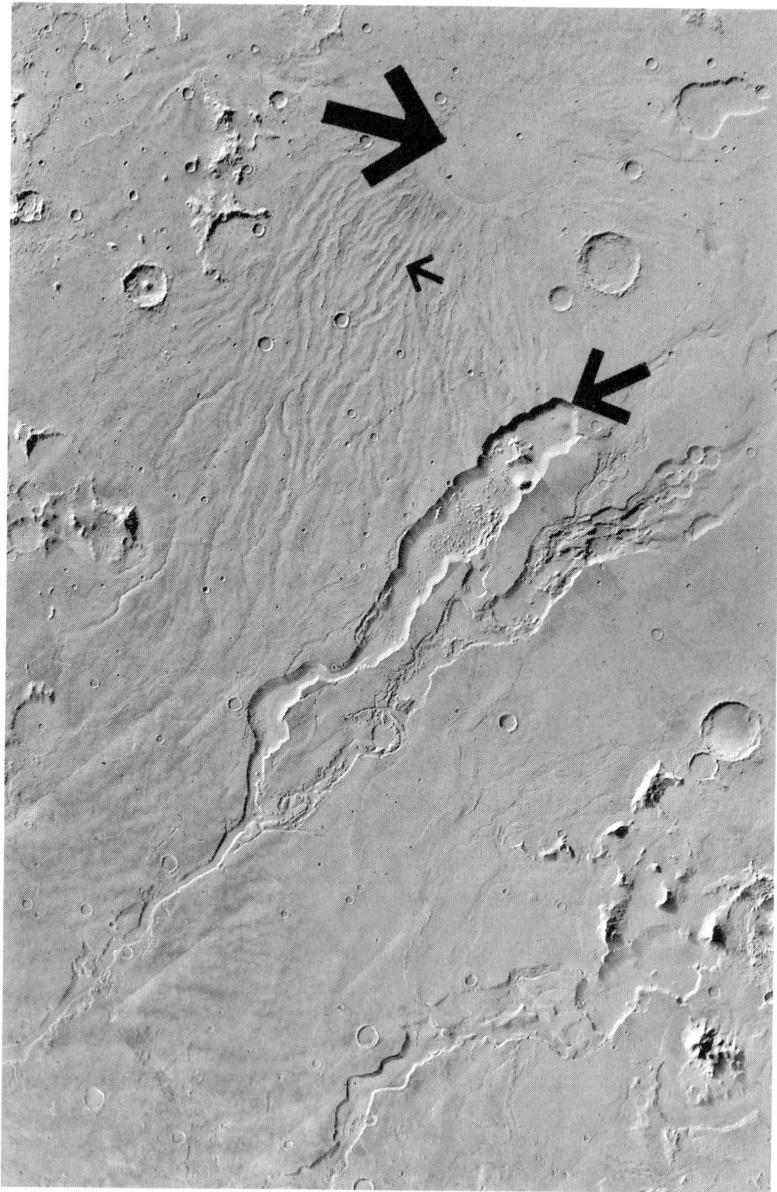

FIG. 1. The ancient Martian volcano, Hadriaca Patera showing summit caldera (~75 km diameter; large arrow), small flank channels (small arrow) and the large outflow channel, Dao Vallis (~45 km wide; medium arrow).

present on the slopes of Apollonaris Patera (Gulick & Baker 1990): (a) narrowly-incised valleys lacking tributaries which originate outside the caldera rim (1 on Fig. 2); (b) shallow, straight channels carved into the lobate deposit described above (2 on Fig. 2); and (c) a valley network on the NE flank of the volcano which possesses short, stubby tributaries (3 on Fig. 2). Interestingly, Robinson & Smith (1995) have interpreted a high albedo feature located near the caldera rim on the NE rim of the volcano (Fig. 2, arrow) to be a mineral deposit formed by past fumerolic activity.

Dao Vallis (33°S, 266°W) is a broad (>45 km wide) outflow channel that originates from an amphitheatre-shaped source area on the southern flank of Hadriaca Patera (Fig. 1, medium arrow). This channel extends southwest a distance of ~500 km where it enters the Hellas Basin, a large impact crater. Squyres et al (1987) attributed these channels to outfloods of subsurface water released when ground ice was melted by localized subsurface heat sources (e.g. shallow magmatic intrusions). Dao Vallis clearly truncates (post-dates) smaller flat-floored channels that originate on the higher flank regions of Hadriaca Patera, although the period separating their formation is problematic. The floor materials of Dao Vallis suggest extensive modification by fluvial processes and headward growth by the collapse of channel walls (Crown et al 1992). The origin of small mounded features on the floor of Dao Vallis cannot be resolved with Viking images, and we presently lack spectral imaging that could help evaluate mineralogy. Thus, floor deposits near the head reaches of Dao Vallis constitute an important target for high resolution imaging and infrared spectrscopy during upcoming orbital missions (Farmer et al 1993).

Caldera floors

Calderas are enlarged craters that form when the summit area of a volcano collapses following large-scale explosive eruptions. Subsidence occurs when magmatic support is withdrawn from below, causing overlying mass of rock to founder along circular fracture systems, or ring faults.

On Earth, caldera floors are often a locus for magmatic-meteoric hydrothermal systems and focused mineralization (see Pirajno 1992). Terrestrial calderas tend to have shallow, laterally confined aquifers, dominated by meteoric recharge (Henley 1996, this volume). Flat convection cells form above relatively massive magma chambers, and because the convecting column is within easy reach of the surface, alkaline-chloride waters enriched in Na, K, chlorides, silica, bicarbonates, fluoride, ammonia, As, Li, Rb, Cs and boric compounds dominate. Boiling produces a two-phase system (liquid plus steam) in the upper part of the convecting column. H_2S becomes concentrated in the vapour phase and produces fumaroles, solfataras and acid sulfate waters at the surface immediately above shallow instrusives. In outlying areas, cooler, chloride-rich waters (pH 5–9) dominate, and because silica solubility is lower for chloride solutions at an alkaline pH, siliceous sinters tend to be deposited along declining temperature gradients. For areas underlain by limestone or other carbonate-rich rocks, springs tend

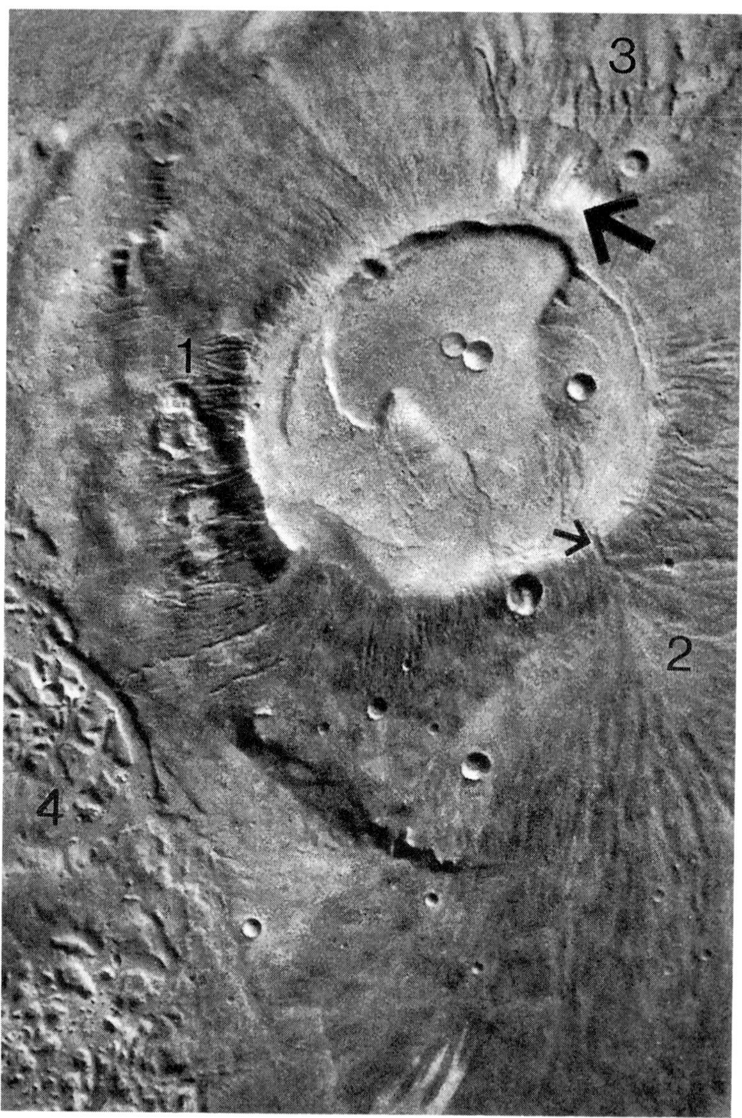

FIG. 2. The ancient Martian volcano, Apollonaris Patera showing summit caldera (~80 km diameter) and high albedo feature (arrow) interpreted to be fumarole deposit. Also visible are several types of channels, including narrowly incised valleys lacking tributaries which originate outside the caldera rim (1), shallow, straight channels carved into a large lobate deposit below a breach in the caldera rim (2), and the upper reaches of a valley network on the NE flank of the volcano which exhibits short, stubby tributaries (3).

to deposit calcareous sinters (travertines), while in mafic volcanic terranes, iron-precipitating spring systems dominate.

Calderas are widespread features of Martian volcanic terranes (Crumpler et al 1996) and they have been identified as high priority targets for exopalaeontology because of their potential for focused hydrothermal mineralization (e.g. Walter & Des Marais 1993, Farmer & Des Marais 1994). An objective of upcoming orbital missions to Mars should be to look for zoned patterns of mineralization on the floors of calderas.

Volcanic fissures

Many ancient Martian volcanoes (e.g. Hadriaca, Apollinaris and Tyrrhena Paterae and Elysium Mons) were apparently dominated by pyroclastic eruptions, suggesting that higher volatile abundances prevailed during earlier periods of Martian volcanic activity. This is in contrast to the dominantly effusive (lava-prone) volcanism of younger Martian volcanoes, such as those at Tharsis. The volcanic plains northwest of Elysium Mons (25°N, 214°W) consist of lobate deposits with well-defined medial channels (Granicus Valles, Fig. 3 small arrows) that originate from a fissure trough on the NW flank of the volcano (Fig. 3, large arrow). Because their estimated volume is 10–100 times larger than the fissure system from which they originated, these flows have been interpreted to be 'lahars', or volcanic mud flows formed when pyroclastic materials incorporated water by melting subsurface ground ice (Christiansen 1989). The surfaces of these pyroclastic flows and the trough fissure systems from which they were erupted are potential targets for hydrothermal and fumarolic activity.

Dark-coloured fractures occur on the Cerberus volcanic plains which are a part of a regional system that encircles the ancient Martian volcano, Elysium Mons (7°N, 200°W). In some cases, down-wind drifts of dark materials appear to originate from fractures. In other cases, symmetrical zones of lighter albedo are observed adjacent to fissures (Fig. 4, arrows). These lighter albedo features could be due either to fumarolic mineralization, or to systematic variations in the grain size of pyroclastics erupted from the fissures. Deciding between these alternatives will require high resolution orbital imaging, accompanied by information about mineralogy and grain size.

Fretted channel terranes

Carruthers (1995) cited evidence for volcano–ground ice interactions in association with an unnamed system of fretted channels and closed canyons located within the southern Ismenius Lacus terrane of Mars (35.5°N, 334°W). The fretted channel segment identified by the arrows in Fig. 5 lies between two features interpreted to be eruptive centres. At one end is a dark-coloured mound that is interpreted to be pyroclastic cone (Fig. 5, bottom arrow). At the other is an irregular, closed depression ~40 km in diameter that is interpreted to be a Maar-type crater (Fig. 5, top arrow). At higher resolution, lobate deposits interpreted to be pyroclastic surge or debris flow deposits are visible on the flanks of the Maar crater. These deposits are cut by 400–800 m wide channels up to 20 km long which are arrayed radially around the cone.

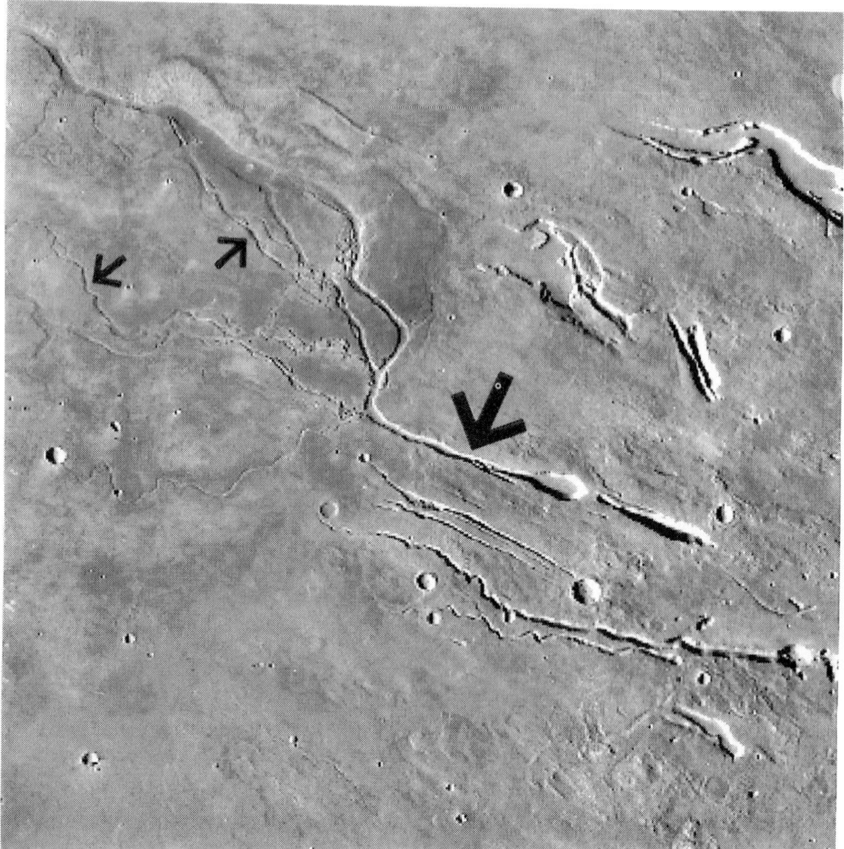

FIG. 3. Sinuous channels of Granicus Valles (small arrows) crossing the surface of lobate deposits that comprise volcanic plains northwest of the Martian volcano, Elysium Mons. These deposits, which originate from a trough fissure system on the NW flank of the volcano (large arrow), are interpreted to be 'lahars', or mud-flows formed when pyroclastics were erupted within an ice-rich terrane. Area shown is about 300 km across.

Small volcanic centres are widespread in many ice-rich, high latitude terranes on Mars (Hodges & Moore 1994), and such features are obvious potential targets for small-scale hydrothermal systems. But the spatial coincidence of small volcanic features with closed canyons, depressions and channel junctions suggests that shallow magmatic processes may have played an important role in the origin of some fretted terranes. Carruthers (1995) suggested a generalized model for fretted channel growth whereby ground ice is melted and vaporized by shallow intrusives, escaping to the surface through fractures and small vents. Channels initiated by phreatic explosions are

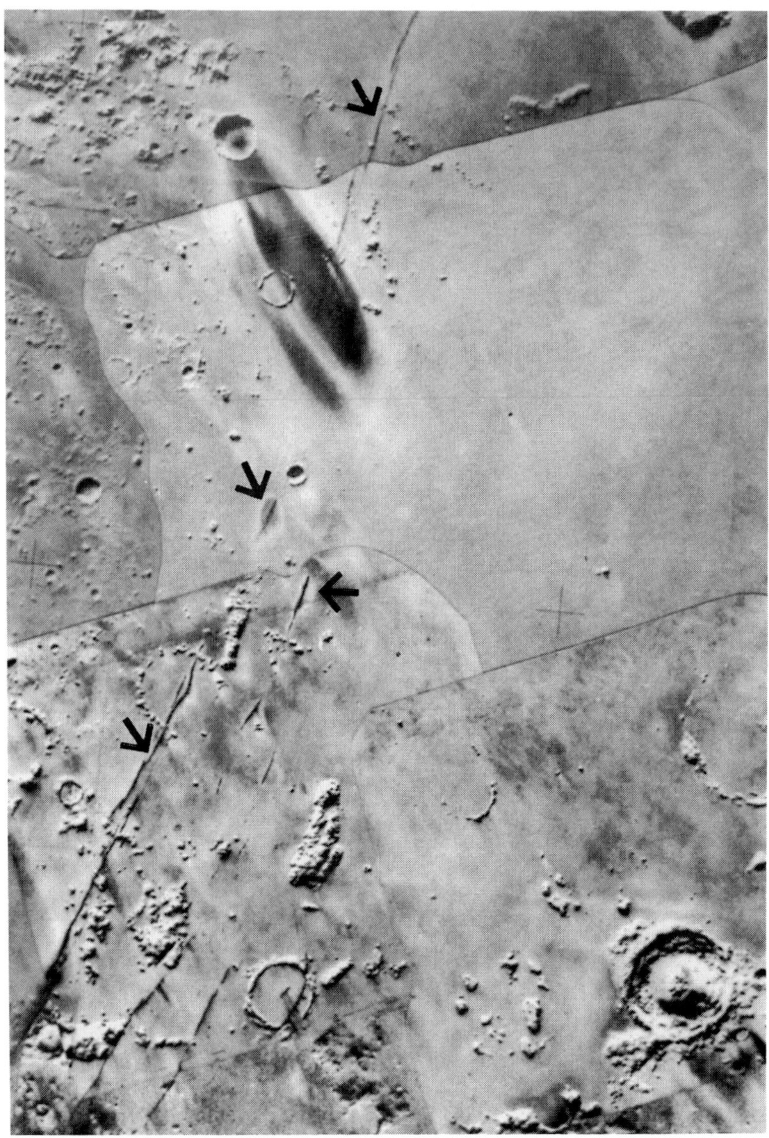

FIG. 4. Cerberus Volcanic Plains located south of Elysium Mons. Dark fractures are a part of the regional fracture system surrounding Elysium Mons. Some fractures exhibit symmetrical zones of lighter albedo (arrows) which could be due either to fumarolic mineralization, or to systematic variations in the grain size of pyroclastics erupted from the fissures. Impact crater at lower right of image is ~50 km in diameter.

FIG. 5. Fretted channels and closed canyons located on southern Ismenius Lacus, Mars. The channel segment lying between the arrows (~100 km long) is bounded by features interpreted to be small eruptive centres, including a dark-coloured mound (bottom arrow) and closed depression believed to be a Maar-type crater (top arrow).

enlarged by subsurface sapping, wall collapse, along with erosion by down-channel ice-rich debris flows. Under such conditions, the floors of fretted channels and closed depressions could have also been sites for hydrothermal mineralization.

Chaotic terranes and associated outflow channels

Most outflow channels on Mars occur in younger Martian terranes around the large volcanic complex at Tharsis. Outflows northeast of Vallis Marineris originate from highly fractured, elongate to circular collapsed zones called chaotic terranes (Baker et al 1992). These regions consist of irregularly-broken and jumbled blocks apparently formed by the withdrawal of subsurface ice and/or water (Mars Channel Working Group 1983). Masursky et al (1986) suggested a thermokarst origin for the chaos source regions of many Martian outflow channels. Outfloods of water would have occurred by the melting of water ice stored in the Martian regolith. The favoured mechanism is heating by shallow intrusives. This suggests that shallow hydrothermal systems could have developed and persisted for some time prior to and following major outflooding events. Small channels on the flanks of circular collapsed terranes, such as Aram Chaos (3°N, 20°W; Fig. 6, arrow) are logical sites for concentrating the surface deposits of hydrothermal systems (Farmer et al 1995). In addition, the walls of collapsed blocks could also expose zones of shallow epithermal mineralization.

Rifted basins

On Earth, rifted basins are often associated with a variety of sediment-hosted (stratiform) and exhalative base metal sulfide deposits (Pirajno 1992). Crustal attenuation during rifting is usually accompanied by an increase in regional heat flow and igneous activity, particularly along basin margins. In such settings, hydrothermal convection transports water from the deeper parts of the basin toward its margins. Hydrothermal fluids usually ascend along basin-bounding normal faults, and invade adjacent aquifers, forming exhalative sulfide deposits at or below the sediment–water interface. In rifted basin lakes, stratiform ore deposits may form either from diagenetic, low temperature, metalliferous brines formed during compactional dewatering, or from high temperature hydrothermal systems associated with basin margin faults (Robbins 1983). Hydrothermal fluids originate from meteoric waters that are heated by intrusives at depth, picking up CO_2, H_2, N_2, and biogenic H_2S and CH_4. In addition, the Na and Cl content of saline lakes in such settings often reflects substantial hydrothermal inputs.

Candor Chasma (5°S, 76°W; Fig. 7) is a semi-enclosed structural trough ~525 km long by ~150 km wide that is part of the Vallis Marineris Canyon system, a large system of steep-walled, canyons formed by extension along east–west trending radial fracture systems associated with the Tharsis volcanic province (Plescia & Saunders 1982). Viking Orbiter multispectral images of the western end of Candor Chasma reveal two 20 km-long depressions of anomalous albedo. These features are located near the margins of interior layered deposits on the chasma floor which have been interpreted to be

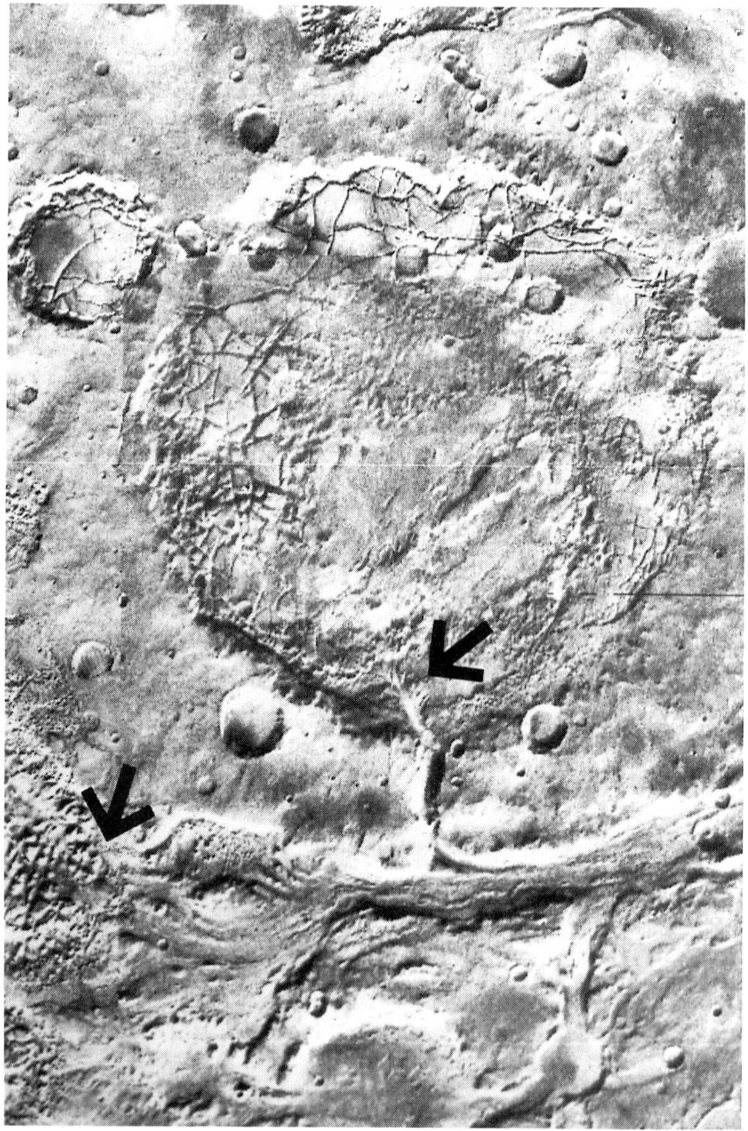

FIG. 6. Aram Chaos is a semi-circular collapse feature (~300 km diameter) located near the upper reaches of Ares Vallis outflow channel. Chaotic terranes are thought to have formed where magmatic intrusions melted ground ice. Under such conditions, near surface hydrothermal systems may have developed near the head reaches of small valleys, or on valley floors (arrows).

FIG. 7. The floor of Candor Chasma, a semi-enclosed structural trough (~150 km wide) that is part of the Vallis Marineris Canyon system. An anomalous spectral unit on the chasma floor is interpreted to be due to the presence of aqueously-deposited ferric oxides, oxyhydroxides and hematite.

ancient lake deposits (Nedell et al 1987). Combined analysis of Viking multispectral data and high spatial resolution infrared reflectance data obtained by the ISM imaging spectrometer on the Russian probe, Phobos 2, suggest a local enrichment in ferric oxides and oxyhydroxides, and a small increase in crystallinity due to hematite (Geissler et al 1993). The anomalous spectral unit is confined to the chasma floor (Fig. 7 arrow), and suggests that mineralization and alteration is restricted to low areas, possibly as a result of groundwater seepage and subaqueous alteration of pre-existing, iron-rich rocks. The tectonic setting and restricted nature of the mineralization suggests that the spectral unit could be hydrothermal in origin, but further evaluation of this hypothesis will require more detailed mineralogical information (Geissler et al 1993).

Channels surrounding impact craters

Newsom (1980) suggested that impact-related hydrothermal systems were widespread on early Mars. Impact melting of volatile-rich crustal materials is likely to have set up hydrothermal systems that would have emerged at the surface along the margins of impact melt sheets. Brakenridge et al (1985) identified two classes of small, sapping valleys associated with impact craters formed just prior to the end of heavy bombardment, ~3.8 Ga. These channels include some of the smallest valleys visible in Viking images — small, flat-floored ravines oriented subparallel to the slopes flanking craters. One group of valleys occur near the margins of impact crater melt sheets and are thought to have been formed by hydrothermal outflows (Fig. 8). The headwall regions of this smaller class of valleys (ravines) adjacent to impact crater melt sheets (e.g. Fig. 8, arrow) are a particularly compelling target to explore for hydrothermal mineralization.

Brakenridge et al (1985) also recognized a second class of larger and more widespread valley systems within inter-crater plains areas. These flat-floored branching valleys are thought to have formed through the interaction of impact, volcanism and erosion.

Given the complex, overlapping events at the end of late bombardment, it is possible that hydrothermal mineralization was widespread within heavily cratered ancient highland terranes. Wilhelms & Baldwin (1989) suggested that magmatism played a role in the origin of small valley systems in some upland terranes of Mars as a result of the widespread emplacement of shallow sills into an ice-rich crust during later periods of Martian history. In terrestrial settings, loosely consolidated sediments and/or pyroclastics marginal to dikes and sills, undergo porosity reduction during magma emplacement, expelling substantial amounts of pore water. For example, it is estimated that an area of 2 km^2 intruded by sills can expel up to 40 × 106 m^3 of water (Einsele et al 1980). Under such conditions, temperatures adjacent to intrusives can exceed 400 °C, and near surface pore waters may reach the boiling point. In permeable country rocks, symmetrically-zoned alteration halos usually form adjacent to tabular plutons as a result of convecting hydrothermal fluids. Water can also be channeled upward along faults and discharge as hot springs at the surface.

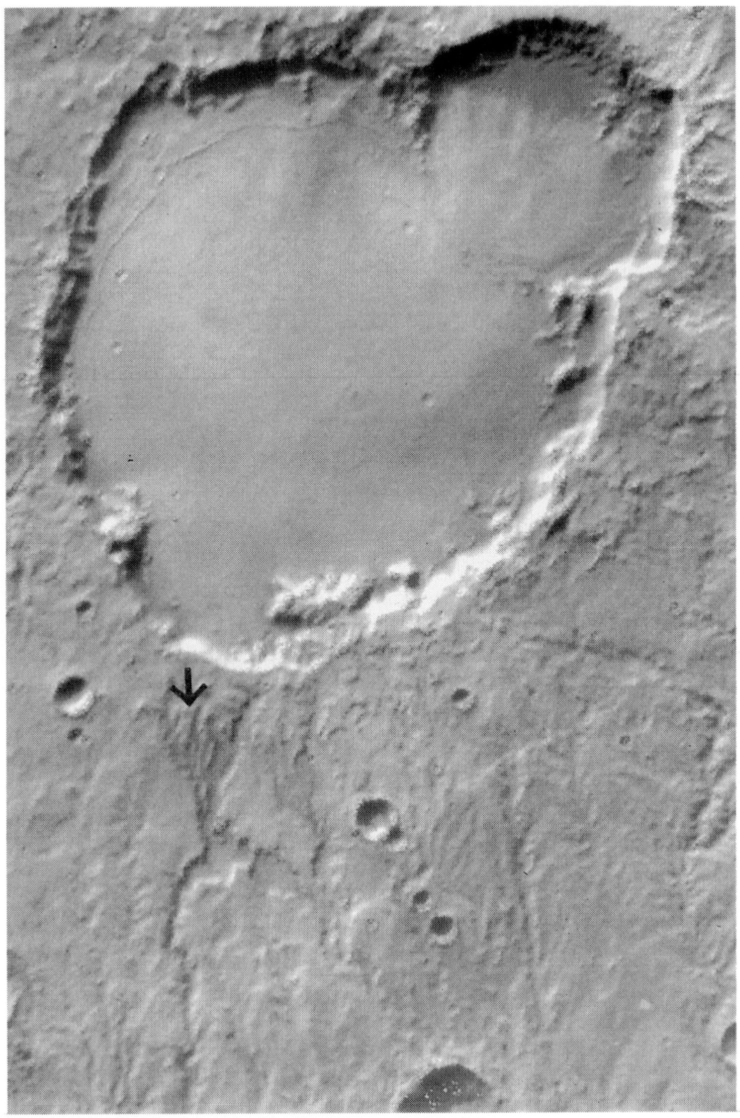

FIG. 8. Small, flat-floored ravines, originating near the margin of an ~70 km diameter impact crater in the Margaritifer Sinus region of Mars. Crater rims have been removed by erosion leaving a flat, smooth crater floor interpreted to be an impact melt sheet. Such valleys often comprise higher order tributaries of small valley networks within ancient cratered highland terranes of Mars and are interpreted by some authors to have formed by surface outflows of hydrothermal systems formed by impact heating.

Summary discussion

The interpretation of geomorphic features on Mars is presently challenged by the lack of spatially-resolved orbital imaging and information about surface mineralogy. Interpretations are also complicated by differences in the present Martian surface environment and uncertainties about how geological and climatic processes have varied through time. But, there are a number of independently-derived observations that are consistent with past hydrothermal processes on Mars and this framework provides the basis for preliminary evaluations.

Thermodynamic models indicate that hydrothermal mineralization could have been an effective means for sequestering atmospheric CO_2 in the early Martian crust. Thus, hydrothermal systems may have played a prominent role in loss of the early atmosphere (Griffith & Shock 1995). This is consistent with isotopic data from the Mars-derived SNC meteorites which indicate there has been an active and ongoing process of crustal–atmosphere exchange throughout Martian history (see McSween 1994). Hydrothermal convection provides the most compelling mechanism for such exchanges (Jakosky & Jones 1994). The young age of the SNCs suggests that hydrothermal systems could still be active on Mars, but given the instability of liquid water at the surface, their expression is likely to be limited to localized anomalies in surface temperature or atmospheric composition.

The selection of potential hydrothermal sites on Mars is presently based on the concurrence of water-carved geomorphic features and potential heat sources (e.g. volcanic centres or suspected magmatic intrusions). In some instances, suspected hydrothermal features are accompanied by albedo or spectral anomalies that suggest focused mineralization. These potential hydrothermal sites are regarded as important for Mars exopalaeontology and should be targeted for high resolution orbital imaging and mineralogical mapping during upcoming missions (see Farmer et al 1993, Farmer & Des Marais 1994).

Perhaps the most convincing geomorphic evidence for hydrothermal activity on Mars is the observation of simple channel systems along the margins of impact craters in ancient cratered highland terranes on Mars (Brakenridge et al 1985, Newsom 1980). In addition, the presence of iron-rich deposits of possible hydrothermal origin observed on the floors of some chasmata (Geissler et al 1993) suggests intriguing similarities to the stratiform ore deposits of many rifted basin settings on Earth. More controversial are numerous examples of small channel systems on the slopes of Martian stratovolcanoes. Although a variety of processes (density-driven pyroclastic flows, sapping processes, and/or surface run-off) may have been involved in their formation, subsurface water appears to have played a key role in many instances (Gulick & Baker 1990).

The visible concentration of channels on volcanic slopes, within terranes where they are otherwise absent, provides important circumstantial evidence for the thermal focusing of hydrological activity. And the presence of anomalous, high albedo features near the crater rims (e.g. Apollinaris Patera; Robinson & Smith 1995) compares favorably to fumarolic deposits found in similar locations on terrestrial volcanoes.

Widespread magma–cryosphere interactions are suggested by chaotic source terranes associated with the large outflow channels surrounding Tharsis (Masursky et al 1986), by volatile-rich pyroclastic fissure eruptions on the plains west of Elysium Mons (Christiansen 1989), and by the association of fretted channels, closed depressions and small volcanic centres in some areas of the northern plains (Carruthers 1995).

Present evidence suggests that classical meteoric systems of the type described for volcanoes on Earth are probably not directly applicable to Mars. Climate models suggest an early cool Mars with surface temperatures below the freezing point. Thus, atmospheric precipitation and meteoric recharge may have been minimal. However, the aquifer recharge needed to sustain channel erosion could have been provided by hydrothermal convection (Squyres & Kasting 1994). In many cases, features of the adjacent terrane suggest that ground ice was widespread prior to volcanism, and that a subsurface cryosphere could have provided a source of water for hydrothermal systems. Numerical modeling of hydrothermal systems developed adjacent to a 500 km^3 intrusion suggest that discharges of water exceeding 1.0×10^{13} m^3 during the cooling history (Gulick 1995). This volume is probably sufficient to carve the small valley networks on Mars even in the presence of a subsurface cryosphere. Planetary-scale hydrological models for Mars support the presence of a global groundwater system that is capable of replenishing long term losses of ice and water to the atmosphere by such surface outflows (Clifford 1993).

In conclusion, under cooler climatic conditions, the shallow portions of hydrothermal systems on early Mars would have been dominated by magma–cryosphere interactions and the effects of declining atmospheric pressure. Given the greater efficiency of volatile outgassing and magmatic cooling afforded by the lower gravity of Mars, the attenuated atmosphere, and a permeable megaregolith, hydrothermal systems are likely to have had lower average surface temperatures than comparable geological settings on Earth. Compositionally speaking, the predominance of basaltic host rocks on Mars suggests hydrothermal fluids and associated deposits relatively enriched in Fe, Mg, Si and Ca, similar to hydrothermal deposits in mafic volcanic terranes on Earth. Thus, surficial hydrothermal deposits are likely to be dominated by lower temperature, mixed iron oxide and carbonate mineralogies.

Acknowledgements

This work was supported by grants from NASA's Exobiology Program.

References

Baker VR 1990 Spring sapping and valley network development. Geol Soc Am Spec Paper 252:235–265

Baker VR, Carr MH, Gulick VC, Williams CR, Marley MS 1992 Channels and valley networks. In: Kieffer HH, Jakosky BM, Snyder CW, Matthews MS (eds) Mars. University of Arizona Press, Tucson, AZ, p 493–554

Brakenridge GR, Newsom HE, Baker VR 1985 Ancient hot springs on Mars: origins and paleoenvironmental significance of small Martian valleys. Geology 13:859–862

Carr MH 1981 The surface of Mars. Yale University Press, New Haven, CT

Carr MH 1996a Water on Mars. Oxford University Press, New York

Carr MH 1996b Water on early Mars. In: Evolution of hydrothermal ecosystems on Earth (and Mars?). Wiley, Chichester (Ciba Found Symp 202) p 249–267

Carruthers MW 1995 Igneous activity and the origin of the fretted channels, Southern Ismenius Lacus, Mars. Lunar Planet Sci 26:217–218

Christensen PR, Anderson DL, Chase SC et al 1992 Thermal emission spectrometer experiment: Mars Observer mission. J Geophys Res 97:7719E–7734E

Christiansen EH 1989 Lahars in the Elysium region of Mars. Geology 17:203–206

Clifford SM 1993 A model for the hydrologic and climatic behavior of water on Mars. J Geophys Res 98:10973–11016

Crown DA, Price KH, Greeley R 1992 Geologic evolution of the East Rim of the Hellas Basin, Mars. Icarus 100:1–25

Crumpler LS, Head JW, Aubele JC 1996 Calderas on Mars: characteristics, structural evolution, and associated flank structures. In: McGuire WC, Jones AP, Neuberg J (eds) Proceedings, conference on volcanic instability on Earth and other planets. Geol Soc Spec Publ 110:307–347

Des Marais DJ 1996 Stable light isotope biogeochemistry of hydrothermal systems. In: Evolution of hydrothermal ecosystems on Earth (and Mars?). Wiley, Chichester (Ciba Found Symp 202) p 83–98

Donahue TM 1995 Evolution of water reservoirs on Mars from D/H ratios in the atmosphere and crust. Nature 374:432–434

Einsele G, Gieskes JM, Curray J et al 1980 Intrusion of basaltic sills into highly porous sediments, and resulting hydrothermal activity. Nature 283:441–445

Farmer JD, Des Marais DJ 1994 Exopaleontology and the search for a fossil record on Mars. Lunar Planet Sci 25:367–368

Farmer JD, Des Marais DJ, Klein H 1993 Mars site selection for Exobiology: criteria and methodology. In: Greeley R, Thomas P Mars landing site catalog. NASA Ref Pub 1238, 2nd edn, p 11–16

Farmer JD, Des Marais DJ, Greeley R 1995 Exopaleontology at the Mars Pathfinder landing site. Lunar Planet Sci 26:393–394

Geissler PE, Singer RB, Komatsu G, Murchie S, Mustard J 1993 An unusual spectral unit west Candor Chasma: evidence for aqueous or hydrothermal alteration in the Martian Canyons. Icarus 106:380–391

Gooding JL 1992 Soil mineralogy and chemistry on Mars: possible clues from salts and clays in SNC meteorites. Icarus 99:28–41

Griffith LL, Shock EL 1995 A geochemical model for the formation of hydrothermal carbonates on Mars. Nature 377:406–408

Gulick VC 1993 Magmatic intrusions and hydrothermal systems: implications for the formation of Martian fluvial valleys. PhD thesis, University of Arizona, Tucson, AZ

Gulick VC 1995 Geological observations and numerical modeling of hydrothermal systems on Mars. Eos, Trans Am Geophys Union 76:330

Gulick VC, Baker VR 1990 Origin and evolution of valleys on Martian volcanoes. J Geophys Res 95:14325–14344

Henley RW 1996 Chemical and physical context for life in terrestrial hydrothermal systems: chemical reactors for the early development of life and hydrothermal ecosystems. In: Evolution of hydrothermal ecosystems on Earth (and Mars?). Wiley, Chichester (Ciba Found Symp 202) p 61–82

Hodges CA, Moore HJ 1994 Atlas of volcanic landforms on Mars. US Geol Survey Prof Paper 1534:1–194

Jakosky BM 1991 Mars volatile evolution: evidence from stable isotopes. Icarus 94:14–31

Jakosky BM, Jones JH 1994 Evolution of water on Mars. Nature 370:328–329

Kasting JF 1991 CO_2 condensation and the climate of early Mars. Icarus 94:1–13
Mars Channel Working Group 1993 Channels and valleys on Mars. Geol Soc Am Bull 94:1035–1054
Masursky H, Chapman MG, Dial AL Jr, Strobell ME 1986 Episodic channeling punctuated by volcanic flows in Mangala Valles region, Mars. NASA Tech Mem 88383:459–461
McSween HY Jr 1994 What we have learned about Mars from SNC meteorites. Meteoritics 29:757–779
McSween HY Jr, Stolper EM, Taylor LA et al 1979 Petrogenetic relationship between the Alan Hills 77005 and other achondrites. Earth Planet Sci Letters 45:275–284
Nedell SS, Squyres SW, Anderson DW 1987 Origin and evolution of the layered deposits in the Vallis Marineris, Mars. Icarus 70:409–411
Neukum G, Hiller K 1981 Martian ages. J Geophys Res 86:3097–3121
Newsom HE 1980 Hydrothermal alteration of impact melt sheets with implications for Mars. Icarus 44:207–216
Pepin RO, Carr MH 1992 Major issues and outstanding questions. In: Kieffer HH, Jakosky BM, Snyder CW, Matthews MS (eds) Mars. University of Arizona Press, Tucson, AZ, p 120–143
Pirajno F 1992 Hydrothermal mineral deposits: principles and fundamental concepts for the exploration geologist. Springer-Verlag, New York
Plescia JB, Saunders RS 1982 The tectonic history of the Tharsis region, Mars. J Geophys Res 87:9775B–9791B
Pollack JB, Kasting JF, Richardson SM, Poliakoff K 1987 The case for a wet, warm climate on early Mars. Icarus 71:203–224
Robbins EI 1983 Accumulation of fossil fuels and metallic minerals in active and ancient rift lakes. Tectonophysics 94:633–658
Robinson M, Smith M 1995 Multi-temporal examination of the Martian Surface in the Apollinaris Patera Region. Eos, Trans Am Geophys Union 76:331
Robinson MS, Mouginis-Mark PJ, Zimbelman JR, Wu SSC, Ablin KK, Howington-Kraus AE 1993 Chrononlogy, eruption duration, and atmospheric contribution of the Martian volcano Apollinaris Patera. Icarus 104:301–323
Romanek CS, Grady MM, Wright IP et al 1994 Record of fluid–rock interactions on Mars from the meteorite ALH84001. Nature 372:655–657
Schubert G, Solomon SC, Turcotte DL, Drake MJ, Sleep NH 1992 Origin and thermal evolution of Mars. In: Kieffer HH, Jakosky BM, Snyder CW, Matthews MS (eds) Mars. University of Arizona Press, Tucson, AZ, p 147–183
Squyres SW, Carr MH 1986 Geomorphic evidence for the distribution of ground ice on Mars. Science 231:249–252
Squyres SW, Kasting JF 1994 Early Mars: how warm and how wet? Science 265:744–749
Squyres SW, Wilhelms DE, Moosman AC 1987 Large-scale volcano–ground ice interactions on Mars. Icarus 70:385–408
Walter MR, Des Marais DJ 1993 Preservation of biological information in thermal spring deposits: developing a strategy for the search for fossil life on Mars. Icarus 101:129–143
Watson LL, Hutcheon ID, Epstein S, Stolper EM 1994 Water on Mars: clues from deuterium/hydrogen and water contents of hydrous phases in SNC meteorites. Science 265:86–90
Wilhelms DE, Baldwin RJ 1989 The role of igneous sills in shaping the Martian uplands. Lunar Planet Sci 19:355–365

DISCUSSION

Jakosky: Why do you think the structure in Fig. 8. is an impact crater?

Farmer: Because of the general circular morphology. The crater rim was removed by erosion, and the flat crater floor is interpreted to be an impact melt sheet. Brakenridge et al (1985) used this as an example in their paper. I wonder if Mike Carr would be willing to comment?

Carr: When you go back to the earliest history of Mars, you see impact craters in all stages of degradation. This one, for example, is a rimless depression, but it is simply an end-member of a continuous series of craters degraded to different degrees.

Jakosky: I have seen many impact craters, and this one looks distinctly different.

Carr: It is obviously a dual crater.

Davies: Has the rim eroded away?

Carr: Yes. There was a dramatic change in erosion rates at the end of heavy bombardment, around 3.8 Ga, such that these old craters look extremely eroded, whereas younger craters are barely eroded at all.

Davies: So was there a much thicker atmosphere then?

Carr: Yes, I think so.

Farmer: How do you feel about the idea of impact-crater-related hydrothermal systems?

Carr: I think that is very plausible. What I do not accept, with respect to the effects of hydrothermal activity, is that under the present climate conditions the water could ever get back in the ground. That's a part of the Gulick and Baker story.

Davies: Where does it all go?

Carr: Various places. Some was lost to space, some was frozen out at the poles, and some formed permanent ice deposits at high latitudes.

Knoll: In years gone by, a paper on this subject would have discussed at length the evidence for standing water: what is the present thinking on this issue?

Farmer: My feeling is that if standing water was present at all, it was probably only for short episodes.

Carr: If you look at the size of some of those large channels, you can calculate the volumes of water must have gone through them. They must have formed small lakes. I see no problem in having small lakes that then freeze over, leaving a permanent ice deposit. My perception is that we're seeing an equilibration process whereby the water at high regions of the planet ends up in low regions where it forms permanent ice deposits. Others have proposed the multiple formation of oceans, but this hypothesis has so little backing and causes so many problems in terms of the mechanisms involved that I can't accept it.

Knoll: That is important, because if what you say is true, hydrothermal systems on Mars would be the only suitable environment for life.

Jakosky: I didn't want to let Mike's comment about the problem of getting water brought to the surface by hydrothermal systems back down again pass completely unchallenged. The problem, of course, is that temperature is too cold for liquid water to be stable at the surface, but I think there are a couple of ways by which you might be able to do it. One is under different climatic conditions with different orbital elements, because the obliquity changes dramatically: if you can put enough CO_2 in the atmosphere, you can raise the temperature enough to let the water dribble back. The other is the exchange near the surface of hydrothermal water with ground ice. It is not clear that either of these will work, but I think it's premature to rule out exchange.

Davies: If there was continued episodic bombardment of Mars, would that raise the temperature sufficiently so that for a short period of time the water could be re-liquified and go back?

Jakosky: I don't know.

Pentecost: Most of Earth's hydrothermal travertines are fault-guided and in many areas you can see alignments of travertine deposits. It may be worth looking for alignments on Mars.

Farmer: We would need much better spatial resolution than we currently have to get to the observational scale that you are talking about. The lineaments along which travertine deposits in Yellowstone National Park are aligned are only hundreds of metres long and would not be visible on the most of the Viking images. The only features I have looked at that might be fissure-related are those I showed on the Cerberus volcanic plains south west of Elysium which have black lineaments with halos of lighter albedo. Those could be due to hydrothermal alteration, or to variations in grain size of pyroclastics erupted from fissures. It is hard to say without additional data.

Pentecost: If one has hot water flowing up from a spring onto a surface, is it possible to model how long that water will remain water, given the discharge and the temperature?

Farmer: Dr Virginia Gulick, a National Research Council Postdoctoral Fellow at Ames, has been doing this kind of model-based work. Perhaps the most interesting models involve the incorporation of a confining layer of near-surface ground ice. The bottom line is that from a small igneous intrusion she can sustain thermal systems for a very long time. Even with the presence of confining ground ice, these systems can apparently be sustained for periods long enough to carve the kinds of channels that we see.

Henley: It occurs to me that whilst those channel ways are impressive, the amount of water that is actually being mobilized compared to the amount of water that over time is convecting within the system is actually very small. So down below, there is a convective system, but only a small amount of water discharging at the surface.

Trewin: I have no experience of the surface of Mars, but it looked as though the rims of the large channel (~45 km wide; Dao Vallis) were raised. The morphology looks very similar to the bottleneck slides that occur on delta fronts in the Mississippi, with slurry and debris flow in the channels. Is it possible that it is not water that was flowing down those channels, but something with viscous strength, for instance a mixture of ice and whatever?

Carr: That has been suggested. With respect to the large flood features, the water interpretation is pretty secure. We have lots of analogies. One can take the largest flood features that we know on earth (the Chuja Basin flood of Siberia and the Channeled Scablands of Eastern Washington) and they match feature for feature these on Mars. There are very plausible mechanisms for generating these large floods. There are many other indicators of ground water, and in my paper I discuss the mechanisms for getting the water out of the ground (Carr 1996, this volume).

Powell: Mars is about half the size of the Earth. On Earth, hydrothermal systems seem to have a standard size. Why are the systems on Mars much larger than their counterparts on Earth, despite its smaller size?

Farmer: It seems as if all the geological features on Mars are large by Earth standards. But if you think of it in terms of integrating a single process over a long time and then not destroying the landform (there's no turnover of the crust), then you can perhaps account for the very large volcanoes and channel systems on Mars.

Nisbet: I lived in Saskatchewan for many years, and so I got quite used to temperatures of −40 °C. Even as cold as this, ice disappears quite fast by ablation, especially if there's some wind. Presumably the Martian ice-covered lake would have disappeared pretty fast, unless it was covered in dust?

Carr: If the lake forms at low latitude, it disappears. But at high latitudes, where the mean annual temperature is well below the frost point, it is permanent. You can calculate what the sublimation rates are. If you simply cover the lake with a thin layer of dust, it lasts forever.

Nisbet: My second question is about finding fluorine, which goes back to Jim Lovelock's old idea of releasing CFCs into the atmosphere (and maybe we could add some methane too), in order to warm up the surface of Mars. Presumably, finding supplies of chlorine and fluorine would be a secondary target of any exploration. I was interested that you mentioned the evidence for maars on Mars. Have you looked for places, such as this sort of volcano, where you might find fluorine and chlorine, particularly fluorine?

Farmer: Not specifically.

Nisbet: Presumably it remains a real, if distant possibility that you could make a million tonnes of CFCs and turn some of the CO_2 into methane by creating ponds and bogs colonized with methanogenic communities. As a long-term option Lovelock has a point: we could think of warming up the planet. It's well within our power, even now. Would it be worth doing? Or ethical? Deep ecologists would prefer it barren of life, I suspect, or would argue that as long as there is a chance of indigenous life, we should not add our own biology.

Huntington: One thing that interests me about these pictures of the surface of Mars is that there's little evidence of any brittle failure. Is this a correct perception, or is it all masked by the ejecta from the craters? There is all this erosion, sapping and these channels, but one never sees intersecting brittle structures.

Carr: You do, it's just the region we're looking at here. There is a vast system of faults; in this particular area you don't see much of them.

Farmer: There are certainly vertical faults; I don't know about lateral shear zones.

Carr: 99% of the faults we see are normal faults, but there are huge fault systems. There is a big bulge on the equator. It is 4000 km across and 10 km high at the centre. Around this bulge is a radial set of fractures that extend perhaps over a third of the planet. There is a complementary set of ridges that are compression features, and they are circumferential around this structure. The fracture pattern and the compressional ridges are consistent with the present load being carried by the lithosphere.

Davies: With the early hydrothermal systems on Earth, the moon would have been closer then and the tides would have been enormous. There are two reasons why this might be important. First, the huge changes in the sea level and, consequently, pressure; second, the dissipation of energy in the Earth's crust, heating it. Is this at all relevant?

Jakosky: Tidal heating is negligible.

Davies: Even with the moon much closer?

Jakosky: Yes, there's just not much energy.

Davies: What I've heard this afternoon doesn't make Mars in the past sound any better suited to life than Mars today. Why did we all think it was a better place for life 3.5 Ga ago than it is now?

Carr: For one thing, the evidence for water: when we go back as far as we can look, about 3.8 Ga, the evidence is that water is far more abundant. There are more channels and valleys. Another factor is that modelling of the heat flow would suggest that when Mars emerged from the end of heavy bombardment about 3.8 Ga, the heat flow was five times as high as at present, so there should have been more volcanism.

Davies: Has all volcanism now ceased? Presumably there is still a heat outflow and processes such as the liberation of water could continue.

Carr: With time, volcanism has become much more restricted to one or two areas of the planet. But early on there was much more extensive volcanism.

Stetter: Is there anything known about the driving forces of volcanism on Mars?

Carr: The source of most of the heat is accretional energy. Schubert et al (1990) have modelled the convective systems that you might expect in a planet the size of Mars. With the appropriate starting points from the current models, they showed a very small number of convective cells which is consistent with the very localized distribution of volcanic action on Mars.

Farmer: Radiogenic element concentrations in the SNC meteorites are comparable to abundances on earth, so radiogenic heating should be an equally effective way of providing internal heat.

Carr: Yes, but the dominant factor is accretional energy. In fact there is somewhat more potassium on Mars than in the Earth, so radiogenic heating may be somewhat larger.

References

Brakenridge GR, Newsom HE, Baker VR 1985 Ancient hot springs on Mars: origins and paleoenvironmental significance of small Martian valleys. Geology 13:859–862

Carr MH 1996 Water on early Mars. In: Evolution of hydrothermal ecosystems on Earth (and Mars?). Wiley, Chichester (Ciba Found Symp 202) p 249–267

Schubert G, Bercovici D, Glatzmaier GA 1990 Mantle dynamics in Mars and Venus: influence of an immobile lithosphere on three-dimensional mantle convection. J Geophys Res 95:4105–4129

General discussion IV

Martian stable isotopes: volatile evolution, climate change and exobiological implications

Jakosky: The Martian climate clearly has undergone change over the last 4 Ga. This is seen most readily in the surface geomorphology: erosion rates prior to about 3.5 Ga ago were about 1000 times what they have been since then. This was sufficient to remove almost all of the impact craters smaller than about 15 km and to substantially degrade those larger than this. In addition, dendritic valley networks appear on these older surfaces, suggesting the possibility of precipitation or of recharge of near-surface groundwater systems. These observations suggest a climate that was warmer and wetter during the earliest epochs, although they do not allow a unique determination of what 'warmer and wetter' means (Squyres & Kasting 1994).

Stable isotopic measurements of the various components of the Martian volatile system provide indications of the nature of the evolution of the volatiles. This is because the processes by which the volatiles can evolve— exchange with the crust, formation of weathering products, or loss to space— all act to fractionate the isotopes of a given element. We can examine the isotopic fractionation of argon, carbon, nitrogen, hydrogen and oxygen to determine the nature of the evolution processes. In turn, these will tell us about the amount and availability of volatiles for affecting the climate and, ultimately, for understanding the nature of any biological systems (Jakosky 1991, Jakosky & Jones 1996).

Measurements of stable isotopes are available for species in the atmosphere, in weathering products that formed from atmospheric species, and from igneous minerals that represent initial starting points for juvenile gases. Most of these measurements actually come from various components of the SNC meteorites. These meteorites are convincingly of Martian origin: most of them are igneous minerals that crystallized within the last 1.3 Ga, requiring formation on one of the larger planets in our solar system. The oxygen isotopes, however, absolutely and uniquely rule out formation on the Earth or Moon, leaving Venus or Mars as the likely sources. It is easier dynamically to eject a Martian rock from the surface by an impact than it is to eject a Venusian rock. However, the most compelling evidence for a Martian origin is that a glassy component of two of the meteorites contains gases trapped within it that are identical in composition to the Martian atmosphere and distinctly different from any other known source of volatiles in the solar system; it has been shown that gas can be trapped like this by the impact event that would have ejected them to space.

Table 1 (*Jakosky*) shows the various isotope measurements as they are thought to apply to the Martian atmosphere (see Jakosky & Jones 1996). All of the ratios show an enrichment relative to terrestrial or to the inferred initial value on Mars. In the case of hydrogen, argon, carbon and nitrogen, the enrichments are substantial. In each case, the only process that both is capable of enriching the isotope ratios to this degree and is plausible is the loss of volatiles to space. Other processes, such as formation of carbonates, condensation of ice, or decarbonation of sediments, cannot produce fractionation of the magnitude seen. In particular, the argon isotopes are notable because only loss to space is capable of fractionating the isotopes at all — once outgassed into the atmosphere, argon is not capable of chemically reacting with other species (Jakosky et al 1994).

Hydrogen is the only gas that can be lost by thermal escape to space. The other species are lost either by photochemical reactions that impart energy to the atoms, or by sputtering at the top of the atmosphere. The fractionation of argon suggests that this latter process is significant. The sputtering occurs by the impact of Martian oxygen ions that are spiralling around the magnetic field of the impinging solar wind at great velocity; these so-called 'pick-up ions' are capable of removing substantial quantities of the Martian atmosphere over time (Luhmann et al 1992). The escape process results in fractionation of the isotopes because, while the loss occurs from the upper atmosphere, the relative abundances of the species are not the same there as in the bulk atmosphere. Rather, the upper atmosphere is not well mixed, and each species will have its own scale height within the atmosphere, determined by its own mass; thus, the lighter species will be more abundant at the altitudes from which loss occurs, and they will be preferentially removed. This leaves the atmosphere enriched in the heavier isotopes (Jakosky 1991).

Models of the structure of the upper atmosphere and the isotopic fractionation induced by escape can be used to estimate the total loss of each species. In fact, the observed fractionation suggests that at least half, and more likely 90 % or more, of each species has been lost to space through time. The fact that four of the five species show this fractionation (oxygen will be discussed below) provides a strong argument that

TABLE 1 *(Jakosky)* Measurements of Martian isotope ratios, as determined primarily from the SNC meteorites and thought to represent the Martian atmosphere

Isotope ratio	*Measured value (relative to terrestrial)*
D/H	5
$^{38}Ar/^{36}Ar$	1.3
$^{13}C/^{12}C$	1.05–1.07
$^{15}N/^{14}N$	1.7
$^{18}O/^{16}O$	1.025

loss has occurred. While a single isotope system can be explained away by one of various processes, it is extremely diffficult to rationalize four separate isotope systems.

The loss rate depends on the properties of the solar wind and on the ultraviolet flux from the sun. Each of these varies with time, and the loss rate has been declining ever since the earliest history of the solar system. As a result, estimates of the abundance of volatiles through Martian history are consistent with the geological inferences: there were more volatiles available during the first Ga of Martian history. An estimated 90 % of the volatiles were lost to space, with the loss being concentrated prior to about 3 Ga ago. Loss is continuing up to the present day, however.

Notice that oxygen does not show a large fractionation. On the basis of its mass, and on the mass difference between the two isotopes, we would expect it to show a larger fractionation than does carbon, yet it is fractionated by much less than carbon. It is not plausible either that the other species have undergone escape to space but oxygen has not or that the lack of oxygen fractionation invalidates the occurrence of escape. Rather, it suggests that oxygen in the atmosphere is mixing with a non-atmospheric reservoir of oxgyen that serves to buffer the isotopic abundances. Because this reservoir does not also buffer the hydrogen or carbon isotopes, and based on estimates of the integrated escape, it probably is neither CO_2 nor H_2O in the Martian environment (Jakosky & Jones 1996). The most available oxygen reservoir then becomes oxygen in silicate minerals in the crust.

If oxygen is exchanging with crustal silicates, it might be doing so through exchange with water in hydrothermal systems. Such exchange occurs quite readily on Earth and should occur on Mars as well. The presence of hydrothermal systems is suspected on Mars on other grounds, including the abundance of regolith or crustal water and of volcanic heat sources, the presence of minerals within the SNC meteorites that are indicative of hydrothermal deposition, and silicate oxygen isotope abundances that suggest alteration (Jakosky & Jones 1996).

There are three key aspects of these results that bear on any potential Martian biota. First, it is clear from both the geological and the geochemical perspectives that the Martian environment has evolved substantially with time. Early Mars appears to have had an active hydrological cycle, to have allowed liquid water at or very near to the surface, and to have had much more abundant volatile species in its environment. Second, exchange of volatiles between the crust and atmosphere has occurred throughout geological time. Such exchange may have occurred in hydrothermal systems but, regardless, must have involved liquid water at or near the surface to facilitate exchange. This liquid water would provide a suitable environment in which any Martian biota could thrive. Third, this exchange must have occurred up through recent geological epochs, in order to provide the appropriate degree of isotopic buffering. This suggests that locations in which liquid water exists could have been present up through recent epochs and, in fact, could still exist.

These results, while apparently robust, suffer from a lack of adequate observations. Most of the atmospheric isotope ratios are inferred rather than measured directly. Clearly, direct measurements of the present-day atmosphere are required. In addition, measurements of the current rate of escape to space, and how it varies throughout the solar cycle, would strongly constrain our models of the loss processes. These would go a long way towards allowing a quantitative understanding of the evolution of Martian volatiles through time. In addition, they provide an understanding of the non-biological evolution of Martian isotopes; such an understanding is required if we are to search for a biological signature in the isotopes.

References

Jakosky BM 1991 Mars volatile evolution: evidence from stable isotopes. Icarus 94:14–31
Jakosky BM, Jones JH 1996 The evolution of martian volatiles. Rev Geophys, in press
Jakosky BM, Pepin RO, Johnson RE, Fox JL 1994 Mars atmospheric loss and isotopic fractionation by solar-wind-induced sputtering and photochemical escape. Icarus 111:271–288
Luhmann JG, Johnson RE, Zhang MHG 1992 Evolutionary impact of sputtering of the martian atmosphere by O^+ pickup ions. Geophys Res Lett 19:2151–2154
Squyres SW, Kasting JF 1994 Early Mars: how warm and how wet? Science 265:744–749

The transfer of viable microorganisms between planets

P. C. W. Davies

Department of Physics and Mathematical Physics, The University of Adelaide, Adelaide, SA 5005, Australia

Abstract. There is increasing acceptance that catastrophic cosmic impacts have played an important role in shaping the history of terrestrial life. Large asteroid and cometary impacts are also capable of displacing substantial quantities of planetary surface material into space. The discovery of Martian rocks on Earth suggests that viable microorganisms within such ejecta could be exchanged between planets. If this conjecture is correct, it will have profound implications for the origin and evolution of life in the solar system.

1996 Evolution of hydrothermal ecosystems on Earth (and Mars?). Wiley, Chichester (Ciba Foundation Symposium 202) p 304–317

The 'panspermia' hypothesis

The conventional scenario for the origin of life on Earth derives from Darwin's 'warm little pond', and involves chemical self-organization in a broth situated on the Earth's surface (Haldane 1929, Oparin 1938). However, the so-called 'panspermia' hypothesis also has a long, if somewhat speculative, history (Arrhenius 1908, Crick & Orgel 1973, Hoyle 1983, Zahnle & Grinspoon 1990). According to this hypothesis, life did not originate on Earth, but arrived from elsewhere in the galaxy in the form of microorganisms propelled through space by some means (e.g. light pressure).

The strongest evidence in favour of the panspermia hypothesis is the fact that life existed on Earth at least 3.6 Ga ago (some estimates suggest >3.8 Ga). The Earth is 4.6 Ga old, and for several hundred million years it would have been inhospitable. The atmosphere was probably poisonous, solar radiation would have been intense and the surface was spasmodically pounded and sterilized (Sleep et al 1989) by intense bombardment by asteroids and planetesimals (see section below). Evidently, life got started here pretty well as soon as conditions permitted. This suggests that either life forms spontaneously and rapidly, or that microorganisms were raining down continually on the new planet and took up residence when conditions became favourable, or both. This hypothesis also implies that life may have established itself on Earth (or Mars) several times, only to be annihilated again by sterilizing cosmic bombardment.

Against the panspermia hypothesis is the absence of a convincing propagation mechanism plus the fact that cosmic radiation and low temperatures would be a deadly hazard for exposed microorganisms in interstellar space.

The ultimate origin of life remains a mystery in the panspermia theory. Increasing the number of potential sites at which the necessary physical and chemical processes might have occurred drastically shortens the odds against the spontaneous assembly of organic molecules of sufficient complexity, but taken at face value these odds are so enormous anyway that this improvement scarcely matters.

There is a hidden cosmological assumption that goes into theories of the origin of life. Conventionally, the universe is assumed to have a finite age of between 10 and 20 Ga. However, an alternative cosmology, the so-called steady state theory, was once popular. According to this theory, the universe had no origin in time — it is infinitely old — and its physical state remains the same (on a large scale) from epoch to epoch. As stars burn out and galaxies gradually die, so new matter and new star systems are continually created to maintain a steady state. One is free (indeed obliged) to hypothesize in the steady state theory that life too must always have existed. Thus the problem of the origin of life is completely circumvented. Life would pervade the universe and propagate by some means or other from old galaxies to new. The astronomical evidence is against the steady state theory at present, but periodically there are attempts to revive it.

The transfer of material between the planets

We are used to thinking of the planets of our solar system as being physically, and hence biologically, isolated. However, there is increasing evidence that material is continually exchanged between the terrestrial planets. Geologists have known for some time that certain meteorites found on Earth originated on the moon. This conclusion is supported by a comparison of isotopic abundances in these meteorites with those from moon rocks recovered during the Apollo missions.

More recently, it has been fairly convincingly demonstrated that a group of nine meteorites known as SNCs (for Shegotty, Nakhla and Chassingny after their places of discovery) originated on Mars (Wright et al 1989, McSween 1994). This conclusion again comes from measurements of isotopic abundances (which do not match those of terrestrial material), radioactive dating of their ages, plus cosmic ray exposure. For example, the material of the Nakhla meteorite, a fragment of which is in a collection at the University of Adelaide, crystallized 1.3 Ga ago on a parent body large enough to trap volatiles, and spent 180 Ma in space before falling to Earth. The material of this meteorite is augite, which suggests it has been subjected to physical processes on a planetary surface, and Mars seems the most plausible candidate.

This raises the question of how material can travel from the moon and Mars to Earth. Only two mechanisms come to mind for the high-speed propulsion of matter from the surface of a planet: vulcanism and cosmic impacts. The former process is almost certainly too feeble to achieve escape velocity (Melosh 1985). A more plausible explanation is that lunar and Martian meteorites consist of detritus from the impact of large comets or asteroids with the lunar and Martian surfaces. The record of such impacts is written in the surface features of these bodies in the form of copious cratering. Although most cosmic bombardment of the planets took place in the early history of the solar system, sporadic (possibly quasi-periodic) impacts occur even today. The recent encounter between comet Shoemaker-Levy and Jupiter served to focus attention on the impact risk.

The possibility that comets might strike the planets, causing massive devastation, was first suggested in the seventeenth century by Edmond Halley (Steel 1995). However, the emphasis on uniformitarianism by geologists and biologists led to a view in which the geological and biological evolution of Earth was treated in terms of slow change and adaptation. In recent years this view has been challenged, with the importance of sudden, catastrophic events that occur episodically, interspersed by epochs of gradual change, slowly gaining acceptance.

A typical impactor will be travelling at a relative speed of 25 km/s. Sizes range from specks of dust up to bodies several kilometres in diameter. It is likely that Mars will be struck by a 1 km object every few million years, and by a 10 km object every few tens-of-millions of years. These events are sufficiently violent to displace material into space with a velocity that exceeds the escape velocity of the planet (5 km/s). Although much of the material of the impactor and the impact site will be vaporized, peripheral material will be ejected more or less intact, i.e. subject to mild shocking, but not necessarily melted (see later).

Material from cosmic impacts will be ejected in many directions and over a range of velocities. Over time, this ejecta will become distributed over a wide volume of interplanetary space. Some fraction of this material will eventually find itself on Earth-intersecting orbits. Over millions or billions of years, the Earth will sweep up a proportion of this debris. It has been estimated that each year the Earth intersects about half a ton of displaced Martian material. Much of this is vaporized on entry into Earth's atmosphere, but larger rocks will survive intact to ground level.

The reciprocal process will also occur, i.e. the transfer of material from Earth to Mars. Earth has a higher escape velocity (11.2 km/s) than Mars, suggesting there will be less displaced material in total. However, its stronger gravitational field will increase the velocity of impact. Impacts sufficiently violent to splash Earth rocks into space are known to occur. It is widely believed that the death of the dinosaurs (and many other species) that occurred at the Cretaceous–Tertiary boundary 65 Ma ago was caused by the impact of one or more bodies at least 10 km in diameter. A layer of iridium of probable extraterrestrial origin (Alvarez et al 1980) and a crater at Chicxulub in Mexico measuring at least 180 km in diameter have been associated with this event.

Sudden mass extinctions of species have occurred several times in the Earth's history, and comic bombardment seems a likely cause. A 10 km impactor will create massive ecological damage, through shock waves, giant tsunamis, atmospheric dust, climate modification and the re-entry heat of displaced detritus (see, for example, Chapman & Morrison 1989, Steel 1995).

Impacts of this magnitude probably occur on average about every 10 Ma. There is some evidence for a 25–30 Ma periodicity, possibly due to the passage of the solar system through the plane of the galaxy, where gravitational perturbations due to nearby stars and gas clouds stir up the Oort cloud, displacing comets into the inner solar system. For a million years or so this 'cometary storm' greatly raises the probability of a comet–planet encounter. (Recently this evidence has been challenged.)

On a smaller scale, 1 km objects strike Earth every 100 000 years or so. Still smaller impactors strike more often, although they are unlikely to deliver enough energy to eject material far into space. Impacts equivalent to 20 megaton explosions occur every 50–300 years, the last well-documented case being that in Tunguska in 1908 (see, for example, Gallant 1994). A one megaton explosion probably occurred in Brazil in 1930 and possibly a still larger event in British Guiana in 1935.

On Earth, small impactors (such as the Tunguska object) explode and vaporize before reaching the ground, so no crater is produced (although surface damage, e.g. to trees, may be extensive). Craters produced by larger impactors suffer rapid erosion, so are not very familiar. The famous Meteor Crater in Arizona (small by the above standards) is the best known. However, many ancient impact sites have been identified by careful surveying. One of these, on the Eyre Peninsula in South Australia, now contains Lake Acraman (Williams 1986, 1994), about 30 km in diameter. Ejected material from this impact was identified by Gostin in the Flinders Ranges, several hundred kilometres away (Gostin et al 1986).

It is thus well established that Earth, the moon and Mars suffer major cosmic impacts from time to time. Venus will not be spared either. The cratering record of Mercury is evidence that it too has been severely bombarded over time. The existence of lunar and Martian material on Earth strongly suggests that there is a continual traffic of rocky debris and dust between the terrestrial planets. Moreover, the cratering record and all plausible cosmogonic models imply that this traffic would have been far greater during the first 1 Ga of the solar system, i.e. during the time that life established itself on Earth.

The deep hot biosphere

One of the most significant advances of recent years concerning the origin and early evolution of terrestrial life was the discovery of complex ecosystems centred around volcanic vents on the sea bed, in total darkness several kilometres below the ocean surface (see, for example, Felbec 1981, Arp & Childress 1981). These life forms, although DNA-based and presumably closely related to surface and near-surface life, are independent of the solar energy chain. Their food–energy chain is based on thermophilic bacteria that process minerals in the hot rocks at the rock–water interface. These processes take place at pressures of several hundred atmospheres and temperatures in excess — possibly well in excess — of 100 °C (see, for example, Brock 1978).

Thermophiles are also known on the Earth's surface, living in hot volcanic springs at temperatures near boiling point (Walter 1976). Many biologists believe that these archaea closely resemble primordial terrestrial microorganisms (Woese 1987).

Since the discovery of deep ocean life, evidence has been steadily accumulating of subsurface bacteria on land, inhabiting porous rocks up to 4 km below ground (Olson et al 1980, Gold 1992, Appenzeller 1992, Stevens & McKinley 1995). According to Gold, the bacterial colonies around ocean vents are just the visible tip of the iceberg, and bacteria that utilise internal (non-solar) energy sources could also inhabit many less accessible locations, within the Earth's crust. Indeed, if bacterial colonies within rocks are widespread, the total biomass below the Earth's surface could rival that on the surface. Gold (1992), using a limiting depth of 10 km for bacteria requiring temperatures below 150 °C, estimates a total subterranean biomass of 2×10^{14} tons. Even if this is a gross over-estimate, it seems increasingly likely that bacteriological action is responsible for large-scale mineral processing, perhaps explaining concentrated mineral deposits such as gold nuggets.

The existence of subsurface microorganisms has a number of implications for the origin of life on Earth and the possibility of life on Mars:

(1) It is possible that life on Earth originated deep underground, and migrated to the Earth's surface only when conditions settled down. Deep rocks have abundant supplies of the necessary hydrocarbons, minerals, fluids and sources of free energy to drive biochemistry. Moreover, photosynthesis, the basis of all surface life on Earth, is a highly complex process that quite probably took a long time to evolve. Therefore the first microorganisms almost certainly were obliged to make a living from non-solar energy sources.

    The deep subterranean zone also constituted a much more stable and equable environment than the surface of the primitive Earth. In particular, it offered a radiation shield and protection against all but the largest cosmic impacts, volcanic eruptions, surface melting, floods, etc. Of course, the deep rocks are characterized by high temperatures and pressures, but the continued existence of microorganisms that can thrive under these conditions proves that this is no obstacle to life. Thus, in place of Darwin's cosy 'warm little pond' we have something more closely resembling Hades as the crucible of terrestrial life!

(2) The existence of subsurface life on Earth raises the question of whether microbial life also exists beneath the surface of Mars. The Martian surface today is probably inhospitable even for bacterial life, but the subsurface zone offers many of the same advantages as its equivalent on Earth: warmth, stability, protection from radiation, presence of water, etc. The Viking space probes searched only for surface life. However, a thorough search for life on Mars requires drilling at least metres and perhaps even kilometres beneath the Martian surface. Of course, it is possible that microfossil evidence for subsurface Martian life may be revealed in the regolith of major cosmic impact events, or in Martian meteorites found on Earth, or in Martian rift valleys (Gold 1992).

(3) The existence of bacteria thriving deep within terrestrial rocks suggests that microorganisms could remain viable within rocky cocoons in outer space. Material displaced from Earth as a result of cosmic impacts will often contain microorganisms.

Could at least some of these organisms survive the journey through space to Mars? If so, could they survive impact with the Martian surface? The evidence suggests the answer 'yes' to both questions (see below). If viable terrestrial organisms were to reach Mars in this manner, a transfer into the Martian subsurface zone would then be possible. Although the physical hazards would be slight, there may be biochemical problems. Nevertheless, the extraordinary fecundity of bacteria and their rapid mutation rate implies that adaptation to the local conditions might be expected. Once local colonization of the deep rocks had occurred, microorganisms could be expected to spread across the planet.

(4) Similar arguments apply to the transfer of viable microorganisms from Mars to Earth. Thus life in the solar system may have originated on or beneath the surface of Mars (or for that matter any of the rocky planets or moons in the solar system) and been transported to Earth in rocky ejecta from cosmic impacts. It could then have rapidly colonized the subsurface and/or surface zones. Finally, life may have formed in several locations (e.g. both Earth and Mars) and subsequently undergone cross-contamination by this process.

(5) The search for life elsewhere in the solar system, and especially Mars, is considerably complicated by the realization that the planets are not quarantined. If, for example, microbes were discovered on Mars based on nucleic acid and proteins, particularly with terrestrial chirality, a common origin may be a more plausible explanation than the independent formation of two ecosystems exploiting a similar biochemistry. As the latter scenario is considerably more significant for fundamental science and philosophy than the former, this suggests that the most interesting search strategy for life on Mars is precisely *not* to seek out Earth-type life there.

Surviving the journey

The hypothesis that viable microorganisms can be transferred from Earth to Mars depends on the organisms concerned overcoming a number of hazards:

(1) Survival of the initial cosmic impact and ejection into space.
(2) Remaining viable for long durations while restricted to a small volume of rock.
(3) Intense UV and cosmic radiation, and cold.
(4) The shock and heat of impact with Mars.
(5) The hostility of the Martian environment.

Similar hazards would face any microorganisms making the journey the other way. Let me consider each of these hazards in turn.

The initial cosmic impact

There is still only a rudimentary understanding of the physics of high-speed cosmic impacts (Melosh 1989). The details depend greatly on the speed and angle of impact, the mass, constitution and density of the impactor, the nature of the planetary atmosphere (if any) and the nature of the surface material. In the case of large impactors (1–10 km), which are likely to avoid incoming fragmentation by the terrestrial or Martian atmospheres, most of the impactor and the surface material immediately beneath it will be vaporized or melted by the sudden concentrated energy release. Much of the splashed solid material will be severely shocked. Any microorganisms in this material will be killed.

There remains the possibility that material near the periphery of the impact site will be accelerated to escape velocity without being shocked or melted. Studies of meteorites of lunar and Martian origin reveal that they have suffered remarkably little violence in this respect.

Theoretical modelling of the physics of impacts (Melosh 1985) may explain this circumstance. The existence of a free surface at essentially zero pressure implies that material near the surface of the impact region cannot be subjected to high compression, even if accelerated rapidly to high speeds. In fact, near-surface material is subjected to both a compressive shock wave and a surface-reflected tensile wave more or less simultaneously. These waves may interfere in such a way as to produce regions of large accelerations with low shock. Plates of material are ejected and then break up into smaller fragments.

Deeper-lying rocks will evidently suffer more compression and heating. These will be ejected at high temperature. Thus there is likely to be a top layer of ejecta in which any microorganisms would probably survive, propelled upwards by a layer of highly shocked and molten material. From Melosh's work it seems as if the total volume of unshocked target material ejected above the escape velocity might be as high as 1% of the impactor volume, representing a considerable mass of material for major impact events.

The survival of microorganisms for long periods in space probably requires their sojourn in rocks of an appreciable size, to protect them from cosmic radiation and to act as a thermal blanket (see below). I would estimate that a minimum size in the range 1–10 m is necessary. Cosmic impacts tend to produce severe fragmentation of high-speed ejecta. Melosh has estimated that a 100 m impactor with an arrival speed of 22 km/s will produce fragments at escape velocity of size only about 1 cm for Mars and less for Earth. However, the fragment size scales in proportion to the size of the impactor, so larger fragments will be produced by rarer, more violent events.

A further consideration is the effect of atmospheric friction and heating on rocks blasted skyward at high speed. This could be important for small impact events and debris fragments, especially in the case of Venus, which has a very thick atmosphere. However, a large impactor ($>$ 10 km) will effectively punch a hole in the atmosphere on its way in, and much of the ejecta could escape into space along this evacuated tunnel. (The escape time is only a few seconds.) In any case, atmospheric heating would probably sterilize only a thin layer of surface material, which would effectively shield the layers beneath.

Melosh has estimated that an impact which produces an initial crater of about 50 km diameter (comparable in size to several known craters on Earth) would eject a million cubic metres of boulders of one metre or more in size from a 'fecund zone' in which microorganisms might survive heating and shock. Events many times larger than this have certainly occurred. Clearly, during the history of the solar system, enormous quantities of terrestrial rocks containing viable microorganisms will have been ejected into space. Even greater quantities of Martian material will have suffered a similar fate.

How long can microorganisms survive in space?

Live bacteria were recovered from the Surveyor 3 spacecraft after three years exposure on the lunar surface. Cocooned within rocks, bacteria would survive for much longer. Three-million-year-old microorganisms have been recovered from the Siberian permafrost. Bacterial spores may remain dormant but viable for enormous durations. Recent reports from California Polytechnic University suggest that spores embedded in amber for up to 130 Ma have been resuscitated, while dehydrated bacteria have apparently survived in New Mexico salt beds for up to 225 Ma (Melosh 1994).

Computer modelling (Melosh 1993) suggests that 5% of Earth ejecta reaches Mars within 150 Ma, while several times that amount is transported in the other direction. A significant fraction of both Earth and Mars ejecta is propelled from the solar system entirely by Jupiter. Sojourn times in space of tens of millions of years seems normal, although significantly shorter transit times will inevitably occur by chance for a small fraction of the ejected material. It is possible, though extremely rare, for a terrestrial rock containing microorganisms to reach Mars in just a few years. For the record, the so-called Farmington meteorite (probably created by fragmentation of a larger body) is thought to have been in space in its final form for as little as a few thousand years. Recent computer studies of the orbits of lunar ejecta in the Earth–moon system (Gladman et al 1996) suggest that ejection is easier, and that transit times may be much shorter, than previously believed: for example, two-thirds of Earth impacts occur within a mere 50 000 years.

A large fraction of Earth ejecta will fall back to Earth eventually, perhaps after some millions or tens of millions of years. We thus have the possibility of ancient microorganisms being returned from space, perhaps after they have become extinct on Earth. If there is a continual or episodic rain of DNA from space in this manner, it may have important consequences for terrestrial biological evolution.

The hostile space environment

Outer space is undoubtedly a hostile environment for life. Nevertheless, a rock would shield microorganisms from the worst hazards. UV radiation may be shielded by a thin layer of solid material, while even cosmic rays would only be a small hazard within a boulder of a metre or more in size. As far as temperature is concerned, a tumbling rock would average out the insolation. In the region between the orbits of Earth and Mars, a fairly stable mean temperature within the rock in the range $-10\,^{\circ}C$ to $-50\,^{\circ}C$ is expected—cold but possibly not fatal.

The impact with Mars

A rock of 1–10 m diameter entering the atmosphere of Mars or Earth would be significantly slowed by friction. Although the outer layer will suffer extreme heating, leading to melting and vaporization, the low thermal conductivity of rock and the short duration of the transit through the atmosphere would ensure that the interior of the rock would be largely unaffected. The exact details depend on the speed and especially the angle of entry into the atmosphere (shallow paths leading to gentler deceleration), and on the size of the projectile. Meteorites of 1–10 m size frequently fragment, and the pieces fall to the ground at terminal velocity slowly enough for there to be negligible shock on impact. Thus any microorganisms that remain viable in the deep interior of the rock at this stage are likely to be exposed and dispersed unharmed on the surface of the host planet.

Duncan Steel has pointed out (personal communication) that Mars will have the highest probability of intersecting ejecta from Earth when it is close to perihelion, and at this time solar heating will vaporize some of the polar caps and release significant quantities of CO_2. In the past, Mars has had perihelia closer to the sun. Its atmosphere could thus have been substantially thicker at just the time that most Earth ejecta were likely to arrive, considerably aiding the chances of a less violent entry.

The hostile Martian environment

Even if terrestrial microorganisms survive their hazardous journey to the surface of Mars, they still face an inhospitable environment. However, in the past Mars had a thicker atmosphere and liquid water, so early colonization is not implausible. Even today, some terrestrial bacteria might be able to survive on Mars. If terrestrial bacteria were able to migrate rapidly into the subsurface Martian rocks, conditions would be more suitable to their continued survival.

Any Martian microorganisms reaching Earth would be presented with the opposite problem: a highly equable environment teeming with indigenous competitive life. Would a Martian bacterium be rapidly attacked and destroyed by its terrestrial competitors, or could a type of co-evolution occur? The answer would depend on the degree of similarity in biochemistry and the utilization of resources.

Finally, it is possible that dust, rather than rocks, may be a more promising mode of conveyance of microorganisms between planets. As pointed out by Steel (1996), small meteoroids represent the lion's share of meteoric mass, and they suffer less violence on re-entry into a planetary atmosphere. They could have played an important part in conveying organic materials to Earth. However, in the case of living organisms, they offer little shielding against cosmic radiation. It is hard to quantify these competing factors in our present state of knowledge.

Conclusion

Theoretical modelling of cosmic impacts suggests that substantial numbers of rocks of one metre diameter and more have been exchanged between the terrestrial planets. It seems likely that terrestrial rocks would harbour microorganisms for some considerable time. If it is established that viable microorganisms can be/have been exchanged between Earth and Mars, it will have major implications for the theories of the origin of life, the theory of evolution and search strategies for life elsewhere in the solar system.

The transfer of microorganisms would imply that Earth and Mars are biologically coupled. One may then identify two regimes: weak and strong coupling. Weak coupling means that the arrival of a viable organism would be a very rare event and/or the host planet would prove inhospitable. In the case that life originated on Earth, Mars may have experienced spasmodic 'infection', possibly of limited duration. There may exist fossil remains of several episodes of unsuccessful colonization, where microorganisms survived only with difficulty and for a relatively short duration. If life arose independently on Earth and Mars, it is not clear what the outcome of an intermingling might be. Our understanding of the origin of life is so poor that we have little idea whether an independently-arising life form would be based on nucleic acids and proteins, or something else entirely. If the two forms were very similar, competition or symbiosis might occur (e.g. mitochondria from Mars invading terrestrial prokaryotic cells). In the (more likely) case of very dissimilar organisms, one might displace the other, or they may peacefully co-exist, perhaps in very different habitats.

Strong coupling implies the regular exchange of organisms between the planets, so that Earth and Mars would need to be considered as a single ecosystem. In this case Martian organisms would be very similar to terrestrial organisms, but presumably largely subterranean in nature. The question of where life originated would then be hard to answer.

Even if Mars turns out to be a totally sterile planet, it is clear that the Earth's ecosystem extends into space, and that the effects of ancient microorganisms returning in ejecta, which has hitherto been overlooked, may well be significant for terrestrial evolution.

Acknowledgements

I thank Prof. Vic Gostin, Prof. Lloyd Hamilton and Dr Duncan Steel for helpful discussions and critical suggestions. I also thank Duncan Steel for supplying many useful references.

References

Alvarez LW, Alvarez W, Asaro F, Michel HV 1980 Extraterrestrial cause for the Cretaceous-Tertiary extinction. Science 208:1095–1107

Appenzeller T 1992 Deep-living microbes mount a relentless attack on rock. Science 258:222

Arp AJ, Childress JJ 1981 Blood function in the hydrothermal vent vestimentiferan tube worm. Science 213:342–344
Arrhenius S 1908 Worlds in the making. Harper Collins, London
Brock TD 1978 Thermophilic microorganisms and life at high temperatures. Springer-Verlag, New York
Chapman CR, Morrison D 1989 Cosmic catastrophes. Plenum, New York
Crick FHC, Orgel LE 1973 Directed panspermia. Icarus 19:341–345
Felbec H 1981 Chemoautotrophic potential of the hydrothermal vent tube worm, *Riftia pachyptila* Jones (Vestimentifera). Science 213:336–338
Gallant RA 1994 Journey to Tunguska. Sky & Telescope 87(6):38
Gladman B, Burns JA, Duncan M, Lee P, Levinson H 1996 The exchange of impact ejecta between terrestrial planets. Science 271:1387–1392
Gold T 1992 The deep, hot biosphere. Proc Natl Acad Sci USA 89:6045–6049
Gostin VA, Haines PW, Jenkins RJF, Compston W, Williams IS 1986 Impact ejecta horizon within late Precambrian shales, Adelaide Geosyncline, South Australia. Science 233:198–200
Haldane JBS 1929 The origin of life. Ration Ann 148:3–10
Hoyle F 1983 The intelligent universe. Michael Joseph, London
McSween HJ 1994 What we have learned about Mars from SNC meteorites. Meteoritics 29:757–779
Melosh HJ 1985 Ejection of rock fragments from planetary bodies. Geology 13:144–148
Melosh HJ 1988 The rocky road to Panspermia. Nature 332:687–688
Melosh HJ 1989 Impact cratering: a geologic process. Oxford University Press, New York
Melosh HJ 1993 Swapping rocks: ejection and exchange of surface material among the terrestrial planets. Meteoritics 28:398
Melosh HJ 1994 Swapping rocks: exchange of surface material among the planets. Planetary Rep 14:16–19
Olson GJ, Dorkins WS, McFeters GA, Iverson WP 1980 Sulfate reducing and methanogenic bacteria from deep aquifers in Montana. Geomicrobiology J 2:327–340
Oparin AI 1938 The origin of life. Macmillan, London
Sleep NH, Zahnle KJ, Kasting JF, Morowitz HJ 1989 Annihilation of ecosystems by large asteroid impacts on the early Earth. Nature 342:139–142
Steel D 1995 Rogue asteroids and doomsday comets. Wiley, New York
Steel D 1996 Cometary impacts on the biosphere. In: Thomas P, McKay C (eds) Comets and the origin and evolution of life. Springer-Verlag, New York, p 1–6
Stevens TO, McKinley JP 1995 Lithotrophic microbial ecosystems in deep basalt aquifers. Science 270:450–454
Walter MR (ed) 1976 Stromatolites. Elsevier, Amsterdam
Williams GE 1986 The Acraman impact structure: source of ejecta in late Precambrian shales, South Australia. Science 233:200–203
Williams GE 1994 Acraman, South Australia: Australia's largest meteorite impact structure. Proc R Soc Victoria 106:105–127
Woese CR 1987 Bacterial evolution. Microbiol Rev 51:221–271
Wright IP, Grady MM, Pillinger CT 1989 Organic materials in a Martian meteorite. Nature 340:220–222
Zahnle K, Grinspoon D 1990 Comet dust as a source of amino-acids at the Cretaceous tertiary boundary. Nature 348:157–160

DISCUSSION

Des Marais: I don't know whether your hypothesis that Earth life has made the journey to Mars (and maybe vice versa) would be a complicating factor in our search strategy for life on Mars, because this search strategy is based on an assumption that Martian life would be similar in basic respect to early life on Earth. However, possible cross-contamination between planets clearly complicates the bigger question about the origin of life.

Pentecost: In order to escape the Earth's atmosphere, fragments of rock would have to accelerate rapidly, placing huge forces on the individual bacteria. The components of bacterial cells are not of equal density: some parts are going to be lighter than others, and serious damage could result. Have any calculations been made on this?

Davies: Not that I'm aware of. The forces of acceleration depend on the mass and the linear size: because the cells are very small I don't think that would be a problem. Although they haven't been done, these are fairly simple calculations.

Shock: You pointed out that Melosh's calculations are model dependent. Which factors would have the greatest effect in these calculations?

Davies: The first problem is that you can't do the experiments. Although some people have tried firing projectiles at targets, it is not possible to achieve the velocities we are talking about here experimentally. The physical models seem to me rather crude. They depend on a number of factors, not least the nature of the material. Of course, the characteristics of an impact on the ocean would be quite different from one on a continent. Melosh's work still seems to be at an elementary stage. However, the principle of having a sort of circular region around the periphery of an impact that can be expelled without being shocked is fairly well established. The amount of material displaced is a woolly figure, but the most suspect figure is the fragment size, which is based on scaling arguments rather than any dynamical models.

Horn: Concerning the fragment size, couldn't we just look at modern meteorite fall-back breccias and see what size they are?

Davies: Yes, but we are interested here in what escapes, not what remains.

Horn: But whatever falls back might give you a size distribution from which the ejected fragment size could be extrapolated.

Davies: One intuitively feels that the little bits would go straight up and the bigger fragments would simply flop to the side.

Carr: One of the major unknowns in the calculations is the state of the materials involved, because we are dealing with such high pressure. But I don't think the uncertainties affect the principle at all, in that you are going to get spallation around the periphery of a large impact and it is going to be minimally shocked, just by virtue of the refraction waves. It is inevitable that pieces of Earth will land on Mars.

However, there is one hazard that you didn't talk about: the effect of solar storms. These major radiation events are a much greater danger than the UV exposure.

Davies: Solar storms produce mainly protons. What will be their penetration depth?

Carr: They are very high energy protons. I remember seeing calculations with respect to the size of shelters that one would need to survive for a manned mission to Mars. You would have to take a storm shelter with you, and elaborate mechanisms have been suggested whereby one could relay the observation of a solar storm so that you could get in the shelter by the time that the radiation arrives. Over tens of millions of years any projectile in space or on the surface of Mars would be exposed to a lot of radiation. An organism would need repair mechanisms to counteract it.

Davies: That is a good point that I had overlooked.

Kuzmin: Another hazard might be temperature variations on the surface of the projectile. The temperature of a rock fragment in space would be highly variable, with a low of about −100 °C. You only talked about the equilibrium temperature; the temperature variation due to illumination and shadowing would be very large.

Davies: But where is the shadowing coming from in interplanetary space?

Kuzmin: During the flight of the projectile in space, all parts of its surface will periodically be illuminated and shadowed because it is rotating. We don't know how rapidly the temperature would change, but the temperature variations between the two extremes might be damaging.

Davies: But unless it is on a very eccentric orbit, I can't see why there would be major temperature variations.

Kuzmin: The temperature variations are a result of the periodic exposure to the sun of the different sides of the projectile. The depth to which the temperature variations penetrate into the projectile depends on the rotation period.

Stetter: You are suggesting that low temperature might be a hazard, but a temperature of −100 °C is ideal for preservation of bacteria.

Kuzmin: Only if this temperature were to be constant. Rapid variations of temperature can be damaging for life forms.

Carr: I think you are speaking at cross purposes. The temperature variations Ruslan Kuzmin is talking about will be right on the surface. For an organism embedded a metre or so within the projectile, the temperature variations won't be very large.

Nisbet: A metre still might not be far enough within—for instance, the seasonal thermal wave where I used to live in Canada, with fluctuations from −40 °C minimum in winter to +40 °C maximum in summer, goes down a metre or more. So in six months, if the projectile is slowly turning, you might easily get a thermal wave changing its temperature by 30–40 °C. That might be a bit much for bacteria if it is going on for 150 Ma. You could just possibly imagine an environment of this sort where frozen bacteria would survive, but the likelihood is that they wouldn't.

Davies: My feeling is that even if there was a temperature fluctuation, the peak would be still well below temperatures where bacteria would be revived, if the projectile is on an orbit between Earth and Mars.

Nisbet: But it seems the most likely time when this sort of process could be possible is very early on, say 3.8 Ga ago or earlier when Mars was potentially fertile. A lot of bombardment was taking place on Earth and there might have been a large number of fragments ejected of which one or two could have been carrying some sort of viable ancestor to life.

Davies: It's clear that this whole process would have been of much greater importance in the past than it is today.

Stetter: Is it possible that a smaller meteorite, which doesn't impact on Earth but just hits the atmosphere, would then land on Mars and take with it some life form from Earth?

Davies: The problem is that the surface temperatures of meteors grazing the atmosphere are enormous. At the sort of layers where there might be live organisms the temperature would be lethal.

Carr: This issue has some very practical implications. I was on a committee a couple of years ago looking at forward contamination of Mars and the levels of sterilization one would need on spacecraft and instruments that landed on its surface. Because of the prospect that Earth has already contaminated Mars, and also the low probability of any multiplication if contaminated material were to land on Mars, we decided that we would no longer sterilize spacecraft going to Mars. The surface is so hostile to life that any contaminating organisms might be able to remain dormant, but the chance of them multiplying and dispersing around the planet is very small.

Parkes: Paul Davies, are you envisaging lumps of rock being bounced into space?

Davies: Yes, but there will of course be a much greater quantity of granules or dust.

Parkes: Bacteria on the surface of dust granules would die; you would need a big lump of rock with the bacteria in the centre for them to survive.

Davies: That's my favoured scenario, but I've discussed this with Duncan Steel and he thinks we should not ignore the smaller grains.

Parkes: There's a paradox here. Bacteria will only be in rocks if there's an energy source: under these conditions bacterial activity will result in weathering and the eventual break-up of the rock. A solid igneous rock without fissures will not have bacteria. Small rock particles may therefore have high bacterial populations, but they are going to be sterilized during interplanetary transport. The big ones may not have the bacteria you need.

Trewin: Do we have pieces of sedimentary rock that have returned to Earth from Earth via space?

Davies: This has got to have happened at some stage, but how would you recognize them?

Carr: You would have to see them fall!

Trewin: My point was that I do not know of records of large chunks of sedimentary rock returning to Earth. They could be recognized (e.g. on Antarctic ice along with meteorites). I feel that if large rocks can be hurled into space in the condition and manner envisaged, we would have evidence on Earth that the process operated.

Henley: Would some of the larger volcanic eruptions have power to blast rocks into space?

Jakosky: The rise of volcanic plumes in the atmosphere is not by the ballistic rise of ejecta. The material comes out of even the largest eruptions at only a few hundred m/s and it rises by the buoyant rise of the plume in the atmosphere. It is simply not possible to put something into orbit, of either Earth or Mars, in this way.

Summing-up

Malcolm R. Walter

School of Earth Sciences, Macquarie University, North Ryde, NSW 2109 and Rix & Walter Pty Ltd, 265 Murramarang Road, Bawley Point, NSW 2539, Australia

I have recently been reading a collection of papers put together by the American Association of Petroleum Geologists called 'Oil is first found in the mind' (Foster & Beaumont 1992). This title aptly expresses the intellectual endeavour we are attempting in searching for the earliest life on Earth and Mars. We are using the same sorts of approaches that the better of the oil explorers use; trying to visualize, predict and 'finesse' our way to the exploration targets. As in petroleum exploration, we will never have all the information we would hope for, or big enough budgets to try every possible exploration strategy.

In 1952 Wallace Pratt, one of the more famous oil explorers, wrote that oil is first found 'in the minds of men. The undiscovered oil field exists only as an idea in the mind of some oil-finder. When no man any longer believes that oil is left to be found, no more oil fields will be discovered. So long as a single oil-finder remains with a mental vision of a new oil field to cherish, along with freedom and incentive to explore, just so long new oil fields may continue to be discovered.' (Introduction, in Foster & Beaumont 1992.) Our meeting, I think, has helped to define and articulate a vision of an exploration strategy that has an excellent chance of revealing significant new information on early life both here and on Mars. In addition to that, one of the most positive messages to come out of this meeting is the very fact that it has occurred at all. In recent years we have seen a remarkable convergence of interests and ideas in the fields of evolutionary biology, palaeobiology and planetary exploration. We are all thinking about the same sorts of issues from very different points of view, and we have seen that mineral explorers have many of the tools that we need to employ in our search. In some ways the meeting self-organized — these ideas converged in different people's minds at about the same time and it seemed appropriate to hold the symposium.

A focus on ancient hydrothermal systems by palaeobiologists seeking evidence of the earliest life on Earth is justified on many grounds, but especially because biologists have recognized that hyperthermophiles cluster around the roots of the tree of extant life. Taking this approach on Mars is justified for the same reasons, and also because hydrothermal systems are sharply defined targets that can be found with techniques perfected by mineral explorers. I doubt that any of us believes that this should be the only search strategy employed on Mars, but it should be given a high priority, especially as climate modelling suggests that liquid water would have been unstable on Mars except in hydrothermal systems.

As I stated in the introduction, in 1982 Carl Sagan published a petition in which he admonished us, particularly with relation to the SETI program, to remember that a priori arguments are of value, but we desperately need experimental observations as well (Sagan 1982). Along those lines I thought it would be useful to list some future directions and important gaps in our knowledge that need to be filled. What follows is my attempt to distil the essence of comments and suggestions made during the meeting. We need:

(1) Research on deep subterranean bacterial ecosystems on Earth, particularly those associated with hydrothermal systems. Rock records of such ecosystems may have a much higher chance of preservation than the thin and ephemeral subaerial expressions of hydrothermal systems; it is feasible that such an ecosystem still exists on Mars.
(2) Baseline surveys of hydrothermal systems—both modern and fossil. Hydrothermal systems need to be thought of as whole systems and we must therefore consider all the factors involved, from the chemistry of the waters to the fossilization of the bacteria.
(3) The continuing census and characterization of extant hyperthermophiles in all three domains of life, including the description of isotope systematics and lipid chemistry.
(4) The development of high pressure and high temperature sampling equipment in order to get the biological samples that are needed.
(5) Uncovering of the fossil record of hyperthermophiles on Earth. Implicit in this is the need to improve our understanding of the taphonomy of these organisms.
(6) Research on the disequilibrium processes that occur in hot springs, to better understand the chemistry of these systems.
(7) Sharpening of our understanding of the climate on Mars 3–4 Ga ago, through techniques such as work on the isotopic composition of air, rock, weathering products and SNC meteorites, the mineralogy of the surface of Mars, and climate modelling.
(8) In terms of exploration of Mars, the problems caused by the very high requirements for data collection could be minimized by the collection of data along traverses, which can then be related to the broader situation of the planet.
(9) The compilation of structural, geophysical, tectonic and mineralogical ore-deposit models for Mars. We have models that we use on Earth; we need to think about how to translate those to the different environment of Mars, and to consider a range of mineralogies, including silica, carbonates, and metal oxides and sulfides.
(10) To locate the hydrothermal systems on Mars.

I'll finish by quoting another oil explorer, who concluded that success in oil exploration 'is a question of exploratory talent. The oil and gas are there to be found, but it [sic] will only be found profitably by those who have competent personnel at all levels. It is a people problem, not a natural resource problem' (J. G. Elam, p 32 in Foster & Beaumont 1992). If diverse groups, such as those of us at this symposium, can continue to work cooperatively, then within our lifetimes we can realistically hope to know not only much more about the history of life on our own planet, but also whether there was once, and maybe still is, life on Mars.

References

Foster NH, Beaumont EA (compilers) 1992 Oil is first found in the mind. Treatise of Petroleum Geology Reprint Series No. 20, American Association of Petroleum Geologists, Tulsa

Sagan C 1982 Extraterrestrial intelligence: an international petition. Science 218:426

Note added in proof

In the 16th August 1996 issue of Science, McKay et al report evidence for the former existence of life on Mars. In a meteorite thought to have come from Mars they have found minerals, microstructures and organic compounds that they consider to be biogenic. They argue that these features are indigenous to the meteorite and 3.6 Ga old, and therefore that there was life on Mars at that time. If this could be confirmed, it would of course be a momentous discovery. The authors concede that individually none of their pieces of evidence is compelling, but consider that taken together they are convincing evidence of former life on Mars. They cite evidence for the former presence of several physiologically disparate forms of nannobacteria such as are presently found in some subterranean environments on Earth. Their study has been meticulous and thorough but suffers from the limitations familiar to all palaeobiologists who work with possible bacterial microfossils: that is, it is frequently difficult to find compelling evidence of biogenicity for such simple structures which might equally plausibly be interpreted as chemical and mineralogical artifacts, or contaminants (see, for instance, Schopf & Walter 1983).

This is not the place to discuss this report in detail, but there is one issue that is particularly troubling. McKay et al (1996) consider that there was a consortium of disparate kinds of bacteria in the subterranean environment from which the meteorite was derived, and that each kind can be interpreted by comparison with counterparts on Earth. So this argument reduces to a truly remarkable conclusion: not only was there life on Mars but there were several forms of bacteria physiologically closely comparable to forms on Earth. This requires either: (1) that there were independant origins of life on Earth and Mars but that evolution was by some mechanism channelled in the same directions; or (2) that life arose on one of these planets and a consortium of bacteria was transferred to the other planet by the mechanism discussed in this book by Paul Davies (which must surely become less likely the more types of bacteria are required to be transferred). Simpler interpretations would be that the features in the meteorite are not biogenic or that they were introduced after the meteorite landed in Antarctica. These explanations seem to me more likely than the two alternatives.

McKay DS, Gibson EK Jr, Thomas-Keprta KL, Vali H, Romanek CS, Clemett SJ, Chillier XDF, Maechling CR, Zare RN 1996 Search for past life on Mars: possible relict biogenic activity in Martian meteorite ALH84001. Science 273:924–930

Schopf JW, Walter MR 1983 Archean microfossils: new evidence of ancient microbes. In: Schopf JW (ed) Earth's earliest biosphere: its origin and evolution. Princeton University Press, p 214–239

Malcolm Walter

Index of contributors

Non-participating co-authors are indicated by asterisks. Entries in bold type indicate papers; other entries refer to discussion contributions.

Indexes compiled by Liza Weinkove

Subject index